Sociology 9e

the essentials

Margaret L. Andersen
University of Delaware

Howard F. Taylor
Princeton University

With
Kim A. Logio
Saint Joseph's University

CENGAGE Learning

Australia • Brazil • Mexico • Singapore • United Kingdom • United States

CENGAGE Learning

Sociology: The Essentials, **Ninth Edition**
Margaret L. Andersen, Howard F. Taylor, and Kim A. Logio

Product Director: Marta Lee-Perriard
Product Manager: Elizabeth Beiting-Lipps
Content Developer: John Chell
Product Assistant: Chelsea Meredith
Marketing Manager: Kara Kindstrom
Content Project Manager: Cheri Palmer
Art Director: Vernon Boes
Manufacturing Planner: Judy Inouye
Production Service: Jill Traut, MPS Limited
Photo Researcher: Lumina Datamatrics
Text Researcher: Lumina Datamatrics
Copy Editor: Heather McElwain
Illustration and Composition: MPS Limited
Text Designer: Jean Calabrese
Cover Designer: Irene Morris
Cover Image: EschCollection/Photonica/Getty Images

WCN: 01-100-101

Library of Congress Control Number: 2015935648

Student Edition:

ISBN: 978-1-305-50308-3

Loose-leaf Edition:

ISBN: 978-1-305-67588-9

Cengage Learning
20 Channel Center Street
Boston, MA 02210
USA

Printed in the United States of America
Print Number: 01 Print Year: 2015

Brief Contents

Contents

PART ONE Introducing the Sociological Imagination

PART TWO Studying Society and Social Structure

5 Social Structure and Social Interaction 103

6 Groups and Organizations 125

7 Deviance and Crime 147

PART THREE Social Inequalities

8 Social Class and Social Stratification 169

9 Global Stratification 201

10 Race and Ethnicity 227

11 Gender 253

12 Sexuality 281

13 Families and Religion 307

14 Education and Health Care 339

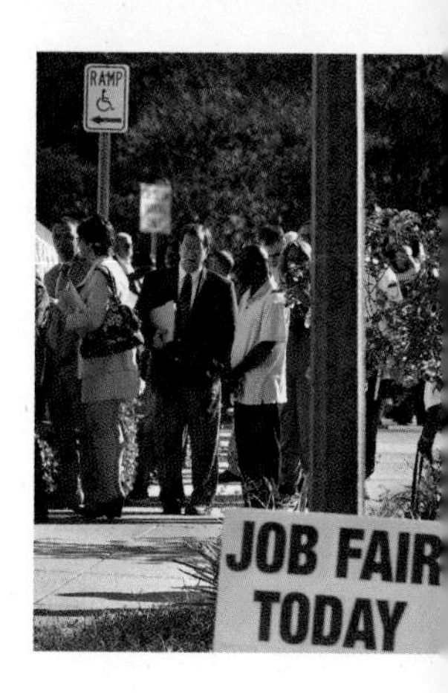
RAMP
JOB FAIR
TODAY

PART FIVE Social Change

16 Environment, Population, and Social Change 393

Features

doing sociological research

what would a sociologist say?

understanding diversity

a sociological eye on the media

maps

Mapping America's Diversity

Viewing Society in Global Perspective

Preface

Sociology: The Essentials is a book that teaches students the basic concepts, theories, and insights of the sociological perspective. With each new edition come new challenges—challenges that stem from new generations of students with different learning styles; challenges that stem from the diversity among students who will study this book; and challenges that stem from the changes that are taking place in society. One of the most important changes taking place today is how students learn and how they are engaged with their course material, often in the form of online learning resources. With that in mind, *Sociology: The Essentials*, ninth edition, takes full advantage of this revolutionary change by having a fully electronic version of the book available, which allows for personalized, fully online digital learning—a platform of content, assignments, and learning resources that will engage students in an interactive mode, while also offering instructors the opportunity to make individualized configurations of course work. Some will want to continue using the printed version of the book, still enhanced with various pedagogical features. Those who want to enhance their curriculum through online resources will be able to utilize the new MindTap Sociology in the way that best suits their course.

However the book is used, we have updated it to reflect the latest social changes and developments in sociological scholarship. We are somewhat amazed, even as sociologists, to see how much change occurs, even in the relatively short period of time between editions. Our book adapts to new research that appears at an amazing pace, as well as addresses the significant changes that occur in society between editions.

In this edition, we have maintained the themes that have been the book's hallmark from the start: a focus on diversity in society, attention to society as both enduring and changing, the significance of social context in explaining human behavior, the increasing impact of globalization on all aspects of society, and a focus on critical thinking and an analysis of society fostered through sociological research and theory. We know that studying sociology opens new ways of looking at the world. As we teach our students, sociology is grounded in careful observation of social facts, as well as analyses of how society operates. For students and faculty alike, studying sociology can be exciting, interesting, and downright fun, even though it also deals with sobering social issues, such as the growing inequality that marks our time, as one example.

In this book, we try to capture the excitement of the sociological perspective, while introducing students to how sociologists do research and how they theoretically approach their subject matter. We know that most students in an introductory course will not become sociology majors, although we hope, of course, that our book and their teacher encourage them to do so. We want to give students, no matter their area of study, a way of thinking about the world that is not immediately apparent. We especially want students to understand how sociology differs from the individualistic and commonsense thinking that tends to predominate. This is showcased in the box feature in every chapter entitled, "What Would a Sociologist Say?" Here, we take a common topic and, with informal writing, briefly discuss how a sociological perspective would approach understanding on that particular issue. We think this feature helps students see the unique ways that sociologists view everyday topics—things as commonplace as the funeral of a superstar, finding a job, or sports in popular culture.

We want our book to be engaging and accessible to undergraduate readers, while also preserving the integrity of sociological research and theory. Our experience in teaching introductory students shows us that students can appreciate the revelations of sociological research and theory if they are presented in an engaging way that connects to their lives. We have kept this in mind throughout this revision and have focused on material that students can understand and apply to their own social worlds.

Critical Thinking and Debunking

We use the theme of *debunking* in the manner first developed by Peter Berger (1963) to look behind the facades of everyday life, challenging the ready-made assumptions that permeate commonsense thinking. Debunking is a way for students to develop their critical thinking, and we use the debunking theme to help students understand how society is constructed and sustained. This theme is highlighted in the **Debunking Society's Myths** feature found throughout each chapter.

We want students to understand the rigor that is involved with sociological research, whether quantitative research or qualitative. The box feature **Doing Sociological Research** presents a diverse array of research studies, presented to students so they can see the question being asked, the method of investigation, the research results, and the study's conclusions. This feature also includes critical thinking questions ("Questions to Consider") to help students think further about the implications of the research presented.

We also include a feature to help students see the relevance of sociology in their everyday lives. The box feature **See for Yourself** allows students to apply a sociological concept to observations from their own lives, thus helping them develop their critical abilities and understand the importance of the sociological perspective.

Critical thinking is a term widely used but often vaguely defined. We use it to describe the process by which students learn to apply sociological concepts to observable events in society. Throughout the book, we ask students to use sociological concepts to analyze and interpret the world they inhabit. This is reflected in the **Thinking Sociologically** feature that is also present in most chapters.

Because contemporary students are so strongly influenced by the media, we also encourage their critical thinking through the box feature called **A Sociological Eye on the Media.** These boxes examine sociological research that challenges some of the ideas and images portrayed in the media. This not only improves students' critical thinking skills but also shows them how research can debunk these ideas and images.

A Focus on Diversity

When we first wrote this book, we did so because we wanted to integrate the then new scholarship on race, gender, and class into the core of the sociological field. We continue to see race, class, and gender—or, more broadly, the study of inequality—as one of the core insights of sociological research and theory. With that in mind, diversity, and the inequality that sometimes results, is a central theme throughout this book. A boxed theme, **Understanding Diversity**, highlights this feature, but you will find that analysis of inequality, especially by race, gender, and class, is woven throughout the book.

Social Change

The sociological perspective helps students see society as characterized both by constant change and social stability. Throughout this book, we analyze how society changes and how events, both dramatic and subtle, influence change. We have added new material throughout the text that shows students how sociological research can help them understand that social changes are influencing their lives, even if students think of these changes as individual problems.

Global Perspective

One of the main things we hope students learn in an introductory course is how broad-scale conditions influence their everyday lives. Understanding this idea is a cornerstone of the sociological perspective. We use a global perspective to examine how global changes are affecting all parts of life within the United States, as well as other parts of the world. This means more than including cross-cultural examples. It means, for example, examining phenomena such as migration and immigration or helping students understand that their own consumption habits are profoundly shaped by global interconnections. The availability of jobs, too, is another way students can learn about the impact of an international division of labor on work within the United States. Our global perspective is found in the research and examples cited throughout the book, as well as in various chapters that directly focus on the influence of globalization on particular topics, such as work, culture, and crime. The map feature **Viewing Society in Global Perspective** also brings a global perspective to the subject matter.

New to the Ninth Edition

We have made various changes to the ninth edition to reflect new developments in sociological research and current social issues. These revisions should make the ninth edition easier for instructors to teach and even more accessible and interesting for students.

Sociology: The Essentials is organized into five major parts: "Introducing the Sociological Imagination" (Chapter 1); "Studying Society and Social Structure" (Chapters 2 through 7); "Social Inequalities" (Chapters 8 through 12); "Social Institutions" (Chapters 13 through 15); and "Social Change" (Chapter 16).

Part I, "Introducing the Sociological Imagination," introduces students to the unique perspective of sociology, differentiating it from other ways of studying society, particularly the individualistic framework students tend to assume. Within this section, **Chapter 1, "The Sociological Perspective,"** introduces students to the sociological perspective. The theme of debunking is introduced, as is the sociological imagination, as developed by C. Wright Mills. This chapter briefly reviews the development of sociology as a discipline, with a focus on the classical frameworks of sociological theory, as well as contemporary theories, including an expanded discussion of feminist theory. There is a stronger discussion of how sociology differs from psychology. The ninth edition adds examples from current events to capture student interest, including new research on growing inequality, the high rate of suicide among veterans, the influence of social media, and new research on how friendship patterns influence the likelihood of pregnancy.

In **Part II, "Studying Society and Social Structure,"** students learn some of the core concepts of sociology. It begins with the study of culture in **Chapter 2, "Culture"** that includes much discussion of social media as a force shaping contemporary culture. This includes research on social media usage both by young and older people. There is new material on the vast growth of digital viewing, but also new work on body images and some of the popular titles that influence young people. Some of the material on ethnocentrism, cultural relativism, and culture shock was reorganized to integrate it better with other chapter material. **Chapter 3, "Doing Sociological Research,"** contains a discussion of the research process and the tools of sociological research—the survey, participant observation, controlled experiments, content analysis, historical research, and evaluation research. The chapter was somewhat reorganized to give better attention to the different types and tools of sociological research. As in the previous edition, we place the chapter on research methods after the chapter on culture as a way of capturing student interest early. **Chapter 4, "Socialization and the Life Course,"** contains material on socialization theory and research, including agents of socialization such as the media, family, and peers. There is new material on military identities, especially in transition to civilian life. More material on child-rearing practices is included, as well as more discussion of the socialization of college students.

Chapter 5, "Social Structure and Social Interaction," emphasizes how changes in the macrostructure of society influence the micro level of social interaction. We do this by focusing on technological changes that are now part of students' everyday lives and making the connection between changes at the societal level in the everyday realities of people's lives. New material on social media usage is included, including how people create identities online and use social media websites to interact with others. The discussion of social interaction includes contemporary examples of romantic relationships, police interviews, and group interactions, such as Comic-Con.

In **Chapter 6, "Groups and Organizations,"** we study social groups and bureaucratic organizations, using sociology to understand the complex processes of group influence, organizational dynamics, and the bureaucratization of society. The chapter includes a discussion of organizational culture, McDonaldization, and the significance of social networks.

Chapter 7, "Deviance and Crime," includes the study of sociological theories and research on deviance and crime. The core material is illustrated with contemporary events, such as police shootings of young, Black men, as well as school rampages. There is new material on gender-based violence, identity theft, human trafficking, and terrorism. The chapter also maintains a focus on race, class, and gender inequality in the criminal justice system, including mass incarceration of Black Americans and Hispanics.

In **Part III, "Social Inequalities,"** each chapter explores a particular dimension of stratification in society. Beginning with the significance of class, **Chapter 8, "Social Class and Social Stratification,"** provides an overview of basic concepts central to the study of class and social stratification. The chapter has a substantial emphasis on growing inequality. New research on extreme poverty and on the connection between poverty and immigration is included. There is updated data throughout and new data on the likelihood of social mobility in the United States compared to other nations.

Chapter 9, "Global Stratification," follows with a particular emphasis on understanding the significance of global stratification, the inequality that has developed among, as well as within, various nations. There is new material on world poverty and the Ebola outbreak, as well as new examples to show students how the clothing they wear is linked to global stratification. Data and examples are updated throughout. **Chapter 10, "Race and Ethnicity,"** is a comprehensive review of the significance of race and ethnicity in society. We have added new material on colorblind racism and the significance of implicit bias, as well as updating examples in this important and growing field of sociological research.

Chapter 11, "Gender," focuses on gender as a central concept in sociology closely linked to systems of stratification in society. This edition was reorganized to better present material on nature–nurture and biological sex differences. There is a more thorough and new discussion of research on transgender people, as well as new work on Black and Latino men's gender identities. More material is included on Title IX and the national concern with sexual assault on college campuses. There is new material on immigrant women, as well. **Chapter 12, "Sexuality,"** treats sexuality as a social construction and a dimension of social stratification and inequality. We have emphasized the influence of feminist theory on the study of sexuality. The chapter also includes new research on pornography and violence against women, as well as the link between rape myths and the sexual double standard. There is new data throughout on topics such as abortion rates, teen pregnancy, and contraception usage.

Part IV, "Social Institutions," includes three chapters, each focusing on basic institutions within society. **Chapter 13, "Families and Religion,"** maintains its inclusion of important topics in the study of families, such as interracial dating, same-sex marriage, fatherhood, gender roles within families, and family violence. We have added new material on women's employment and divorce rates, gender and housework sharing, as well as the impact of economic stress on families. **Chapter 14, "Education and Health Care,"** has been substantially reorganized to emphasize inequality. There is updated information on school segregation, including the impact of choice and charter schools on segregation. In the section on health, details about the Affordable Care Act have been included, including the increased usage. Data on both education and health care is updated throughout. **Chapter 15, "Economy and Politics,"** analyzes the state, power, authority, and bureaucratic government. It also contains a detailed discussion of theories of power in addition to coverage of the economy seen globally and characteristics of the labor force. The new edition includes more information on Native American unemployment, as well as new research on LGBT experiences in the workplace. The section on politics was substantially revised to show the influence of super-PACs and the Citizens United court case on political elections, as well as more emphasis on the influence of power elites in politics.

Part V, "Social Change," includes **Chapter 16, "Environment, Population, and Social Change."** This chapter has been substantially revised for this edition so that a sociological analysis of environmental issues frames the chapters. The chapter focuses on sustainability and climate change. There is an updated discussion of population growth as well as recent examples from disasters such as Hurricane Sandy. The social movements section includes an illustration from the "Black Lives Matter" movement that followed the police shootings in Ferguson, Missouri, and other places.

Features and Pedagogical Aids

The special features of this book flow from its major themes: diversity, current theory and research, debunking and critical thinking, social change, and a global perspective. The features are also designed to help students develop critical thinking skills so that they can apply abstract concepts to observed experiences in their everyday life and learn how to interpret different theoretical paradigms and approaches to sociological research questions.

Critical Thinking Features

The feature **Thinking Sociologically** takes concepts from each chapter and asks students to think about these concepts in relationship to something they can easily observe in an exercise or class discussion. The feature **Debunking Society's Myths** takes certain common assumptions and shows students how the sociological perspective would inform such assumptions and beliefs.

See for Yourself

The feature **See for Yourself** provides students with the chance to apply sociological concepts and ideas to their own observations. This feature can also be used as the basis for writing exercises, helping students improve both their analytic skills and their writing skills.

An Extensive and Content-Rich Map Feature

We use the map feature that appears throughout the book to help students visualize some of the ideas presented, as well as to learn more about regional and international diversity. One map theme is **Mapping America's Diversity** and the other is **Viewing Society in Global Perspective.** These maps have multiple uses for instructional value, beyond instructing students about world and national geography. The maps have been designed primarily to show the differentiation by county, state, and/or country on key social facts.

High-Interest Theme Boxes

We use high-interest themes for the box features that embellish our focus on diversity and sociological research throughout the text. **Understanding Diversity** boxes further explore the approach to diversity taken throughout the book. In most cases, these box features provide personal narratives or other information designed to teach students about the experiences of different groups in society.

Because many are written as first-person narratives, they can invoke students' empathy toward groups other than those to which they belong—something we think is critical to teaching about diversity. We hope to show students the connections between race, class, and other social groups that they otherwise find difficult to grasp.

The box feature **Doing Sociological Research** is intended to show students the diversity of research questions that form the basis of sociological knowledge and, equally important, how the questions researchers ask influence the methods used to investigate the questions.

We see this as an important part of sociological research—that how one investigates a question is determined as much by the nature of the question as by allegiance to a particular research method. Some questions require a more qualitative approach; others, a more quantitative approach. In developing these box features, we ask: What is the central question sociologists are asking? How did they explore this question using sociological research methods? What did they find? What are the implications of this research? We deliberately selected questions that show the full and diverse range of sociological theories and research methods, as well as the diversity of sociologists. Each box feature ends with **Questions to Consider** to encourage students to think further about the implications and applications of the research.

What Would a Sociologist Say? boxes take a topic of interest and examine how a sociologist would likely interpret this subject. The topics are selected to capture student interest, such as a discussion of veteran suicides, hip-hop culture, and sex and popular culture. We think this box brings a sociological perspective to commonplace events.

The feature **A Sociological Eye on the Media,** found in several chapters, examines some aspect of how the media influence public understanding of some of the subjects in this book. We think this is important because sociological research often debunks taken-for-granted points of view presented in the media, and we want students to be able to look at the media with a more critical eye. Because of the enormous influence of the media, we think this is increasingly important in educating students about sociology. In addition to the features just described, we offer an entire set of learning aids within each chapter that promotes student mastery of the sociological concepts.

In-Text Learning Aids

Learning Objectives. We have added learning objectives to this edition, which appear near the beginning of every chapter. Matched to the major chapter headings, these objectives identify what we expect students to learn from the chapter. Faculty may choose to use these learning objectives to assess how well students comprehend the material. We tried to develop the learning objectives based on different levels of understanding and analysis, recognizing the various paths that students take in how they learn material.

Chapter Outlines. A concise chapter outline at the beginning of each chapter provides students with an overview of the major topics to be covered.

Key Terms. Key terms and major concepts appear in bold when first introduced in the chapter. A list of the key terms is found at the end of the chapter, which makes study more effective. Definitions for the key terms are found in the glossary.

Theory Tables. Each chapter includes a table that summarizes different theoretical perspectives by comparing and contrasting how these theories illuminate different aspects of different subjects.

Chapter Summary in Question-and-Answer Format. Questions and answers highlight the major points in each chapter and provide a quick review of major concepts and themes covered in the chapter.

A **Glossary** and complete **References** for the whole text are found at the back of the book.

MindTap Sociology: The Personal Learning Experience

MindTap Sociology for *Sociology: The Essentials*, ninth edition, powered by Knewton from Cengage Learning represents a new approach to a highly personalized, online learning platform. A fully online learning solution, MindTap Sociology combines all of a student's learning tools—readings, multimedia, activities, and assessments—into a singular learning path that guides students through an introduction to sociology course. Instructors personalize the experience by customizing the presentation of these learning tools for their students, even seamlessly introducing their own content into the learning path via "apps" that integrate into the MindTap platform. Learn more at **www.cengage.com/mindtap**.

MindTap Sociology for *Sociology: The Essentials*, ninth edition, powered by Knewton, is easy to use and saves instructors' time by allowing them to:

- Seamlessly deliver appropriate content and technology assets from a number of providers to students, as they need them.
- Break course content down into movable objects to promote personalization, encourage interactivity, and ensure student engagement.
- Customize the course—from tools to text—and make adjustments "on the fly," making it possible to intertwine breaking news into their lessons and incorporate today's teachable moments.
- Bring interactivity into learning through the integration of multimedia assets (apps from Cengage Learning and other providers) and numerous in-context exercises and supplements; student engagement will increase, leading to better student outcomes.
- Track students' use, activities, and comprehension in real time, which provides opportunities for early intervention to influence progress and outcomes. Grades are visible and archived so students and instructors always have access to current standings in the class.
- Assess knowledge throughout each section: after readings, in activities, homework, and quizzes.
- Automatically grade all homework and quizzes.
- MindTap Sociology for *Sociology: The Essentials*, ninth edition, features Aplia assignments, which help students learn to use their sociological imagination through compelling content and thought-provoking questions. Students complete interactive activities that encourage them to think critically in order to practice and apply course concepts. These valuable critical thinking skills help students become thoughtful and engaged members of society.

Instructor Resources

Sociology: The Essentials, ninth edition, is accompanied by a wide array of supplements prepared to create the best learning environment inside as well as outside the classroom for both instructors and students. All the continuing supplements for *Sociology: The Essentials*, ninth edition, have been thoroughly revised and updated. We invite you to take full advantage of the teaching and learning tools available to you.

Instructor's Resource Manual. This supplement offers instructors brief chapter outlines, student learning objectives, American Sociological Association recommendations, key terms and people, detailed chapter lecture outlines, lecture/discussion suggestions, student activities, chapter worksheets, video suggestions, video activities, and Internet exercises. The ninth edition also includes a syllabus to help instructors easily organize learning tools and create lesson plans.

Cengage Learning Testing Powered by Cognero. This flexible, online system allows teachers to author, edit, and manage test bank content from multiple Cengage Learning solutions, create multiple test versions in an instant, and deliver tests from your LMs, your classroom, or wherever you want.

PowerPoint Slides. Preassembled Microsoft® PowerPoint® lecture slides with graphics from the text make it easy for you to assemble, edit, publish, and present custom lectures for your course.

Acknowledgments

We relied on the comments of many reviewers to improve the book, and we thank them for the time they gave in developing very thoughtful commentaries on the different chapters. Thanks to Mark Beisler, Rowan-Cabarrus Community College; Andrew Butz, Portland Community College; Nancy Dimonte, SUNY-Farmingdale; Maureen Erickson, Cayuga Community College; Kathryn J. Fox, University of Vermont; Jamie Gusrang, Community College of Philadelphia; Caroll Hodgson, Rowan-Cabarrus Community College; Traci Sullivan, Lakeland Community College; Nicole Vadino, Community College of Philadelphia; Stan Weeber, McNeese State University; Gailynn White, Citrus College; and Porscha Orndof of Asheville Buncombe Technical College.

We appreciate the efforts of many people who make this project possible. We are fortunate to be working with a publishing team with great enthusiasm for this project. We thank all of the people at Cengage Learning who have worked with us on this and other projects, especially John Chell who shepherded this edition through important revisions. His attention to detail is especially appreciated. We were also fortunate to have the guidance of Marta Lee-Perriard during a transition to a new editor. Cheri Palmer, once again, oversaw the many aspects of production that are critical to the book's success. We especially thank Jill Traut of MPS Limited and Heather McElwain for their work in the production process. Thanks to Nazveena Begum Syed for photographic research. Finally, our special thanks also go to our spouses Richard Morris Rosenfeld, Patricia Epps Taylor, and Jim Rau for their ongoing love and willingness to put up with us when we are frazzled by the project details!

About the Authors

Courtesy of Margaret Andersen

Margaret L. Andersen is the Edward F. and Elizabeth Goodman Rosenberg Professor of sociology at the University of Delaware where she also holds joint appointments in women's studies and Black American studies. She is the author of *Thinking about Women: Sociological Perspectives on Sex and Gender; Race, Class and Gender* (with Patricia Hill Collins); *Race and Ethnicity in Society: The Changing Landscape* (with Elizabeth Higginbotham); *On Land and On Sea: A Century of Women in the Rosenfeld Collection;* and *Living Art: The Life of Paul R. Jones, African American Art Collector.* She is a recipient of the American Sociological Association's Jessie Bernard Award and the Merit Award of the Eastern Sociological Society. She is the former vice president of the American Sociological Association, former president of the Eastern Sociological Society, and a recipient of the University of Delaware's Excellence in Teaching Award and the College of Arts and Sciences Award for Outstanding Teaching.

Courtesy of Howard Taylor

Howard F. Taylor was raised in Cleveland, Ohio. He graduated Phi Beta Kappa from Hiram College and has a Ph.D. in sociology from Yale University. He has taught at the Illinois Institute of Technology, Syracuse University, and Princeton University, where he is presently professor of sociology and former director of the Center for African American studies. He has published over fifty articles in sociology, education, social psychology, and race relations. His books include *The IQ Game* (Rutgers University Press), a critique of hereditarian accounts of intelligence; *Balance in Small Groups* (Van Nostrand Reinhold), translated into Japanese; and the forthcoming *The SAT Triple Whammy: Race, Gender, and Social Class Bias.* He has appeared widely before college, radio, and TV audiences, including ABC's *Nightline.* He is past president of the Eastern Sociological Society, and a member of the American Sociological Association and the Sociological Research Association, an honorary society for distinguished research. He is a winner of the DuBois-Johnson-Frazier Award, given by the American Sociological Association for distinguished research in race and ethnic relations, and the President's Award for Distinguished Teaching at Princeton University. He lives in Pennington, New Jersey, with his wife, a corporate lawyer.

Courtesy of Kim A. Logio

Kim A. Logio is currently associate professor and chair of sociology at Saint Joseph's University in Philadelphia, Pennsylvania. She has been interviewed for local television and National Public Radio for her work on body image and race, class, and gender differences in nutrition and weight control behavior. She teaches research methods and data analysis courses, guiding students through the completion of their undergraduate thesis projects. She has been awarded a teaching award at Saint Joseph's University. She lives in Delaware County, Pennsylvania, with her husband and three children.

Charles Cook/Getty Images

The Sociological Perspective

in this chapter, you will learn to:

- Illustrate what is meant by saying that human behavior is shaped by social structure
- Question individualistic explanations of human behavior
- Describe the significance of studying diversity in contemporary society
- Explain the origins of sociological thought
- Compare and contrast the major frameworks of sociological theory

Imagine you had been switched with another infant at birth. How different would your life be? What if your accidental family was very poor ... or very rich? How might this have affected the schools you attended, the health care you received, and the possibilities for your future career? If you had been raised in a different religion, would this have affected your beliefs, values, and attitudes? Taking a greater leap, what if you had been born another sex or a different race? What would you be like now?

We are talking about changing the basic facts of your life—your family, social class, education, religion, sex, and race. Each has major consequences for who you are and how you will fare in life. These factors play a major part in writing your life script. Your social location (meaning a person's place in society) establishes the limits and possibilities of a life.

Consider this:

- The people least likely to attend college are those most likely to benefit from it (Brand and Xie 2010).
- In the past, marriages in which wives had more education than their husbands were more likely than other marriages to end in divorce. This is no longer true (Schwartz and Han 2014).
- Fourteen percent of households in the United States (18 million households) are considered "food insecure," meaning that they do not have the money for an adequate amount of food (Piontak and Schulman 2014).
- Gender and racial diversity in for-profit business organizations is associated with increased sales revenues, more customers, and higher profits (Herring 2009).

These conclusions, drawn from current sociological research, describe some consequences of particular social locations in society. Although we may take our place in society for granted, our social location has a profound effect on our chances in life. The power of sociology is that it teaches us to see how society influences our lives and the lives of others, and it helps us explain the consequences of different social arrangements.

Sociology also has the power to help us understand the influence of major changes on people. Currently, rapidly developing technologies, increasing globalization, a more diverse population in the United States, and changes in women's roles are affecting everyone, although in different ways. How are these changes affecting your life? Perhaps you rely on social media to keep in touch with friends. Maybe your community is witnessing an increase in immigrants from other places. Perhaps you see women and men trying hard to manage the demands of both work and family life. All of these are issues that guide sociological questions. Sociology explains some of the causes and consequences of these changes.

"Actually, Lou, I think it was more than just my being in the right place at the right time. I think it was my being the right race, the right religion, the right sex, the right socioeconomic group, having the right accent, the right clothes, going to the right schools . . ."

Warren Miller/The New Yorker Collection/Cartoonbank.com

Although society is always changing, it is also remarkably stable. People generally follow established patterns of human behavior, and you can often anticipate how people will behave in certain situations. You can even anticipate how different social conditions will affect different groups of people in society. This is what sociologists find so interesting: Society is marked by both change and stability. Societies continually evolve, creating the need for people to adapt to change while still following generally established patterns of behavior.

What Is Sociology?

Sociology is the study of human behavior in society. Sociologists are interested in the study of people and have learned a fundamental lesson: Human behavior, even when seemingly "natural" or taken for granted, is shaped by social structures—structures that have their origins beyond the immediately visible behaviors of everyday life. In other words, *all human behavior occurs in a social context.* That context—the institutions and culture that surround us—shapes what people do and think. In this book, we will examine the dimensions of society and analyze the elements of social context that influence human behavior.

Sociology is a scientific way of thinking about society and its influence on human groups. Observation, reasoning, and logical analysis are the tools of sociologists. Sociology is inspired by the fascination people have for observing people, but it goes far beyond casual observations. It builds from objective analyses that others can validate as reliable.

David Grossman/Alamy

Sociology is the study of human behavior. What social behaviors do you see here?

Every day, the media in their various forms (television, film, video, digital, and print) bombard us with social commentary. Media commentators provide endless

Key Sociological Concepts

As you build your sociological perspective, you must learn certain key concepts to begin understanding how sociologists view human behavior. Social structure, social institutions, social change, and social interaction are not the only sociological concepts, but they are fundamental to grasping the sociological perspective.

Social Interaction. Sociologists see **social interaction** as behavior between two or more people that is given meaning. Through social interaction, people react and change, depending on the actions and reactions of others. Because society changes as new forms of human behavior emerge, change is always in the works.

Social Structure. We define **social structure** as the organized pattern of social relationships and social institutions that together constitute society. Social structure is not a "thing," but refers to the fact that social forces not always visible to the human eye guide and shape human behavior. Acknowledging that social structure exists does not mean that humans have no choice in how they behave, only that those choices are largely conditioned by one's location in society.

Social Institutions. In this book, you will also learn about the significance of **social institutions**, defined as established and organized systems of social behavior with a particular and recognized purpose. Family, religion, marriage, government, and the economy are examples of major social institutions. Social institutions confront individuals at birth and transcend individual experience, but they still influence individual behavior.

Social Change. As you can tell, sociologists are also interested in the process of **social change**, the alteration of society over time. As much as sociologists see society as producing certain outcomes, they do not see society as fixed, nor do they see humans as passive recipients of social expectations. Sociologists view society as stable but constantly changing.

As you read this book, you will see that these key concepts—social interaction, social structure, social institutions, and social change—are central to the sociological imagination.

opinion about the various and sometimes bizarre forms of behavior in society. Sociology is different. Sociologists often appear in the media, and they study some of the same subjects that the media examine, such as crime, violence, or income inequality, but sociologists use specific research techniques and well-tested theories to explain social issues. Indeed, sociology can provide the tools for testing whether the things we hear about society are actually true. Much of what we hear in the media and elsewhere about society, although delivered with perfect earnestness, is misstated and sometimes completely wrong, as you will see in some of the "Debunking Society's Myths" examples featured throughout this book.

THINKING Sociologically

Q: What do the following people have in common?

First Lady Michelle Obama
Robin Williams (actor, comedian)
Ronald Reagan (former president)
Reverend Martin Luther King, Jr.
Regis Philbin (TV personality)
Reverend Jesse Jackson
Saul Bellow (novelist; Nobel Prize recipient)
Joe Theismann (former football player and TV personality)
Congresswoman Maxine Waters (from California)
Senator Barbara Mikulski (from Maryland)

A: They were all sociology majors!

Source: Compiled by Peter Dreier, Occidental College.

The subject matter of sociology is everywhere. This is why people sometimes wrongly believe that sociology just explains the obvious. Sociologists bring a unique perspective to understanding social behavior and social change. Even though sociologists often do research on familiar topics, such as youth cultures or racial inequality, they do so using particular research tools and specific frames of analysis (known as sociological theory). Psychologists, anthropologists, political scientists, economists, social workers, and others also study social behavior, although each has a different perspective or "angle" on people in society.

Students often wonder what makes sociology different from psychology. After all, both study people and both identify some of the social forces that shape our lives. There is, however, a difference. Psychologists study groups. Research in psychology can inform some sociological analyses, but the focus in psychology is more on individuals—what makes individuals do what they do and how individual minds and emotions work. Increasingly, psychology is also influenced by the studies of the brain that are emerging from the techniques of neuroscience. Sociology, on the other hand, though it can learn from psychological research, is more interested in the broader social forces that shape society as a whole and the people within it. (See the box "What Would a Sociologist Say?" for an example.) Together, these various social sciences provide compelling, though different, views of human behavior.

what would a sociologist say?

Getting Pregnant: A Very Social Act

When does a woman get pregnant? Simple, you might think—it's biological. Of course, you can think of pregnancy from a biological perspective, explaining the process of fertilization. Or, you might think of pregnancy from a psychological perspective, analyzing the desire to have a child as deeply rooted in emotion and individual decision-making processes. You might even think about pregnancy from a cross-cultural or historical perspective, analyzing childbirth in different cultural contexts or analyzing historical changes in how pregnancy is managed by the medical profession. But, what would a sociologist say about getting pregnant?

From a sociological perspective, pregnancy is deeply social behavior. There would be many sociological angles for studying pregnancy. An example from recent research reveals the power of sociological thinking. Sociological researchers have found that the likelihood of becoming pregnant increases significantly in the two years following a friend's having had a child. As the researchers conclude, even such personal decisions as the decision to have a child result from the *web of social relationships in which people are embedded* (Balbo and Barban 2014). Pregnancy may seem like a very personal decision, but it is fertile ground for sociological study. What other social forces do you think might influence the likelihood of getting pregnant?

The Sociological Perspective

Think back to the chapter opening where we asked you to imagine yourself growing up under different circumstances. Our goal in that passage was to make you feel the stirring of the *sociological perspective*—the ability to see societal patterns that influence individual and group life. The beginnings of the sociological perspective can be as simple as the pleasures of watching people or wondering how society influences people's lives. Indeed, many students begin their study of sociology because they are "interested in people." Sociologists convert this curiosity into the systematic study of how society influences different people's experiences within it.

C. Wright Mills (1916–1962) was one of the first to write about the sociological perspective in his classic book, *The Sociological Imagination* (1959). He wrote that the task of sociology was to understand the relationship between individuals and the society in which they live. He defined the **sociological imagination** as the ability to see the societal patterns that influence the individual as well as groups of individuals. Sociology should be used, Mills argued, to reveal how the context of society shapes our lives. He thought that to understand the experience of a given person or group of people, one had to have knowledge of the social and historical context in which people lived.

Think, for example, about the time and effort that many people put into their appearance. You might ordinarily think of this as merely personal grooming or an individual attempt to "look good," but this behavior has significant social origins. When you stand in front of a mirror, you are probably not thinking about how society is present in your reflection. As you look in the mirror, though, you are seeing how others see you and are very likely adjusting your appearance with that in mind, even if not consciously.

This seemingly individual behavior is actually a very social act. If you are trying to achieve a particular look, you are likely doing so because of social forces that establish particular ideals. These ideals are produced by industries that profit enormously from the products and services that people buy, even when people do so believing they are making an individual choice. Some industries suggest that you should be thinner or curvier, your pants should be baggy or straight, your breasts should be minimized or maximized—either way, you need more products. Maybe you should have a complete makeover! Many people go to great lengths to try to achieve a constantly changing beauty ideal, one that is probably not even attainable (such as flawless skin, hair always in place, perfectly proportioned body parts). Sometimes trying to meet these ideals can even be hazardous to your physical and mental health.

The point is that the alleged standards of beauty are produced by social forces that extend far beyond an individual's concern with personal appearance. Beauty ideals, like other socially established beliefs and practices, are produced in particular social and historical contexts. People may come up with all kinds of personal strategies for achieving these ideals: They may buy more products, try to lose more weight, get a Botox treatment, or even become extremely depressed and anxious if they think their efforts are failing. These personal behaviors may seem to be only individual issues, but they have basic social causes. The sociological imagination permits us to see that something as seemingly personal as how you look arises from a social context, not just individual behavior.

Sociologists are certainly concerned about individuals, but they are attuned to the social and historical

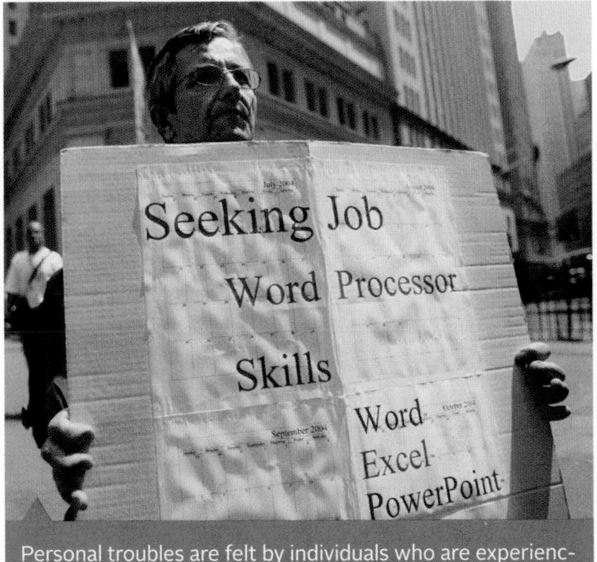

Spencer Platt/Getty Images News/Getty Images

Personal troubles are felt by individuals who are experiencing problems; social issues arise when large numbers of people experience problems that are rooted in the social structure of society.

context that shapes individual and group experiences. The sociological imagination distinguishes between *troubles* and *issues*. **Troubles** are privately felt problems that spring from events or feelings in a person's life. **Issues** affect large numbers of people and have their origins in the institutional arrangements and history of a society (Mills 1959). This distinction is the crux of the difference between individual experience and **social structure**, defined as the organized pattern of social relationships and social institutions that together constitute society. Issues shape the context within which troubles arise. Sociologists employ the sociological perspective to understand how issues are shaped by social structures.

Mills used the example of unemployment to explain the meaning of troubles versus issues—an example that still has resonance given people's concerns about finding work. When an individual person becomes unemployed—or cannot find work—he or she has a personal trouble, such as the worry that many college graduates have experienced in trying to find work following graduation. The personal trouble unemployment brings may include financial problems as well as the person feeling a loss of identity, becoming depressed, or having to uproot a family and move. College students may have to move back home with parents after graduation.

The problem of unemployment, however, is deeper than the experience of any one person. Unemployment is rooted in the structure of society; this is what interests sociologists. What societal forces cause unemployment? Who is most likely to become unemployed at different times? How does unemployment affect an entire community (for instance, when a large plant shuts down) or an entire nation (such as when recessions hit)? Sociologists know that unemployment causes personal troubles, but understanding unemployment is more than understanding one person's experience. It requires understanding the social structural conditions that influence people's lives.

THINKING Sociologically

Troubles and Issues

Personal troubles are everywhere around us: alcohol abuse or worries about money or even being upset about how you look. At an individual level, these things can be deeply troubling, and people sometimes need personal help to deal with them. But most personal troubles, as C. Wright Mills would say, also have their origins in societal arrangements. Take the example of alcohol abuse.

What are some of the things about society—not just individuals—that might influence this personal trouble? Is there a culture of drinking on your campus that generates peer pressure to drink? Do people drink more when they are unemployed? Is drinking more common among particular groups or at different times in history? Who profits from people's drinking? Thinking about these questions can help you understand the distinction that Mills makes between *personal troubles* and *social issues*.

The specific task of sociology, according to Mills, is to comprehend the whole of human society—its personal and public dimensions, historical and contemporary—and its influence on the lives of human beings. Mills had an important point: People often feel that things are beyond their control, meaning that people are shaped by social forces larger than their individual lives. Social forces influence our lives in profound ways, even though we may not always know how. Consider this: Sociologists have noted a current trend, popularly labeled "the boomerang generation" or "accordion families" (Newman 2012). These terms refer to the pattern whereby many young people, after having left their family home to attend college, are returning home after graduation. Although this may seem like an individual decision to save money on housing or live "free" while paying off student loans, when a whole generation experiences this living arrangement, there are social forces at work that extend beyond individual decisions. In other words, people feel the impact of social forces in their personal lives, even though they may not always know the full dimensions of those forces. This is where sociology comes into play—revealing the *social structures* that shape the different dimensions of our day-to-day lives. Social structure is a lot like air: You cannot directly "see" it, but it is essential to living our lives.

Sociologists see social structures through careful and systematic observation. This makes sociology an **empirical** discipline. Empirical refers to careful observation, not just conjecture or opinion. In this way, sociology

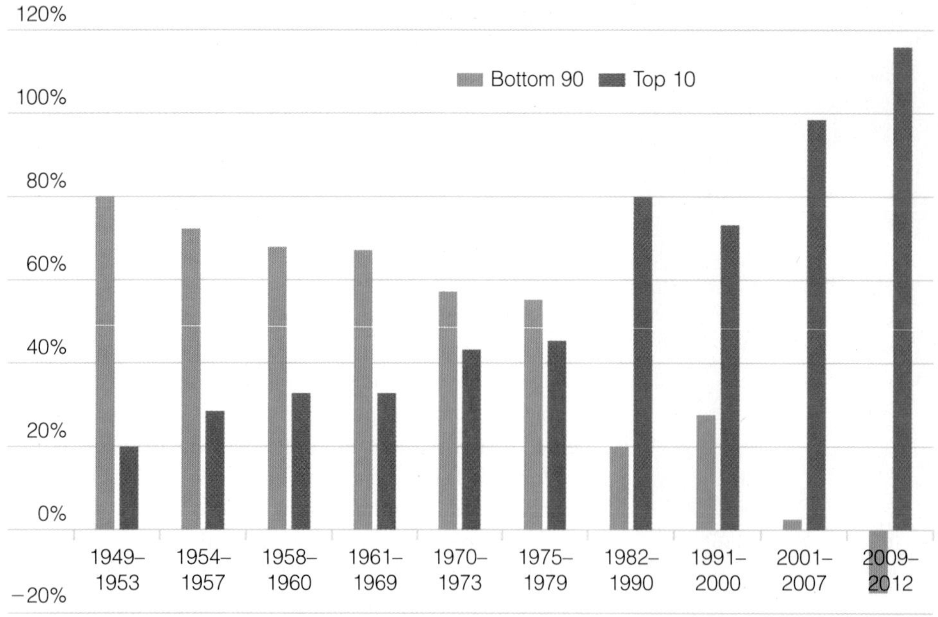

▲ **Figure 1.1**
Distribution of Average Income Growth during Economic Expansions
This figure shows how the bottom 90 percent and top 10 percent of the population experience change in their income during periods of economic expansion. What trends do you see here and how might they be affecting people's personal troubles and social issues?

Source: Tcherneva, Pavlina R. 2014. *Growth for Whom?* Levy Economics Institute of Bard College. Retrieved April 1, 2015. **www.levyinstitute.org/pubs/op_47.pdf**

is very different from common sense. For empirical observations to be useful to others, they must be gathered and recorded rigorously. Sociologists are also obliged to reexamine their assumptions and conclusions constantly. Although the specific methods that sociologists use to examine different problems vary, as we will see in Chapter 3 on sociological research methods, the empirical basis of sociology is what distinguishes it from mere opinion or other forms of social commentary.

Discovering Unsettling Facts

In studying sociology, it is crucial to examine the most controversial topics and to do so with an open mind, even when you see the most disquieting facts. The facts we learn through sociological research can be "inconvenient" because the data can challenge familiar ways of thinking. Consider the following:

- Many think of the Internet as promoting more impersonal social interaction. Sociological research, however, finds that people with Internet access are actually *more likely* to have romantic partners because of the ease of meeting people online (Rosenfeld and Thomas 2012).
- Despite the widespread idea promoted in the media that well-educated women are opting out of professional careers to become "stay-at-home moms," the proportion of college-educated White women who stay home with children has actually declined; those who opt out of work do so more typically because of frustration with how they are treated at work (Stone 2007).
- The number of women prisoners has increased at almost twice the rate of increase for men; two-thirds of women and half of men in prison are parents (Glaze and Maruschak 2008; Sabol and Couture 2008).

These facts provide unsettling evidence of persistent problems in the United States, *problems that are embedded in society, not just in individual behavior*. Sociologists try to reveal the social factors that shape society and determine the chances of success for different groups. Some never get the chance to go to college; others are unlikely to ever go to jail. These divisions persist because of people's placement within society.

▲ Figure 1.1 provides graphic evidence of how changes in society might determine the opportunities for success of different groups. This image shows what percentage of income growth went to the top 10 percent and the bottom 90 percent of the U.S. population since World War II. This was a period of great economic expansion in the United States. How was income growth distributed over this time period and who benefitted? As you can see in this image, since 2000, the bottom 90 percent of the population has actually experienced a rather dramatic decline in income growth. How does this affect opportunity for people like you? How might it help explain the growing concern with class inequality? We will discuss these changes more in Chapter 8, but for now, perhaps you can begin to understand how sociologists study the broad social forces that shape people's life chances. Something as simple as being born in a particular generation can shape the course of your lifetime.

DEBUNKING Society's Myths

Myth: Anyone who works hard enough in the United States can get ahead.

Sociological Research: There are periods in society when some groups are able to move ahead. As examples, the Black middle class expanded following changes in civil rights laws in the 1960s; the White middle class also grew in the post-World War II period as the result of such things as GI benefits for returning vets and government support for home ownership. However, although there are exceptions, most people do not change their social class position from that in which they were born. As Figure 1.1 shows you, at times groups may even fall further behind as the result of conditions in society (Piketty 2014; Noah 2013).

Sociologists study not just the disquieting side of society. Sociologists may also study questions that affect everyday life, such as how young boys and men are affected by changing gender roles (Kimmel 2008), worker–customer dynamics in nail salons (Kang 2010), or the expectations that young women and men have for combining work and family life (Gerson 2010). There are also many intriguing studies of unusual groups, such as cyberspace users (Kendall 2002), strip clubs and dancers (Price-Glynn 2010; Barton 2006), or competitive eaters (Ferguson 2014). The subject matter of sociology is vast. Some research illuminates odd corners of society; other studies address urgent problems of society that may affect the lives of millions.

Debunking in Sociology

The power of sociological thinking is that it helps us see everyday life in new ways. Sociologists question actions and ideas that are usually taken for granted. Peter Berger (1963) calls this process "debunking." **Debunking** refers to looking behind the facades of everyday life—what Berger called the "unmasking tendency" of sociology (1963: 38). In other words, sociologists look at the behind-the-scenes patterns and processes that shape the behavior they observe in the social world.

Take schooling, for example: We can see how the sociological perspective debunks common assumptions about education. Most people think that education is primarily a way to learn and get ahead. Although this is true, a sociological perspective on education reveals something more. Sociologists have concluded that more than learning takes place in schools; other social processes are at work. Social cliques are formed where some students are "insiders" and others are excluded "outsiders." Young schoolchildren acquire not just formal knowledge but also the expectations of society and people's place within it. Race and class conflicts are often played out in schools (Lewis 2003). Poor children seldom have the same resources in schools as middle-class or elite children, and they are often assumed to be incapable of doing schoolwork and are treated accordingly. The somber reality is that schools may actually stifle the opportunities of some children rather than launch all children toward success (Kozol 2006).

Debunking is sometimes easier to do when looking at a culture or society different from one's own. Consider how behaviors that are unquestioned in one society may seem positively bizarre to an outsider. For a thousand years in China, it was usual for the elite classes to bind the feet of young girls to keep the feet from growing bigger—a practice allegedly derived from a mistress of the emperor. Bound feet were a sign of delicacy and vulnerability. A woman with large feet (defined as more than 4 inches long!) was thought to bring shame to her husband's household. The practice was supported by the belief that men were highly aroused by small feet, even though men never actually saw the naked foot. If they had, they might have been repulsed, because a woman's actual foot was U-shaped and often rotten and covered with dead skin (Blake 1994). Outside the social, cultural, and historical context in which it was practiced,

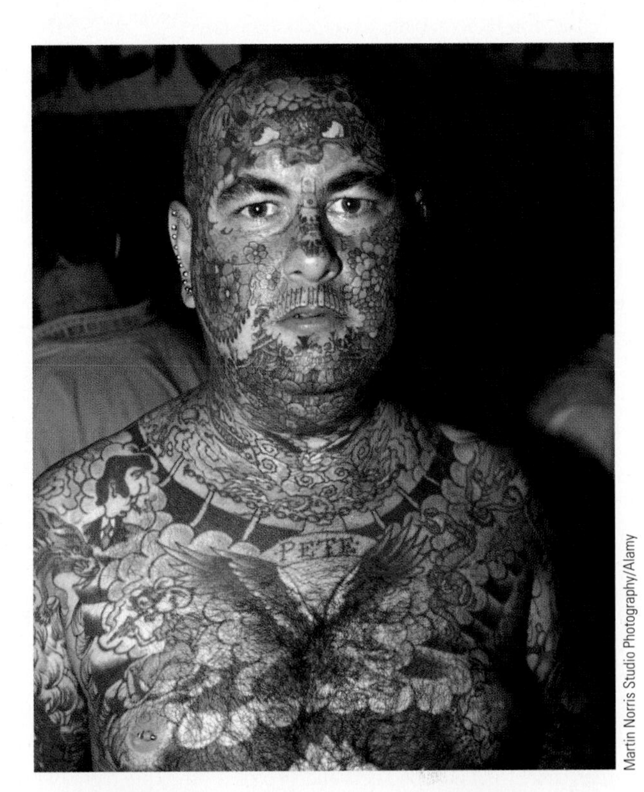

Martin Norris Studio Photography/Alamy

Lindsay Hebberd/Encyclopedia/Corbis

Cultural practices that seem bizarre to outsiders may be taken for granted or defined as appropriate by insiders.

doing sociological research

Debunking the Myths of Black Teenage Motherhood

Research Question: Sociologist Elaine Bell Kaplan knew that there was a stereotypical view of Black teen mothers that they had grown up in fatherless households where their mothers had no moral values and no control over their children. The myth of Black teenage motherhood also depicts teen mothers as unable to control their sexuality, as having children to collect welfare checks, and as having families who condone their behavior. Is this true?

Research Method: Kaplan did extensive research in two communities in the San Francisco Bay area—East Oakland and Richmond—both communities with a large African American population and typical of many inner-city, poor neighborhoods. Once thriving Black communities, East Oakland and Richmond are now characterized by high rates of unemployment, poverty, inadequate schools, crime, drug-related violence, and high numbers of single-parent households. Having grown up herself in Harlem, Kaplan knew that communities like those she studied have not always had these problems, nor have they condoned teen pregnancy. She spent several months in these communities, working as a volunteer in a community teen center that provided educational programs, day care, and counseling to teen parents, and "hanging out" with a core group of teen mothers. She did extensive interviews with thirty-two teen mothers, supplementing them when she could with interviews with their mothers and, sometimes, the fathers of their children.

Research Results: Kaplan found that teen mothers adopt strategies for survival that help them cope with their environment, even though these same strategies do not help them overcome the problems they face. Unlike what the popular stereotype suggests, she did not find that the Black community condones teen pregnancy; quite the contrary, the teens felt embarrassed and stigmatized by being pregnant and experienced tension and conflict with their mothers, who saw their pregnancy as disrupting the hopes they had for their daughters' success. These conclusions run directly counter to the public image that such women do not value success and live in a culture that promotes welfare dependency.

Conclusions and Implications: Instead of simply stereotyping these teens as young and tough, Kaplan sees them as struggling to develop their own gender and sexual identity. Like other teens, they are highly vulnerable, searching for love and aspiring to create a meaningful life. Often locked out of the job market, the young women's struggle to develop an identity is compounded by the disruptive social and economic conditions in which they live.

Kaplan's research is a fine example of how sociologists debunk some of the commonly shared myths that surround contemporary issues. Carefully placing her analysis in the context of the social structural changes that affect these young women's lives, Kaplan provides an excellent example of how sociological research can shed new light on some of our most pressing social problems.

Questions to Consider

1. Suppose that Kaplan had studied middle-class teen mothers. What similarities and differences would you predict in the experiences of middle-class and poor teen mothers? Does race matter? In what ways does your answer debunk myths about teen pregnancy?
2. Make a list of the challenges you would face were you to be a teen parent. Having done so, indicate those that would be considered personal troubles and those that are social issues. How are the two related?

Source: Kaplan, Elaine Bell. 1996. *Not Our Kind of Girl: Unraveling the Myths of Black Teenage Motherhood*. Berkeley, CA: University of California Press.

foot binding seems bizarre, even dangerous. Feminists have pointed out that Chinese women were crippled by this practice, making them unable to move about freely and more dependent on men (Chang 1991).

This is an example of outsiders debunking a practice that was taken for granted by those within the culture. Debunking can also call into question practices in one's own culture that may normally go unexamined. Strange as the practice of Chinese foot binding may seem to you, how might someone from another culture view wearing shoes that make it difficult to walk? Or piercing one's tongue or eyebrow? Many take these practices of contemporary U.S. culture for granted, just as they do Chinese foot binding. Until these cultural processes are debunked, seen as if for the first time, they might seem normal.

DEBUNKING Society's Myths

Myth: Email scams promising to deliver a large sum of cash from some African bank if you contact the email deliverer prey on people who are just stupid or old.
Sociological Research: Studies of these email scams indicate that Americans and Brits are especially susceptible to such scams because they play on widely held cultural stereotypes about Africa (that these are economically unsophisticated nations in which people are unable to manage money). These scams also exploit the American cultural belief that it is possible to "get rich quick"—reflecting a belief in individualism and the belief that anyone who tries hard enough can get ahead (Smith 2009).

Establishing Critical Distance

Debunking requires critical distance—that is, being able to detach from the situation at hand and view things with a critical mind. The role of critical distance in developing a sociological imagination is well explained by the early sociologist **Georg Simmel** (1858–1918). Simmel was especially interested in the role of *strangers* in social groups. Strangers have a position both inside and outside social groups. They are part of a group without necessarily sharing the group's assumptions and points of view. Because of this, the stranger can sometimes see the social structure of a group more readily than can people who are thoroughly imbued with the group's worldview. Simmel suggests that the sociological perspective requires a combination of nearness and distance. One must have enough critical distance to avoid being taken in by the group's definition of the situation, but be near enough to understand the group's experience.

◆ **Table 1.1** U.S. Population Projections, 2010–2050

	2010	2020	2030	2040	2050
White	79.5%	78.0%	76.6%	75.3%	74.0%
Black	12.9%	13.0%	13.1%	13.0%	13.0%
American Indian and Alaskan Native	1.0%	1.1%	1.2%	1.2%	1.2%
Asian	4.6%	5.5%	6.3%	7.1%	7.8%
Native Hawaiian and Other Pacific Islander	0.2%	0.2%	0.2%	0.3%	0.3%
Two or more races	1.8%	2.1%	2.7%	3.2%	3.7%

Note: The U.S. census counts race and Hispanic ethnicity separately. Thus, Hispanics may fall into any of the race categories. Those who identified themselves as Hispanic were 16 percent of the total U.S. population in the 2010 census.

Source: U.S. Census Bureau. 2012. *National Population Projections: Summary Table.* Washington, DC: U.S. Department of Commerce, **www.census.gov**

Sociologists are not typically strangers to the society they study. You can acquire critical distance through a willingness to question the forces that shape social behavior. Often, sociologists become interested in things because of their own experiences. The biographies of sociologists are rich with examples of how their personal lives informed the questions they asked. Among sociologists are former ministers and nuns now studying the sociology of religion, women who have encountered sexism who now study the significance of gender in society, rock-and-roll fans studying music in popular culture, and sons and daughters of immigrants now analyzing race and ethnic relations (see the box "Understanding Diversity: Becoming a Sociologist").

The Significance of Diversity

The analysis of diversity is a central theme of sociology. Differences among groups, especially differences in the treatment of groups, are significant in any society, but they are particularly compelling in a society as diverse as that in the United States.

Defining Diversity

Today, the United States includes people from all nations and races. In 1900, one in eight Americans was not White; today, racial and ethnic minority groups (including African Americans, Hispanics, American Indians, Native Hawaiians, Asian Americans, and people of more than one race) represent 27 percent of Americans, and that proportion is growing (see ◆ Table 1.1 and ■ Map 1.1).

Perhaps the most basic lesson of sociology is that people are shaped by the social context around them. In the United States, with so much cultural diversity, people will share some experiences, but not all. Experiences not held in common can include some of the most important influences on social development, such as

AP Images/Jae C. Hong

In an increasingly diverse society, valuing and understanding diversity is a part of fully understanding society.

understanding **diversity**

Become a Sociologist

Individual biographies often have a great influence on the subjects sociologists choose to study. The authors of this book are no exception. Margaret Andersen, a White woman, now studies the sociology of race and women's studies. Howard Taylor, an African American man, studies race, social psychology, and especially race and intelligence testing. Here, each of them writes about the influence of their early experiences on becoming a sociologist.

Margaret Andersen As I was growing up in the 1950s and 1960s, my family moved from California to Georgia, then to Massachusetts, and then back to Georgia. Moving as we did from urban to small-town environments and in and out of regions of the country that were very different in their racial character, I probably could not help becoming fascinated by the sociology of race. Oakland, California, where I was born, was highly diverse; my neighborhood was mostly White and Asian American. When I moved to a small town in Georgia in the 1950s, I was ten years old, but I was shocked by the racial norms I encountered. I had always loved riding in the back of the bus—our major mode of transportation in Oakland—and could not understand why this was no longer allowed. Labeled by my peers as an outsider because I was not southern, I painfully learned what it meant to feel excluded just because of "where you are from."

When I moved again to suburban Boston in the 1960s, I was defined by Bostonians as a southerner and was ridiculed. Nicknamed "Dixie," I was teased for how I talked. Unlike in the South, where Black people were part of White people's daily lives despite strict racial segregation, Black people in Boston were even less visible. In my high school of 2500 or so students, Black students were rare. To me, the school seemed not much different from the strictly segregated schools I had attended in Georgia. My family soon returned to Georgia, where I was an outsider again; when I later returned to Massachusetts for graduate school in the 1970s, I worried about how a southerner would be accepted in this "Yankee" environment. Because I had acquired a southern accent, I think many of my teachers stereotyped me and thought I was not as smart as the students from other places.

Courtesy of Margaret L. Andersen

These early lessons, which I may have been unaware of at the time, must have kindled my interest in the sociology of race relations. As I explored sociology, I wondered how the concepts and theories of race relations applied to women's lives. So much of what I had experienced growing up as a woman in this society was completely unexamined in what I studied in school. As the women's movement developed in the 1970s, I found sociology to be the framework that helped me understand the significance of gender and race in people's lives. To this day, I write and teach about race and gender, using sociology to help students understand their significance in society.

Howard Taylor I grew up in Cleveland, Ohio, the son of African American professional parents. My mother, Murtis Taylor, was a social worker and the founder and then president of a social work agency called the Murtis H. Taylor Human Services Center in Cleveland, Ohio. She is well known for her contributions to the city of Cleveland and was an early "superwoman," working days and nights, cooking, caring for her two sons, and being active in many professional and civic activities. I think this gave me an early appreciation for the roles of women and the place of gender in society, although I surely would not have articulated it as such at the time.

My father was a businessman in a then all-Black life insurance company. He was

Courtesy of Howard Taylor

also a "closet scientist," always doing physics experiments, talking about scientific studies, and bringing home scientific gadgets. He encouraged my brother and me to engage in science, so we were always experimenting with scientific studies in the basement of our house. In the summers, I worked for my mother in the social service agency where she worked, as a camp counselor, and in other jobs. Early on, I contemplated becoming a social worker, but I was also excited by science. As a young child, I acquired my father's love of science and my mother's interest in society. In college, the one field that would gratify both sides of me, science and social work, was sociology. I wanted to study human interaction, but I also wanted to be a scientist, so the appeal of sociology was clear.

At the same time, growing up African American meant that I faced the consequences of race every day. It was always there, and like other young African American children, I spent much of my childhood confronting racism and prejudice. When I discovered sociology, in addition to bridging the scientific and humanistic parts of my interests, I found a field that provided a framework for studying race and ethnic relations. The merging of two ways of thinking, coupled with the analysis of race that sociology has long provided, made sociology fascinating to me.

Today, my research on race, class, gender, and intelligence testing seems rooted in these early experiences. I do quantitative research in sociology and see sociology as a science that reveals the workings of race, class, and gender in society.

map 1.1

Mapping America's Diversity: A Changing Population

The nation is becoming increasingly diverse, but the distribution of minority groups differs in various regions of the country. Looking at this map, what factors do you think influence the distribution of the population?

Data: U.S. Census Bureau. 2010. **www.census.gov**

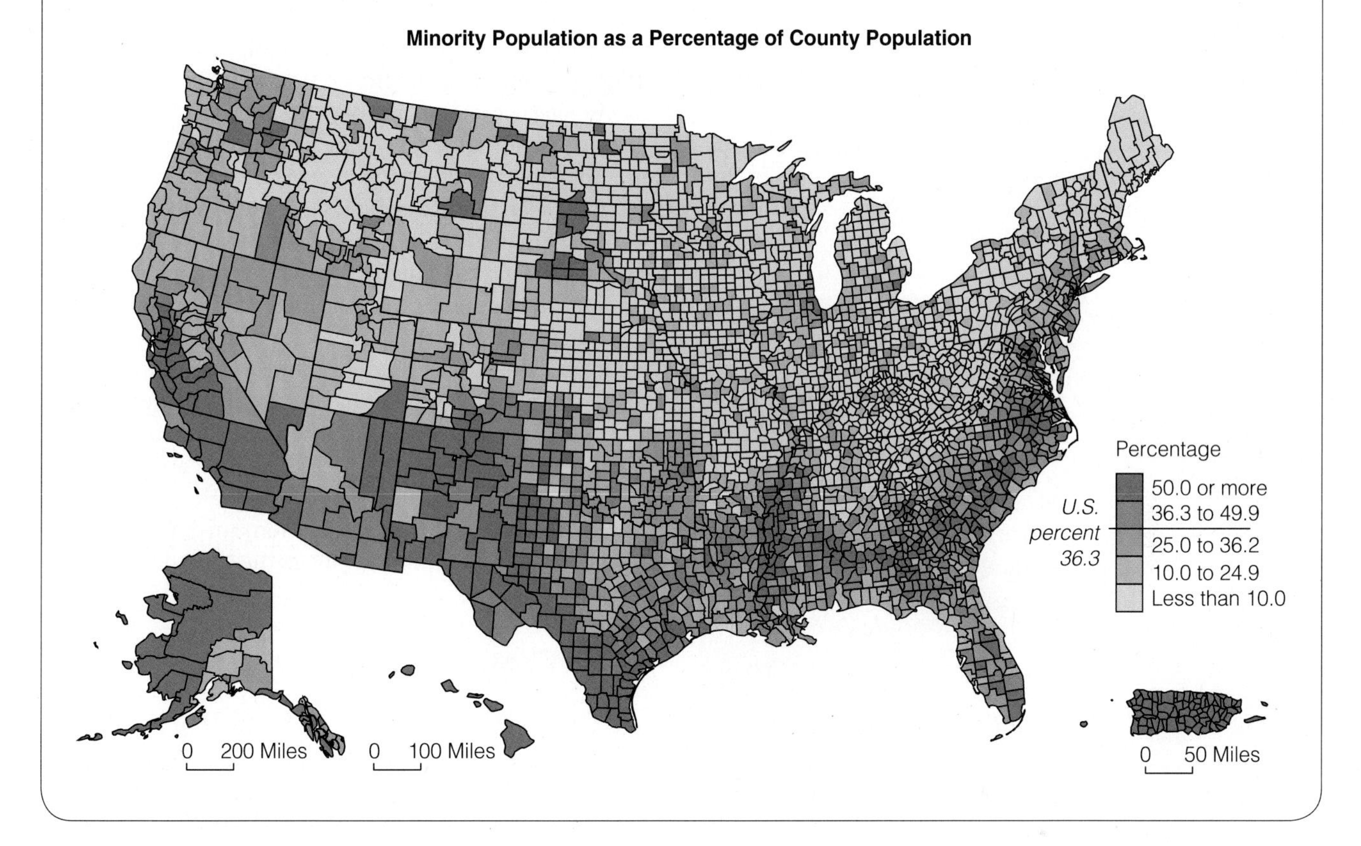

language, religion, and the traditions of family and community. Understanding diversity means recognizing this diversity and making it central to sociological analyses.

In this book, we use the term **diversity** to refer to the variety of group experiences that result from the social structure of society. Diversity is a broad concept that includes studying group differences in society's opportunities, the shaping of social institutions by different social factors, the formation of group and individual identity, and the process of social change. Diversity includes the study of different cultural orientations, although diversity is not exclusively about culture.

Understanding diversity is crucial to understanding society because fundamental patterns of social change and social structure are increasingly patterned by diverse group experiences. There are numerous sources of diversity, including race, class, gender, and others as well. Age, nationality, sexual orientation, and region of residence, among other factors, also differentiate the experience of diverse groups in the United States. As the world is increasingly interconnected through global communication and a global economy, the study of diversity also encompasses a global perspective—that is, an understanding of the international connections existing across national borders and the impact of such connections on life throughout the world.

THINKING Sociologically

What are some of the sources of *diversity* on your campus? How does this diversity affect social relations on campus?

Society in Global Perspective

No society can be understood apart from the global context that now influences the development of all societies. The social and economic system of any one society is increasingly intertwined with those of other nations. Coupled with the increasing ease of travel and telecommunication, a global perspective is necessary to understand change both in the United States and in other parts of the world.

To understand globalization, you must look beyond the boundaries of your own society to see how patterns in any given society are increasingly being shaped by the connections between societies. Comparing and contrasting societies across different cultures is valuable. It helps you see patterns in your own society that you might otherwise take for granted, and it enriches your appreciation of the diverse patterns of culture that mark human society and human history. A global perspective, however, goes beyond just comparing different cultures; it also helps you see how events in one society or community may be linked to events occurring on the other side of the globe.

AP Images/Eugene Hoshiko

Globalization brings diverse cultures together, but it is also a process by which Western markets have penetrated much of the world.

For instance, return to the example of unemployment that C. Wright Mills used to distinguish between troubles and issues. One man may lose his job in Peoria, Illinois, and a woman in Los Angeles may employ a Latina domestic worker to take care of her child while she pursues a career. On the one hand, these are individual experiences for all three people, but they are linked in a pattern of globalization that shapes the lives of all three. The Latina domestic may have a family whom she has left in a different nation so that she can afford to support them. The corporation for which the Los Angeles woman works may have invested in a new plant overseas that employs cheap labor, resulting in the unemployment of the man in Peoria. The man in Peoria may have seen immigrant workers moving into his community. One of his children may have made a friend at school who speaks a language other than English.

Such processes are increasingly shaping many of the subjects examined in this book—work, family, education, politics, just to name a few. Without a global perspective, you would not be able to fully understand the experience of any one of the people just mentioned, much less how these processes of change and global context shape society. Throughout this book, we will use a global perspective to understand some of the developments shaping contemporary life in the United States.

The Development of Sociological Theory

Like the subjects it studies, sociology is itself a social product. Sociology first emerged in western Europe during the eighteenth and nineteenth centuries. In this period, the political and economic systems of Europe were rapidly changing. Monarchy, the rule of society by kings and queens, was disappearing, and new ways of thinking were emerging. Religion as the system of authority and law was giving way to scientific authority. At the same time, capitalism grew. Contact between different societies increased, and worldwide economic markets developed. The traditional ways of the past were giving way to a new social order. The time was ripe for a new understanding.

The Influence of the Enlightenment

The **Enlightenment** in eighteenth- and nineteenth-century Europe had an enormous influence on the development of modern sociology. Also known as the Age of Reason, the Enlightenment was characterized by faith in the ability of human reason to solve society's problems. Intellectuals believed that there were natural laws and processes in society to be discovered and used for the general good. Modern science was gradually supplanting traditional and religious explanations

the higher social classes earn more money because they are more important (functional) to society. Critics disagree, saying that functionalism is too accepting of the status quo. From a functionalist perspective though, inequality serves a purpose in society: It provides an incentive system for people to work and promotes solidarity among groups linked by their common social standing.

Conflict Theory

Conflict theory emphasizes the role of coercion and power in society and the ability of some to influence and control others. Functionalism emphasizes cohesion within society. Conflict theory emphasizes strife and friction. Conflict theory pictures society as comprised of groups that compete for social and economic resources. Social order is maintained not by consensus but by domination, with power in the hands of those with the greatest political, economic, and social resources. When consensus exists, according to conflict theorists, it is attributable to people being united around common interests, often in opposition to other groups (Dahrendorf 1959; Mills 1956).

According to conflict theory, inequality exists because those in control of a disproportionate share of society's resources actively defend their advantages. The masses are not bound to society by their shared values but by coercion at the hands of the powerful. In conflict theory, the emphasis is on social control, not on consensus and conformity. Those with the most resources exercise power over others; inequality and power struggles are the result. Conflict theory gives great attention to class, race, gender, and sexuality in society because these are seen as the grounds of the most pertinent and enduring struggles in society.

Conflict theorists see inequality as inherently unfair, persisting only because groups who are economically advantaged use their social position to their own betterment. Their dominance even extends to the point of shaping the beliefs of other members of the society by controlling public information and holding power in institutions such as education and religion that shape what people think and know. From the conflict perspective, power struggles between conflicting groups are the source of social change. Those with the greatest power are typically able to maintain their advantage at the expense of other groups.

Conflict theory has been criticized for neglecting the importance of shared values and public consensus in society while overemphasizing inequality. Like functionalist theory, conflict theory finds the origins of social behavior in the structure of society, but it differs from functionalism in emphasizing the importance of power.

Symbolic Interaction

The third major framework of sociological theory is **symbolic interaction**. Instead of thinking of society in terms of abstract institutions, symbolic interaction emphasizes immediate social interaction as the place where "society" exists. Because of the human capacity for reflection, people give meaning to their behavior. The creation of meaning is how they interpret the different behaviors, events, or things that happen in society.

As its name implies, symbolic interaction relies extensively on the symbolic meaning that people develop and employ in the process of social interaction. Symbolic interaction theory emphasizes face-to-face interaction and thus is a form of microsociology, whereas functionalism and conflict theory are more macrosociological.

Derived from the work of the Chicago School, symbolic interaction theory analyzes society by addressing the subjective meanings that people impose on objects, events, and behaviors. Subjective meanings are important because, according to symbolic interaction, people behave based on what they *believe*, not just on what is objectively true. Symbolic interaction sees society as socially constructed through human interpretation (Blumer 1969; Berger and Luckmann 1967; Shibutani 1961). Social meanings are constantly modified through social interaction.

People interpret one another's behavior; these interpretations form social bonds. These interpretations are called the "definition of the situation." For example, why would young people smoke cigarettes even though all objective medical evidence points to the danger of doing so? The answer is in the definition of the situation that people create. Studies find that teenagers are well informed about the risks of tobacco, but they also think that "smoking is cool," that they themselves will be safe from harm, and that smoking projects an image—a positive identity for boys as a "tough guy" and for girls as fun-loving, mature, and glamorous. Smoking is also defined by young women as keeping you thin—an ideal constructed through dominant images of beauty. In other words, the symbolic meaning of smoking overrides the actual facts regarding smoking and risk.

THINKING Sociologically

Think about the example given about smoking, and using *symbolic interaction*, how would you explain other risky behaviors, such as steroid use among athletes or eating disorders among young women?

Careers in Sociology

Now that you understand a bit more what sociology is about, you may ask, "What can I do with a degree in sociology?" This is a question we often hear from students. There is no single job called "sociologist" like there is "engineer" or "nurse" or "teacher," but sociology prepares you well for many kinds of jobs, whether with a bachelor's degree or a postgraduate education. The skills you acquire from your sociological education are useful for jobs in business, health care, criminal justice, government agencies, various nonprofit organizations, and other job venues.

For example, the research skills one gains through sociology can be important in analyzing business data or organizing information for a food bank or homeless shelter. Students in sociology also gain experience working with and understanding those with different cultural and social backgrounds; this is an important and valued skill that employers seek. Also, the ability to dissect the different causes of a social problem can be an asset for jobs in various social service organizations.

Some sociologists have worked in their communities to deliver more effective social services. Some are employed in business organizations and social services where they use their sociological training to address issues such as poverty, crime and delinquency, population studies, substance abuse, violence against women, family social services, immigration policy, and any number of other important issues. Sociologists also work in the offices of U.S. representatives and senators, doing background research on the various issues addressed in the political process.

These are just a few examples of how sociology can prepare you for various careers. A good way to learn more about how sociology prepares you for work is to consider doing an internship while you are still in college.

For more information about careers in sociology, see the booklet, "21st Century Careers with an Undergraduate Degree in Sociology," available through the American Sociological Association (www.asanet.org).

Critical Thinking Exercise

1. Read a national newspaper over a period of one week and identify any experts who use a sociological perspective in their commentary. What does this suggest to you as a possible career in sociology? What are some of the different subjects about which sociologists provide expert information?
2. Identify some of the students from your college who have finished degrees in sociology. What different ways have they used their sociological knowledge?

Flake/Alamy

Symbolic interaction theory can help explain why people might do things that otherwise seem contrary to what one might expect.

Symbolic interaction interprets social order as constantly negotiated and created through the interpretations people give to their behavior. In observing society, symbolic interactionists see not simply facts but "social constructions," the meanings attached to things, whether those are concrete symbols (like a certain way of dress or a tattoo) or nonverbal behaviors. In symbolic interaction theory, society is highly subjective—existing in the minds of people, even though its effects are very real.

Feminist Theory

Contemporary sociological theory has been greatly influenced by the development of **feminist theory**. Prior to the emergence of second-wave feminism (the feminist movement emerging in the 1960s and 1970s), women were largely absent and invisible within most sociological work—indeed, within most academic work. When seen, they were strongly stereotyped in traditional roles as wives and mothers. Feminist theory developed to understand the status of women in society and with the purpose of using that knowledge to better women's lives.

Feminist theory has created vital new knowledge about women and has also transformed what is

◆ **Table 1.4 Comparing Sociological Theories**

Basic Questions	Functionalism	Conflict Theory	Symbolic Interaction	Feminist Theory
What is the relationship of individuals to society?	Individuals occupy fixed social roles.	Individuals are subordinated to society.	Individuals and society are interdependent.	Women and men are bound together in a system of gender relationships that shape identities and beliefs.
Why is there inequality?	Inequality is inevitable and functional for society.	Inequality results from a struggle over scarce resources.	Inequality is demonstrated through the importance of symbols.	Inequality stems from the matrix of domination that links gender, race, class, and sexuality.
How is social order possible?	Social order stems from consensus on public values.	Social order is maintained through power and coercion.	Social order is sustained through social interaction and adherence to social norms.	Patriarchal social orders are maintained by the power that men hold over women.
What is the source of social change?	Society seeks equilibrium when there is social disorganization.	Change comes through the mobilization of people struggling for resources.	Change evolves from an ever-evolving set of social relationships and the creation of new meaning systems.	Social change comes from the mobilization of women and their allies on behalf of women's liberation.
Major Criticisms				
	This is a conservative view of society that underplays power differences among and between groups.	The theory understates the degree of cohesion and stability in society.	There is little analysis of inequality, and it overstates the subjective basis of society.	Feminist theory has too often been anchored in the experiences of White, middle-class women.

understood about men. Feminist scholarship in sociology, by focusing on the experiences of women, provides new ways of seeing the world and contributes to a more complete view of society.

Feminist theory takes gender as a primary lens through which to view society. Beyond that, feminist theory makes the claim that without considering gender in society, one's analysis of any social behavior is incomplete and, thus, incorrect. At the same time, feminist theory purports to analyze society with an eye to improving the status of women. Men are not excluded from feminist theory. In fact, feminist theory, as we will see in various chapters that follow also argues that men are gendered subjects too. We cannot understand society without understanding how gender is structured in society and in women's and men's lives.

Feminist theory is a now vibrant and rich perspective in sociology, and it has added much to how people understand the sociology of gender—and its connection to other social factors, such as race, sexuality, age, and class. Along with the classical traditions of sociology, feminist theory is included throughout this book in the context of particular topics.

Functionalism, conflict theory, symbolic interaction, and feminist theory are by no means the only theoretical frameworks in sociology. For some time, however, they have provided the most prominent general explanations of society. Each has a unique view of the social realm. None is a perfect explanation of society, yet each has something to contribute. Functionalism gives special weight to the order and cohesion that usually characterizes society. Conflict theory emphasizes the inequalities and power imbalances in society. Symbolic interaction emphasizes the meanings that humans give to their behavior. Feminist theory takes gender as a primary lens through which to understand society, especially in relation to other structures of inequality. Together, these frameworks provide a rich, comprehensive perspective on society, individuals within society, and social change (see ◆ Table 1.4).

Whatever the theoretical framework used, theory is evaluated in terms of its ability to explain observed social facts. The sociological imagination is not a single-minded way of looking at the world. It is the ability to observe social behavior and interpret that behavior in light of societal influences.

Chapter Summary

What is sociology?

Sociology is the study of human behavior in society. The *sociological imagination* is the ability to see societal patterns that influence individuals. Sociology is an *empirical* discipline, relying on careful observations as the basis for its knowledge.

What is debunking?

Debunking in sociology refers to the ability to look behind things taken for granted, looking instead to the origins of social behavior.

Why is diversity central to the study of sociology?

One of the central insights of sociology is its analysis of social diversity and inequality. Understanding *diversity* is critical to sociology because it is necessary to analyze *social institutions* and because diversity shapes most of our social and cultural institutions.

When and how did sociology emerge as a field of study?

Sociology emerged in western Europe during the *Enlightenment* and was influenced by the values of critical reason, humanitarianism, and positivism. *Auguste Comte*, one of the earliest sociologists, emphasized sociology as a positivist discipline. *Alexis de Tocqueville* and *Harriet Martineau* developed early and insightful analyses of American culture.

What are some of the basic insights of classical sociological theory?

Emile Durkheim is credited with conceptualizing society as a social system and with identifying *social facts* as patterns of behavior that are external to the individual. *Karl Marx* showed how capitalism shaped the development of society. *Max Weber* sought to explain society through cultural, political, and economic factors. W.E.B. DuBois saw racial inequality as the greatest challenge in U.S. society.

What are the major theoretical frameworks in sociology?

Functionalism emphasizes the stability and integration in society. *Conflict theory* sees society as organized around the unequal distribution of resources and held together through power and coercion. *Symbolic interaction* emphasizes the role of individuals in giving meaning to social behavior, thereby creating society. *Feminist theory* is the analysis of women and men in society and is intended to improve women's lives.

Key Terms

conflict theory **20**
debunking **9**
diversity **13**
empirical **7**
Enlightenment **14**
feminist theory **22**
functionalism **19**
issues **7**
positivism **15**
social change **5**
social facts **15**
social institution **5**
social interaction **5**
social structure **7**
sociological imagination **6**
sociology **4**
symbolic interaction **20**
troubles **7**
verstehen **17**

2

Collage/Corbis

Culture

In one contemporary society known for its technological sophistication, people—especially the young—walk around with plugs in their ears. The plugs are connected to small wires that are themselves coated with a plastic film. These little plastic-covered wires are then connected to small devices made of metal, plastic, silicon, and other modern components, although most people who use them have no idea how they are made. When turned on, these devices put music into people's ears or, in some cases, show pictures and movies on a screen not much larger than a bar of soap. Some people who use these devices wouldn't even consider walking around without them. It is as if the devices shield them from other elements of their culture.

The same people who carry these devices around have other habits that, when seen from the perspective of someone unfamiliar with this culture, might seem peculiar and certainly highly ritualized. Apparently, when young people in this society go away to school, most take a large number of various technological devices along with them. Many sleep with one of these devices turned on all night. They look like a large box—some square, others flat—and project pictures and sound when users click buttons on another small device that, though detached from the bigger box, can be placed anywhere in the room. If you click the buttons on this portable device, the pictures and sound from the larger box will change possibly hundreds of times, revealing a huge assortment of images that seem to influence what people in this culture believe and, in many cases, how they behave. They say that in over 40 percent of the households in this culture, this device is turned on 24 hours a day (Gitlin 2002)! Indeed, it seems that everything these young people do involves looking at some kind of screen, enough so that one of the authors of this book has labeled their generation "screenagers."

Not everyone in this culture has access to all of these devices, although many want them. Indeed, having more devices seems to be a mark of one's social status, that is, how you are regarded in this culture, but very few people know where the devices are made, what they are made of, or how they work. The young also often ridicule older people for not understanding how the devices work or why they are so important to young people. From outside the culture, these practices seem strange, yet few within the culture think the behaviors associated with these devices are anything but perfectly ordinary.

in this chapter, you will learn to:

- Define culture
- Recall the elements of culture
- Explain the significance of cultural diversity
- Relate the influence of the mass media and popular culture
- Compare and contrast theoretical explanations of culture and the media
- Discuss the components of cultural change

M. BURGESS/ClassicStock/Alamy; © Jinga/Shutterstock.com

Cultural practices may seem strange to outsiders, but may be taken for granted by those within the culture. How might some contemporary cultural practices in the United States look strange to people from a very different culture?

You have surely guessed that the practices described here are taken from U.S. culture: iPads, smartphones, television/video use. These are such daily practices that they practically define modern American culture. Unless they are somehow interrupted, most people do not think much about their influence on society, on people's relationships, or on people's definitions of themselves.[1]

When viewed from the outside, cultural habits that seem perfectly normal often seem strange. Take an example from a different culture. The Tchikrin people—a remote culture of the central Brazilian rain forest—paint their bodies in elaborate designs. Painted bodies communicate to others the relationship of the person to his or her body, to society, and to the spiritual world. The designs and colors symbolize the balance the Tchikrin people think exists between biological powers and the integration of people into the social group. The Tchikrin also associate hair with sexual powers; lovers get a special thrill from using their teeth to pluck an eyebrow or eyelash from their partner's face (Sanders and Vail 2008; Turner 1969). To Tchikrin people, these practices are no more unusual or exotic than the daily habits we practice in the United States.

To study culture, to analyze it and measure its significance in society, we must separate ourselves from judgments such as "strange" or "normal." We must see a culture as insiders see it, but we cannot be completely taken in by that view. We should know the culture as insiders and understand it as outsiders.

Defining Culture

Culture is the complex system of meaning and behavior that defines the way of life for a given group or society. It includes beliefs, values, knowledge, art, morals, laws, customs, habits, language, and dress, among other things. Culture includes ways of thinking as well as patterns of behavior. Observing culture involves studying what people think, how they interact, and the objects they use.

In any society, culture defines what is perceived as beautiful and ugly, right and wrong, good and bad. Culture helps hold society together, giving people a sense of belonging, instructing them on how to behave, and telling them what to think in particular situations.

Culture is both material and nonmaterial. **Material culture** consists of the objects created in a given society—its buildings, art, tools, toys, literature, and other tangible objects, such as those discussed in the chapter opener. In the popular mind, material artifacts constitute culture because they can be collected in museums or archives and analyzed for what they represent. These objects are significant because of the meaning they are given. A temple, for example, is not merely a building, nor is it only a place of worship. Its form and presentation signify the religious meaning system of the faithful.

Nonmaterial culture includes the norms, laws, customs, ideas, and beliefs of a group of people. Nonmaterial culture is less tangible than material culture,

[1]This introduction is inspired by a classic article on the "Nacirema"—American, backward—by Horace Miner (1956). But it is also written based on essays students at the University of Delaware wrote regarding the media blackout exercise described later in this chapter. Students have written that, without access to their usual media devices, they "had no personality" and that the period of the blackout was the "worst forty-eight hours of my life!"

but it has an equally strong, if not stronger, presence in social behavior. Nonmaterial culture is found in patterns of everyday life. For example, in some cultures, people eat with utensils; in others, people do not. The eating utensils are part of material culture, but the belief about whether to use them is nonmaterial culture.

Cultural patterns make humans interesting. Some animal species develop what we might call culture. Chimpanzees, for example, learn behavior through observing and imitating others, a point proved by observing different eating practices among chimpanzees in the same species but raised in different groups (Whiten et al. 1999). Elephants have been observed picking up and fondling bones of dead elephants, perhaps evidence of grieving behavior (Meredith 2003). Dolphins have a complex auditory language. Most people also think that their pets communicate with them. Apparently, humans are not unique in their ability to develop systems of communication. Are human beings different from animals? Scientists generally conclude that animals lack the elaborate symbol-based forms of knowing and communication that are common in human societies—in other words, culture.

Understanding culture is critical to knowing how human societies operate. Culture can even shape the physical and biological characteristics of human beings. Nutrition, for instance, is greatly influenced by the cultural environment. Cultural eating habits will shape the body height and weight of a given population, even though height and weight are also biological phenomenon. Without understanding culture, you cannot understand such things as changes in idealized images of beauty over time, as the photos on this page show.

In the 1920s, the ideal woman was portrayed as curvaceous with an emphasis on her reproductive characteristics—wide, childbearing hips and large

JT Vintage/Glasshouse Images/Alamy

Justin de Villeneuve/Getty Images

Press2000/Alamy

Body size ideals have changed dramatically since the 1950s. Jayne Mansfield was a major star and sex symbol in the 1950s; she was a size 4. Marilyn Monroe was a size 8. When Twiggy became the ideal in the 1960s, she was the equivalent of a size triple zero! Kate Moss, considered now to be "average" size would wear a size 4 dress. In reality, not the ideal, the average American woman wears a size 14!

breasts. In more recent years, idealized images of women have become increasingly thin. Body mass index (BMI) is a measure of relative size, using height and weight. In the 1950s, the body mass index of idolized women, such as Marilyn Monroe, was 20. Now models have a body mass index in the mid-teens, far below the average BMI for U.S. adult women, which is 28. The point is that the media communicate that only certain forms of beauty are culturally valued. These ideals are not "natural"; they are created within a society's culture.

THINKING Sociologically

Celebrating Your Birthday!

Birthday cake, candles, friends singing "Happy birthday to you!" Once a year, you feel like the day is yours, and your friends and family gather to celebrate with you. Some people give you presents, send cards, and maybe a drinking ritual is associated with turning a particular age. If you are older, say turning forty or fifty, perhaps people kid you about "being over the hill" and decorate your office in black crepe paper. Such are the cultural rituals associated with birthdays in the United States.

What if you had been born in another culture? Traditionally in Vietnam everyone's birthday is celebrated on the first day of the year, and few really acknowledge the day they were born. In Russia, you might get a birthday pie, not a cake, with a birthday message carved into the crust. In Newfoundland, you might get ambushed and have butter rubbed on your nose for good luck—the butter is considered too greasy for bad luck to catch you. Many of these cultural practices are being changed by the infusion of Western culture, but they show how something as seemingly "normal" as celebrating your birthday has strong cultural roots.

What are the norms associated with birthday parties that you have attended? How do these reflect the values in U.S. culture?

The Power of Culture: Ethnocentrism, Cultural Relativism, and Culture Shock

Would you dice a jellyfish and serve it as a delicacy? Roll a cabbage through your house on New Year's Day to ensure good luck in the year ahead? Peculiar or revolting as these examples may seem, from within particular cultures, each seems perfectly normal. Because culture tends to be taken for granted, it can be difficult for people within a culture to see their culture as anything but "the way things are." Seen from outside the culture, everyday habits and practices can seem bizarre, certainly unusual or quirky. Such reactions show just how deeply influential culture is.

We take our own culture for granted to such a degree that it can be difficult to view other cultures without making judgments based on one's own cultural views. **Ethnocentrism** is the habit of seeing things only from the point of view of one's own group. An ethnocentric perspective prevents you from understanding the world as others experience it, and it can lead to narrow-minded conclusions about the worth of diverse cultures.

Any group can be ethnocentric. Ethnocentrism can be extreme or subtle—as in the example of social groups who think their way of life is better than that of any other group. Is there such a ranking among groups in your community? Fraternities and sororities often build group rituals around such claims; youth groups see their way of life as superior to adults; urbanites may think their cultural habits are more sophisticated than those of groups labeled "country hicks." Ethnocentrism is a powerful force because it combines a strong sense of group solidarity with the idea of group superiority.

Ethnocentrism can build group solidarity, but it can limit intergroup understanding (see, for example, ▲ Figure 2.1). Taken to extremes, ethnocentrism can lead to overt political conflict, war, terrorism, even *genocide*, the mass killing of people based on their membership in a particular group. You might wonder how people could believe so much in the righteousness of their religious faith that they would murder people. Ethnocentrism is a key part of the answer. Understanding ethnocentrism does not excuse or fully explain such behavior, but it helps you understand how such murderous behavior can occur.

Contrasting with ethnocentrism is cultural relativism. **Cultural relativism** is the idea that something can be understood and judged only in relation to the cultural context in which it appears. This does not make every cultural practice morally acceptable, but it suggests that without knowing the cultural context, it is impossible to understand why people behave as they do.

Mores (pronounced "more-ays") are strict norms that control moral and ethical behavior. Mores provide strict codes of behavior, such as the injunctions, legal and religious, against killing others and committing adultery. Mores are often upheld through **laws**, the written set of guidelines that define right and wrong in society. Basically, laws are formalized mores. Violating mores can bring serious repercussions. When any social norm is violated, the violator is typically punished.

Social sanctions are mechanisms of social control that enforce folkways, norms, and mores. The seriousness of a social sanction depends on how strictly the norms or mores are held. **Taboos** are those behaviors that bring the most serious sanctions. Dressing in an unusual way that violates the folkways of dress may bring ridicule but is usually not seriously punished. In some cultures, the rules of dress are strictly interpreted, such as the requirement by Islamic fundamentalists that women who appear in public have their bodies cloaked and faces veiled. It would be considered a taboo for women in this culture to appear in public without being veiled. The sanctions for doing so can be as severe as whipping, branding, banishment, even death.

Sanctions can be positive or negative, that is, based on rewards or punishment. When children learn social norms, for example, correct behavior may elicit positive sanctions; the behavior is reinforced through praise, approval, or an explicit reward. Early on, for example, parents might praise children for learning to put on their own clothes. Later, children might get an allowance if they keep their rooms clean. Bad behavior earns negative sanctions, such as getting spanked or grounded. In society, negative sanctions may be mild or severe, ranging from subtle mechanisms of control, such as ridicule, to overt forms of punishment, such as imprisonment, physical coercion, or death.

One way to study social norms is to observe what happens when they are violated. Once you become aware of how social situations are controlled by norms, you can see how easy it is to disrupt situations where adherence to the norms produces social order. **Ethnomethodology** is a theoretical approach in sociology based on the idea that you can discover the normal social order through disrupting it. As a technique of study, ethnomethodologists often deliberately disrupt social norms to see how people respond, thus revealing the ordinary social order (Garfinkel 1967).

In a famous series of ethnomethodological experiments, college students were asked to pretend they were boarders in their own homes for a period of fifteen minutes to one hour. They did not tell their families what they were doing. The students were instructed to be polite and impersonal, to use formal terms of address, and to speak only when spoken to. After the experiment, two of the participating students reported that their families treated the experiment as a joke; another's family thought the daughter was being extra nice because she wanted something. One family believed that the student was hiding some serious problem. In all the other cases, parents reacted with shock, bewilderment, and anger. Students were accused of being mean, nasty, impolite, and inconsiderate; the parents demanded explanations for their sons' and daughters' behavior. Through this experiment, the student researchers were able to see that even the informal norms governing behavior in one's home are carefully structured. By violating the norms of the household, the norms were revealed (Garfinkel 1967).

Ethnomethodological research teaches us that society proceeds on an "as if" basis. That is, society exists because people behave as if there were no other way to do so. Usually, people go along with what is expected of them. Culture is actually "enforced" through the social sanctions applied to those who violate social norms. Usually, specific sanctions are unnecessary because people have learned the normative expectations. When the norms are violated, their existence becomes apparent (see also Chapter 5).

Beliefs

As important as social norms are the beliefs of people in society. **Beliefs** are shared ideas held collectively by people within a given culture about what is true. Shared beliefs are part of what binds people together in society. Beliefs are also the basis for many norms and values of a given culture. In the United States, belief in God or a higher power is widely shared.

Some beliefs are so strongly held that people find it difficult to cope with ideas or experiences that contradict them. Someone who devoutly believes in God may find atheism intolerable; those who believe in magic may seem merely superstitious to those with a more scientific and rational view of the world.

Whatever beliefs people hold, they orient us to the world. They provide answers to otherwise imponderable questions about the meaning of life. Beliefs provide a meaning system around which culture is organized. Whether belief stems from religion, myth, folklore, or science, it shapes what people take to be possible and true. Although a given belief may be logically impossible, it nonetheless guides people through their lives.

Values

Deeply intertwined with beliefs are the values of a culture. **Values** are the abstract standards in a society or group that define ideal principles. Values define what is desirable and morally correct, determining what is considered right and wrong, beautiful and ugly, good and bad. Although values are abstract, they provide

understanding **diversity**

The Social Meaning of Language

Language reflects the assumptions of a culture. This can be seen and exemplified in several ways:

- **Language affects people's perception of reality.**
  Example: Researchers have found that using male pronouns, even when intended to be gender neutral, produces male-centered imagery and ideas (Switzer 1990; Hamilton 1988).
- **Language reflects the social and political status of different groups in society.**
  Example: A term such as *woman doctor* suggests that men are the standard and women the exception. Ask yourself what the term *working man* connotes and how this differs from *working woman*.
- **Groups may advocate changing language that refers to them as a way of asserting a positive group identity.**
  Example: Advocates for the disabled challenge the term *handicapped*, arguing that it stigmatizes people who may have many abilities, even if they are physically distinctive.
- **Language emerges in specific historical and cultural contexts.**
  Example: The naming of so-called races comes from the social and historical processes that have defined different groups as inferior or superior. The term *Caucasian*, for example, was coined in the seventeenth century when racist thinkers developed alleged scientific classification systems to rank different societal groups. Alfred Blumenbach used the label *Caucasian* to refer to people from the Caucasus of Russia whom he thought were more beautiful and intelligent than any other people in the world.
- **Language can distort actual group experience.**
  Example: The terms *Hispanic* and *Latino* lump together Mexican Americans, island Puerto Ricans, U.S.-born Puerto Ricans, as well as people from Honduras, Panama, El Salvador, and other Central and South American countries. *Hispanic* and *Latino* point to the shared experience of those from Latin cultures, but like the terms *Native American* and *American Indian*, the terms obscure the experiences of unique groups, such as the Lakota, Nanticoke, Cherokee, Yavapai, or Navajo.
- **Language shapes people's perceptions of groups and events in society.**
  Example: Following Hurricane Katrina in New Orleans, African American people taking food from abandoned stores were described as "looting" and White people as "finding food."
- **Terms used to define different groups change over time and can originate in movements to assert a positive identity.**
  Example: In the 1960s, *Black American* replaced the term *Negro* because the civil rights and Black Power movements inspired Black pride and the importance of self-naming (Smith et al. 1992). Earlier, *Negro* and *colored* were used to define African Americans. Currently, it is popular to refer to all so-called racial groups as "people of color." This term is meant to emphasize the common experiences of groups as diverse as African Americans, Latinos/as*, Asian Americans, and American Indians. Some people find the use of "color" in this label offensive because it harkens back to the phrase "colored people," a phrase originating in the racist treatment of African Americans.

In this book, we have tried to be sensitive to the language used to describe different groups. We recognize that language is fraught with cultural and political assumptions and that what seems acceptable now may be offensive later. The best way to solve this problem is for different groups to learn as much as they can about one another, becoming more aware of the meaning and nuances of naming and language. Greater sensitivity to the language used to describe different groups is an important step in promoting better intergroup relationships.

**Latina* is the feminine form in Spanish and refers to women; *Latino*, to men.

a general outline for behavior. Freedom, for example, is a value held to be important in U.S. culture, as is democracy. Values are ideals forming the abstract standards for group behavior, but they are also ideals that may not be realized in every situation.

Values can be a basis for cultural cohesion, but they can also be a source of conflict. Some of our most contested issues can often be traced to value conflicts. Should sex education be taught in schools? Should public schools allow school prayer? Should women have the right to choose to terminate a pregnancy? These and numerous other examples you can likely identify are matters of great debate—debates made more heated by the value conflicts that lie at the core of these public issues.

Values guide the behavior of people in society; they also shape social norms. An example of the impact that values have on people's behavior comes from an American Indian society known as the Kwakiutl (pronounced "kwa-kee-YOO-tal"), a group from the coastal region of southern Alaska, Washington State, and British Columbia. The Kwakiutl developed a practice known as *potlatch*, in which wealthy chiefs would periodically the cultural framework. On a college campus, for example, even with a strong system of fraternities and sororities, the number of students belonging to the Greek system may be a numerical minority of the total student body. Still, the campus culture may be dominated by Greek life. In a society as complex as the United States, it is hard to isolate a single dominant culture, although there is a widely acknowledged "American" culture that is considered to be the dominant one. Stemming from middle-class values, habits, and economic resources, this culture is strongly influenced by the mass media, the fashion industry, and Anglo-European traditions. It includes diverse elements such as fast food, Christmas shopping, and professional sports. It is also a culture that emphasizes achievement and individual effort—a cultural tradition that we will later see has a tremendous impact on how many in the United States view inequality (see Chapter 8).

Subcultures

Subcultures are the cultures of groups whose values and norms of behavior differ to some degree from those of the dominant culture. Members of subcultures tend to interact frequently with one another and share a common worldview. They may be identifiable by their appearance (style of clothing or adornments) or perhaps by language, dialect, or other cultural markers. You can view subcultures along a continuum of how well they are integrated into the dominant culture. Subcultures typically share some elements of the dominant culture and coexist within it, although some subcultures may be quite separated from the dominant one. This separation occurs because they are either unwilling or unable to assimilate into the dominant culture, that is, to share its values, norms, and beliefs (Dowd and Dowd 2003).

Rap and hip-hop music first emerged as a subculture as young African Americans developed their own style of dress and music to articulate their resistance to the dominant White culture. Now, rap and hip-hop have been incorporated into mainstream youth culture. Indeed, they are now global phenomena, as cultural industries have turned hip-hop and rap into a profitable industry. Even so, rap still expresses an oppositional identity for Black and White youth and other groups who feel marginalized by the dominant culture (Morgan 2010, 2009).

Some subcultures retreat from the dominant culture, such as the Amish, some religious cults, and some communal groups. In these cases, the subculture is actually a separate community that lives as independently from the dominant culture as possible. Other subcultures may coexist with the dominant society, and members of the subculture may participate in both the subculture and the dominant culture.

AP Images/Amy Sancetta

The Amish people form a subculture in the United States, although preserving their traditional way of life can be a challenge in the context of contemporary society.

Subcultures also develop when new groups enter a society. Puerto Rican immigration to the U.S. mainland, for example, has generated distinct Puerto Rican subcultures within many urban areas. Although Puerto Ricans also partake in the dominant culture, their unique heritage is part of their subcultural experience. Parts of this culture are now entering the dominant culture, such as salsa music. The themes in salsa mix the musical traditions of other Latin music, including rumba, mambo, and cha-cha. As with other subcultures, the boundaries between the dominant culture and the subculture are permeable, resulting in cultural change as new groups enter society.

THINKING Sociologically

Identify a group on your campus that you would call a *subculture*. What are the distinctive norms of this group? Based on your observations of this group, how would you describe its relationship to the dominant culture on campus?

Kyodo/Newscom

Cultural diffusion is occurring as U.S. culture is being exported to other nations, as well as the other way around. This photo shows the Old Navy store that opened in Tokyo, Japan.

Countercultures

Countercultures are subcultures created as a reaction against the values of the dominant culture. Members of the counterculture reject the dominant cultural values, often for political or moral reasons, and develop cultural practices that explicitly defy the norms and values of the dominant group. Nonconformity to the dominant culture is often the hallmark of a counterculture. Youth groups often form countercultures. Why? In part, they do so to resist the culture of older generations, thereby asserting their independence and identity. Countercultures among youth, like other countercultures, usually have a unique way of dress, their own special language, perhaps even different values and rituals.

Some countercultures directly challenge the dominant society. The white supremacist movement is an example. People affiliated with this movement have an extreme worldview, one that is in direct opposition to dominant values. White supremacist groups have developed a shared worldview, one based on extreme hostility to racial minorities, gays, lesbians, and feminists. Because of their self-contained culture—one focused on hate—they can be very dangerous (Ferber 1998).

Countercultures may also develop in situations where there is political repression and some groups are forced "underground." Under a dictatorship, for example, some groups may be forbidden to practice their religion or speak their own language. In Spain, under the dictator Francisco Franco, people were forbidden to speak Catalan—the language of the region around Barcelona. When Franco died in 1975 and Spain became more democratic, the Catalan language flourished—both in public speaking and in the press.

The Globalization of Culture

The infusion of Western culture throughout the world seems to be accelerating as the commercialized culture of the United States is marketed worldwide. One can go to quite distant places in the world and see familiar elements of U.S. culture, whether it is McDonald's in Hong Kong, Old Navy in Japan, or Disney products in western Europe. From films to fast food, the United States dominates, largely through the influence of capitalist markets. The diffusion of a single culture throughout the world is referred to as **global culture**. Despite the enormous diversity of cultures worldwide, U.S. markets increasingly dominate fashion, food, entertainment, and other cultural values, thereby creating a more homogenous world culture. Global culture is increasingly marked by capitalist interests, squeezing out the more diverse folk cultures that have been common throughout the world (Steger 2009).

Does increasing globalization of culture change traditional cultural values? Some worry that globalization imposes Western values on non-Western cultures, thus eroding long-held cultural traditions. Global economic change can also introduce more tolerant values to cultures that might have had a narrower worldview previously. As globalization occurs, *both* economic changes *and* traditional cultural values shape the emerging national culture of different societies.

The conflict between traditional and more commercial values is now being played out in world affairs. Some of the conflicts in international relations are rooted in a struggle between the values of the consumer-based, capitalist Western culture and the traditional values

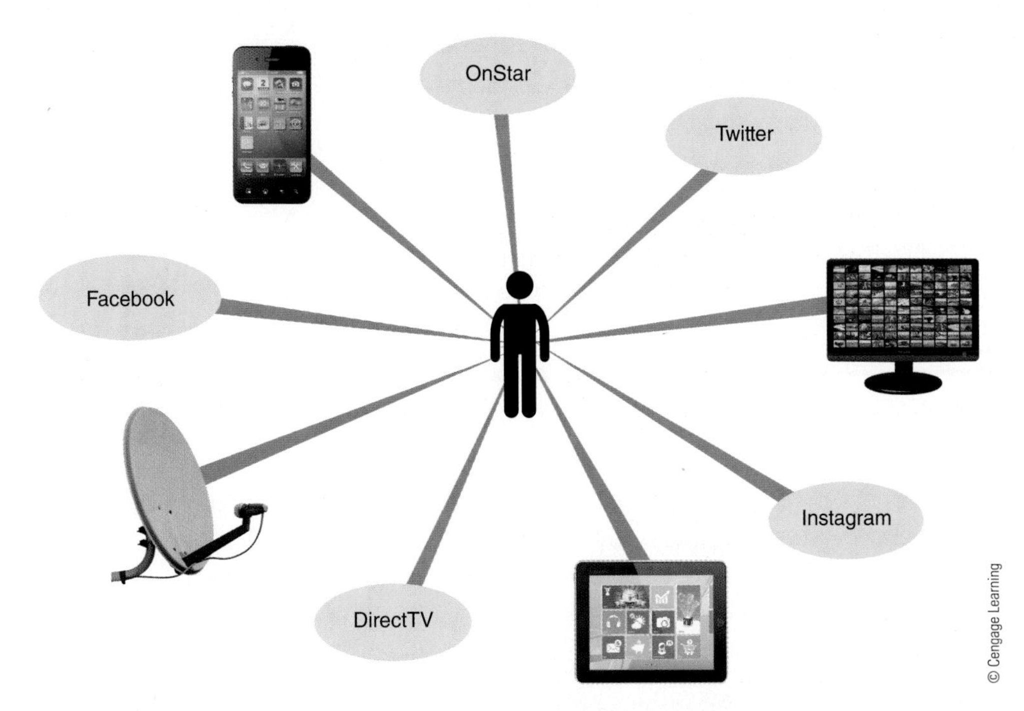

You can see how strong cultural monopolies have become if you just imagine how surrounded you are, even as an individual, by various devices (many of them owned by the same company) that deliver culture to you.

Photos: Phone, Maxx-Studio/Shutterstock.com; Satellite Dish, Roobcio/Shutterstock.com; LCD, Oleksiy Mark/Shutterstock; Tablet, Telnov Oleksii/Shutterstock.com

of local communities. As some people resist the influence of market-driven values, movements to reclaim or maintain ethnic and cultural identity can intensify, such as seen among extremist groups in the Middle East, even while pro-democratic movements also exist there.

The Mass Media and Popular Culture

Increasingly, culture in the United States and around the world is dominated and shaped by the mass media. Indeed, the culture of the United States is so infused by the media that, when people think of U.S. culture, they are likely thinking of something connected to the media—television, film, video, and so forth. The term **mass media** refers to the channels of communication that are available to wide segments of the population—the print, film, and electronic media.

The mass media have extraordinary power to shape culture, including what people believe and the information available to them. If you doubt this, observe how much the mass media affect your everyday life. A YouTube video "goes viral." Friends may talk about last night's episode of a particular show or laugh about the antics of their favorite sitcom character. You may have even met your partner or spouse via electronic media. Your way of dressing, talking, and even thinking has likely been shaped by the media, despite the fact that most people deny this, claiming "they are just individuals."

You can find the mass media everywhere—in living rooms, airports, classrooms, bars, restaurants, and doctor's offices. Even entering an elevator in a hotel, you might find CNN or the Weather Channel on twenty-four hours a day. You may even be born to the sounds and images of television, because they are turned on in many hospital delivery rooms. Television is now so ever-present in our lives that 42 percent of all U.S. households are called "constant television households"—that is, households where television is on most of the time (Gitlin 2002). For many families, TV and video are the "babysitters." The average person consumes some form of media sixty-eight hours a week—more time than is likely spent in school or at work; thirty-one of these hours are spent watching television (U.S. Census Bureau 2012a). More than half (59 percent) of young Americans (those aged 18 to 29) even report that they spend too much time on their cell phones and on the Internet (Newport 2012c). With the growth of digital viewing, the time people spend with media is increasing, with the greatest growth among people over 35 and among Asian and African Americans (Nielsen Report 2014).

One of the truly powerful communicators of culture is television. For most Americans, television consumes half of all leisure time (U.S. Bureau of Labor Statistics 2012a). Even with all of the channels and choices available, television portrays a very homogeneous view of culture because in seeking the widest possible audience, networks and sponsors find the most common ground and take few risks. The mass media also shape our understanding of social problems by determining the range of opinion or information that is defined as legitimate and by deciding which experts will be called on to elaborate an issue (Gitlin 2002). Turn on a news talk show, for example, and ask yourself who gets to lead the public discussion of current events. Are the diverse groups in society represented at the table? Do some perspectives seem off-limits or outside the boundaries of

a sociological eye on the media

Death of a Superstar

When Whitney Houston, fabulous superstar, extraordinary singer, and beautiful woman, tragically died in February 2012, millions of people grieved her passing. Her funeral was broadcast live on several major national television networks, with over 14 million people tuning in, far exceeding the usual number of television viewers during that time of day (*The New York Times*, February 21, 2012). How can people be so moved by someone's death, even when they do not know her personally?

We live in a celebrity culture, one in which the public seems endlessly fascinated by the lives of stars, especially those from the world of entertainment and popular culture. If you look at media coverage of the deaths of superstars, you will likely see a common tale told through the media coverage: the tragic and premature loss of someone with enormous talent who rose from common origins to soaring heights of wealth, popularity, and power. The very lyrics in one of Whitney Houston's songs, "Didn't we almost have it all?" reverberate in the cultural tale relayed through the media—that is, the American dream that one can rise from humble beginnings to "having it all." As sociologist Karen Sternheimer writes, "Celebrity and fame are unique manifestations of our sense of American social mobility; they provide the illusion that material wealth is possible for anyone" (2011: xiii).

don Emmert/AFP/Getty Images/Newscom

Emile Durkheim would say (as would functionalist theorists) that celebrity funerals have a sociological dimension. That is, they produce the *collective consciousness*, thus binding us together in a cultural system and reaffirming our collective beliefs and values.

Whitney Houston's death is not the first time that the public has grieved over a superstar (think of Michael Jackson, Elvis Presley, Marilyn Monroe, James Dean), nor will it be the last. But you don't have to wait for a tragic death to see the cultural ideal of the American dream retold through the media. Observe celebrity culture with a sociological perspective and ask yourself where, when, and how you see the American dream replayed through various media reports.

the media discourse? What age, race, gender, and social class are those who seem to get the most time on air?

With the advent of smartphone technology, the public's viewing habits are also changing. Almost two-thirds (58 percent) of Americans now have a smartphone. Ninety percent have cell phones. Two-thirds of cell phone owners say they check their phones for messages, alerts, and calls—even when the phone has not rung. As recent as this technology is, many think they cannot now live without it (Smith 2012; Pew Research Internet Project 2014; see ▲ Figure 2.2)!

The widespread availability of Internet-based blogs, chat groups, and social networks is, however, radically changing how people communicate, including about current events. Young people, especially, spend more time using computers for games and other leisure activities than they use for reading (U.S. Bureau of Labor Statistics 2012a). Facebook, Twitter, Snapchat, and other electronic networks have become such a common form of interaction that they are now referred to as **social media**—the term used to refer to the vast networks of social interaction that new media have inspired (See ▲ Figure 2.3). Such usage increases the possibility of democratic participation by allowing the open discussion and transmittal of information (Ferdinand 2000). At the same time, however, these forms of communication can mean increased surveillance, both by governments and by hackers. As with other forms of culture, how these networks are used and controlled is a social process.

Despite the vast reach of the mass media, many—including you, perhaps—believe that it has little effect on their beliefs and values, no matter how much they

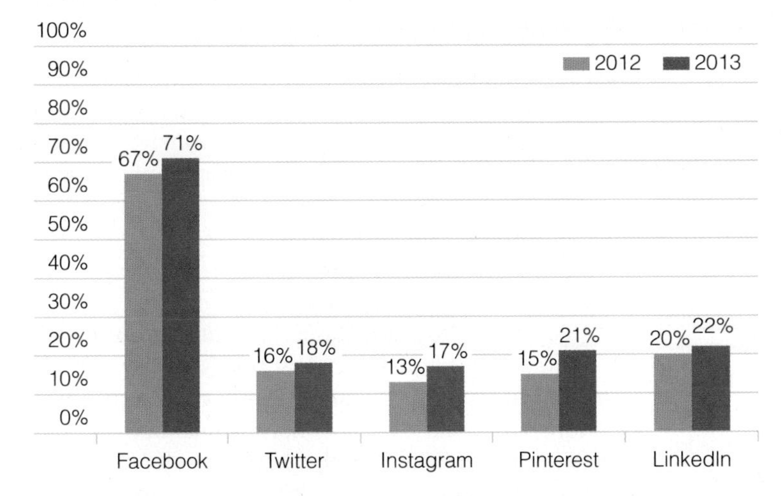

▲ **Figure 2.2** Social Media Use among U.S. Adults As you can see, the use of social media by adults in the United States is both extensive and changing rapidly. These data show the change in just one year, indicating how rapidly culture can change. How might the data change over the next five years? Ten?

Source: Pew Internet Project. 2014. **www.pewinternet.org**

enjoy it. The influence of the mass media is made apparent by trying to do without it—even for a brief period of time. Simply getting away from all of the forms of media that permeate daily life may be extremely difficult to do, as you will see if you try the experiment in the "See for Yourself: Two Days without the Media" box later in this chapter. Turn it all off for a short period of time, and see if you feel suddenly "left out" of society. Then ask yourself how the mass media influence your life, your opinions, your values, and even how you look!

The Organization of Mass Media

Mass media are not only a pervasive part of daily life, but they are also a huge business. On average, consumers spend $900 per year on media consumption, most of which is for television. That may not seem like much until you realize that the television industry (including cable) is a multibillion-dollar industry that is organized by powerful economic interests (U.S. Census Bureau 2012a)!

Increasingly, the media are owned by a small number of companies—companies that form huge media monopolies. This means that a few very powerful groups—media conglomerates—are the major producers and distributors of culture. A single corporation can control a huge share of television, radio, newspapers, music, publishing, film, and the Internet. As the production of popular culture becomes concentrated in the hands of just a few, there may be less diversity in the content.

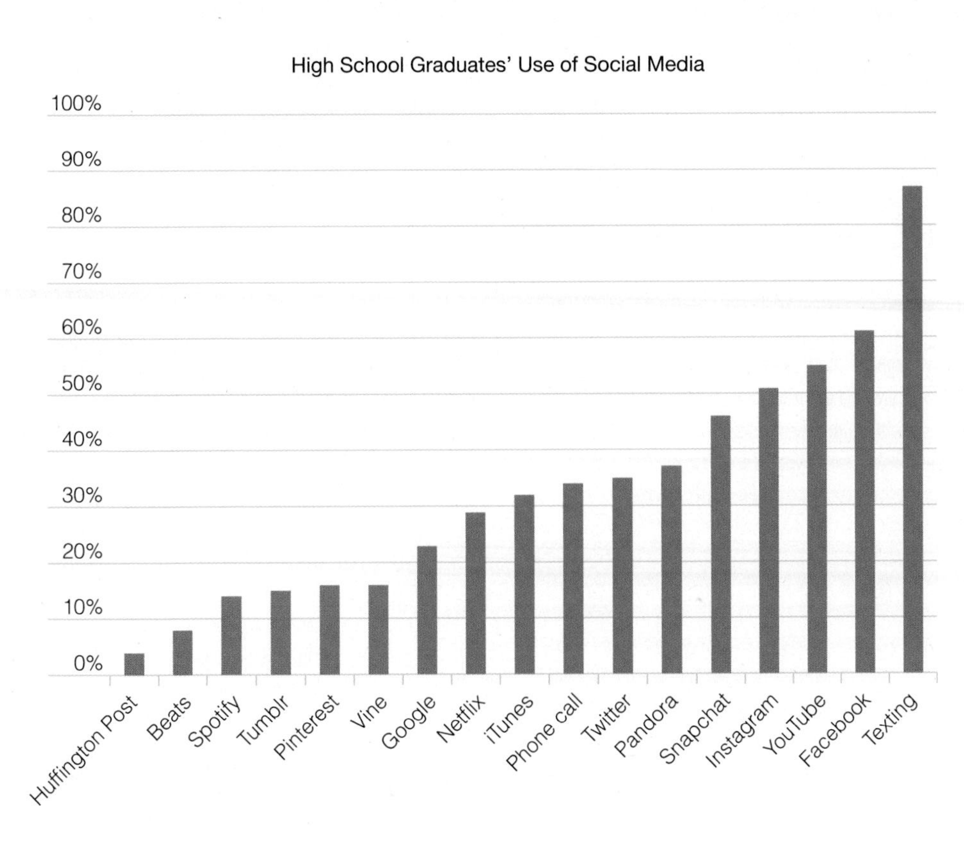

▲ **Figure 2.3** Young People's Use of Social Media Youth are often the first to pick up new social media. Comparing young people to the data in Figure 2.3, what differences do you see? How would you explain this? Do these patterns hold for your peers?

Source: Thompson, Derek. 2014. "The Most Popular Social Network for Young People? Texting." *The Atlantic*, June 19. **www.theatlantic.com**

The organization of the mass media as a system of economic interests means that there is enormous power in the hands of a few to shape the culture of the whole society. Sociologists refer to the concentration of cultural power as **cultural hegemony** (pronounced "heh-JeM-o-nee"), defined as the pervasive and excessive influence of one culture throughout society. Cultural hegemony means that people may conform to cultural patterns and interests that benefit powerful elites, even without those elites overtly forcing people into conformity. Although there seems to be enormous choice in what media forms people consume, the cultural messages are largely homogenous (meaning "same"). Cultural monopolies are then a means by which powerful groups gain the assent of those they rule. The concept of cultural hegemony implies that culture is highly politicized, even if it does not appear to be so. Those who control cultural institutions can control people's political awareness by creating cultural beliefs that make the rule of those in power seem inevitable and right. As a result, political resistance to the dominant culture is blunted (Gramsci 1971). We explore this idea further in the discussion on sociological theories of culture.

DEBUNKING Society's Myths

Myth: Teens are addicted to social media, isolating them from face-to-face interaction.

Sociological Perspective: People tend to misuse the term *addiction* by referring to activities that people enjoy and engage in frequently. Teens say they spend more time on social media than they would like, but policies that prevent teens from gathering in public places push them on to social media more than they actually say they would like (Boyd 2014).

The Media and Popular Culture

Because the mass media pervade the whole society, the media influence such things as popular styles, language, and value systems. **Popular culture** refers to the beliefs, practices, and objects that are part of everyday traditions, such as music and films, mass-marketed books and magazines, newspapers, and Internet websites. Popular culture is produced for the masses and thus has a huge impact on the nations' culture.

Popular culture is distinct from *elite culture*, which is shared by only a select few but is highly valued. Unlike elite culture (sometimes referred to as "high culture"), popular culture is mass-consumed and has enormous significance in the formation of public attitudes and values. Popular culture is also supported by mass consumption, as the many objects associated with popular culture are promoted and sold to a consuming public.

The distinction between popular and elite culture means that various segments of the population consume culture in different ways. This is affected by patterns of social class, race, and gender in the society. Although popular culture may be widely available and relatively cheap for consumers, some groups derive their cultural experiences from expensive theater shows or opera performances where tickets may cost hundreds of dollars. Meanwhile, millions of "ordinary" citizens get their primary cultural experience from television and, increasingly, the Internet. Even something as seemingly common as Internet usage reflects patterns of social class differences in society, as you can see in ▲ Figure 2.4. The **digital divide** is a term used to refer to persistence of inequality in people's access to electronic information. This inequality has led many to advocate for free wireless service in some cities to make Internet access more democratic.

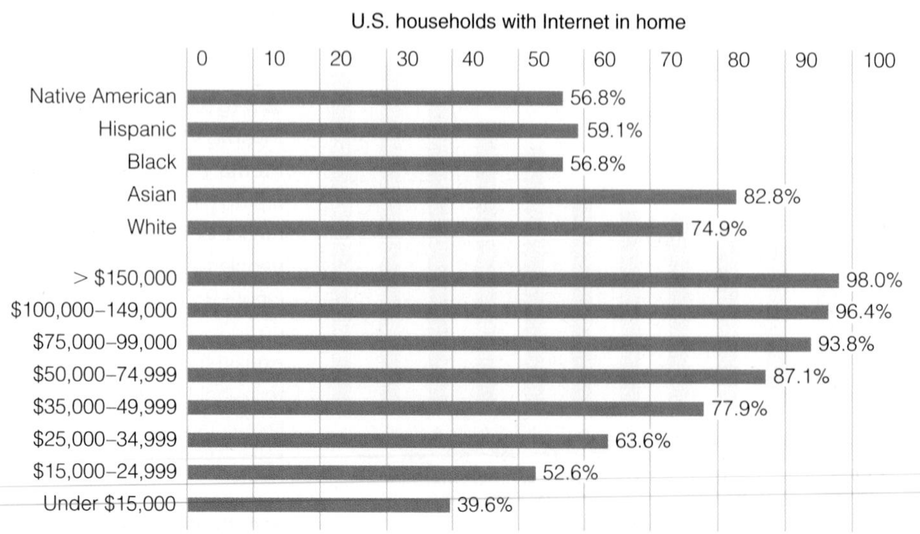

▲ **Figure 2.4** The Digital Divide Even with the widespread availability of the Internet, there are still significant social class and race differences in who has household access. What difference do you think this makes in the daily lives of those in different income brackets and in different racial/ethnic groups? What social policies might you suggest for remedying this so that there is less of a *"digital divide"*?

Data: U.S. Census Bureau. 2012. *Statistical Abstract 2012*. Washington, DC: U.S. Department of Commerce, p. 723. **www.census.gov**

Race, Gender, and Class in the Media

Many sociologists argue that the mass media can promote narrow definitions of who people are and what they can be. Even though you may think books you read, movies you see, and so forth are "just for fun," they can relay powerful messages about gender roles, race relations, and class ideals. Take the popular *Twilight* series of books, widely read by teen girls. Sociological study of the books finds that, as entertaining as they are, they reproduce stereotypes of girls as weak, passive, and needing protection. The men in the books are strong, violent, and dominating. Native Americans are portrayed as animalistic werewolves (Hayes-Smith 2011). Most likely, young girls reading these novels do not think about the messages being projected, but when popular culture is replete with such images, you cannot help but be influenced. This is why it is important for alternative images to be presented, especially to young people. The trilogy *Hunger Games* provides an example. Here the central female character is strong and self-reliant. Unlike most popular heroines, she does not wait for men to rescue her. Such images can provide young women with new models for their own leadership (McCabe et al. 2011).

Images of beauty in the media send similar messages. Youth is defined as beautiful; aging, not. Light skin is promoted as more beautiful than dark skin, although being tan is seen as more beautiful than being pale. Models in African American women's magazines are often those with Anglo features of light skin, blue eyes, and straight or wavy hair. European facial features are also pervasive in the images of Asian women and Latinas appearing in popular culture.

Content analyses of the media (a research method discussed in the following chapter) show distinct patterns of how race, gender, and class are depicted in various media forms. On prime-time television, men are still a large majority of the characters shown. Over the years, there has been an increase in how much women and people of color are depicted in professional jobs. Still, these portrayals typically depict professional women as young (suggesting that career success comes early), thin, and beautiful. In music videos, women wear sexy and skimpy clothing and are more often the object of another's gaze than is true for their male counterparts; music videos are especially represented in sexualized ways (Coy 2014; Collins 2004).

SEE for YOURSELF

Two Days without the Media

Suppose that you lived for a few days without use of the mass media that permeate our lives. How would this affect you? In an intriguing experiment, Charles Gallagher (a sociologist at La Salle University) has developed a research project for students in which he asks them to stage a media blackout in their lives for just forty-eight hours. You can try this yourself.

Begin by keeping a written log for forty-eight hours of exactly how much time you spend with some form of media. Include all time spent watching television, on the Internet, reading books and magazines, listening to music, viewing films, even using smartphones—any activity that can be construed as part of the media monopoly on people's time.

Next, eliminate all use of the media, except for that required for work and school, for a forty-eight-hour period. Keep a log as you go of what happens, what you are thinking, what others say, and how people interact with you. *Warning: If you try the media blackout, be sure to have some plan in place for having your family and/or friends contact you in case of an emergency!* When one of the authors of this book (Andersen) had her students do this experiment, they complained even before starting that they wouldn't be able to do it! But, they had to try. What happened?

First, Andersen's students had help: The week of the assignment came during a hurricane on the East Coast when many were without power for several days. This did not deter the students from thinking they just *had to have* their DVD players, music, TV, and cell phones! Many of the students said they could not stand being without access to the media—even for a few hours. Most could not go the full two days without using the media.

Most reported that they felt isolated during the media exercise, not just from information, but also mostly from other people. They were excluded from conversations with friends about what happened on a given television episode or about film characters or movie stars profiled in magazines and from playing video games. One even wrote that without the media, she felt that she had no personality! Without their connection to the media, students felt alienated, isolated, and detached, although most also reported that they studied more without the distraction of the media. A most interesting finding was that several reported that they were much more reflective during this time and had more meaningful conversations with friends.

After trying this experiment, think about the enormous influence that the mass media have in shaping everyday life, including your self-concept and your relationship with other people. What does this exercise teach you about *cultural hegemony*? The role of the mass media in shaping society? How would each of the following theoretical frameworks explain what happened during your media blackout: functionalism, conflict theory, feminist theory, or symbolic interaction?

Source: Personal correspondence, Charles Gallagher, La Salle University.

African Americans, who watch more television than do White people (see ▲ Figure 2.5), are generally confined to a narrow variety of character types in the

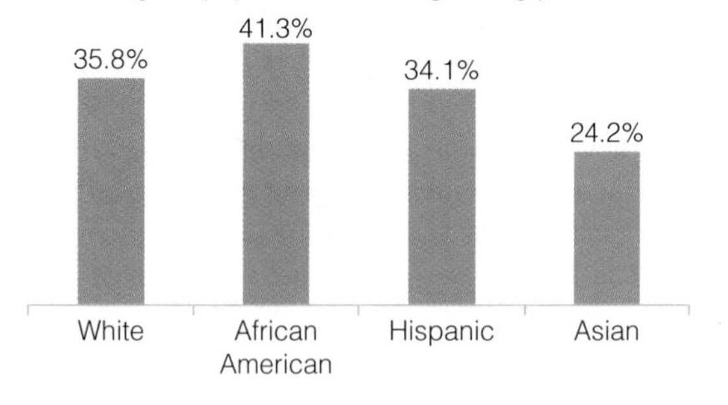

▲ **Figure 2.5** Prime-Time Television Usage by Race and Ethnicity

Source: The Nielsen Company. 2009. *Ethnic Trends in Media.* **www.blog.nielsen.com**

media. In recent years, the number of African American characters shown in television has come to match their proportion in the population, but largely because of their casting in situation comedies and in programs that are mostly minority. A recent report has found that, even though Latinos have increased as a share of the U.S. population, their presence in film and television, especially as leading actors and actresses, is shockingly low—even less than in the past. When shown, Latinos are often stereotyped as criminals, law enforcers, cheap labor, and hypersexualized or comic figures (Negrón-Muntaner 2014).

In a similar vein, African American men are most often seen as athletes and sports commentators, criminals, or entertainers. Women who work as football sports commentators are typically on the sidelines, reporting not so much on the play of the game as on human interest stories or injury reports—suggesting that women's role in sports is limited to that of nurturer. It is difficult to find a single show where Asians are the principal characters—usually they are depicted in silent roles, as sidekicks, domestic workers, or behind-the-scenes characters. Native Americans make occasional appearances, where they usually are depicted as mystics or warriors. Jewish women are generally invisible on popular TV programming, except when they are ridiculed in stereotypical roles. Arab Americans are likewise stereotyped, depicted as terrorists, rich oil magnates, or in the case of women, as perpetually veiled and secluded (Read 2003; Mandel 2001).

The popular show, *The Bachelor*, provides a good example of how race and gender stereotypes merge in the mass media. Supposedly, the women all have an equal chance at being selected as the bachelor's mate, but analyst Dubrofsky (2006) shows, women of color are never chosen as the bachelor's mate; they are, in fact, typically eliminated early from the competition. Equally revealing, Dubrofsky shows how the show's set suggests a harem-like quality—multiple women available to one man, women lounging around on plush furniture, assembled to resemble a stereotypical harem—with plush, overstuffed cushions, lush gardens, and often Middle Eastern tapestries on the walls, thereby producing stereotypes about the supposed sexual excess and availability of Middle Eastern women. Research documents numerous examples of stereotyped portrayals in the media—stereotypes you will see for yourself if you step outside of the taken-for-granted views with which you ordinarily observe the media.

SEE for YOURSELF

Watch a particular kind of television show (situation comedy, sports broadcast, children's cartoon, or news program, for example) and make careful written notes on the depiction of different groups in this show. How often are women and men or boys and girls shown?

How are they depicted? You could also observe the portrayal of Asian Americans, Native Americans, African Americans, or Latinos. What do your observations tell you about the cultural ideals that are communicated through *popular culture*?

Class stereotypes abound in the media and popular culture as well, with working-class men typically portrayed as being ineffectual, even buffoonish (Dines and Humez 2002; Butsch 1992). This has been demonstrated in research by sociologist Laura Grindstaff, who spent six months working on two popular talk shows. She did careful participant observation and interviewed the production staff and talk show guests. She found that to get airtime, guests had to enact social class stereotypes, acting vulgar and loud. She concluded that, although these popular talk shows give ordinary people a place to air their problems and be heard, the shows exploit the working class, making a spectacle of their troubles (Grindstaff 2002; Press 2002).

Even a brief glance at popular television sitcoms reveals rampant homophobic joking. Recently, however, representation of gays and lesbians has increased in the media, after years of being virtually invisible or only the subject of ridicule. As advertisers have sought to expand their commercial markets, they are showing more gay and lesbian characters on television. This makes gays and lesbians more visible, although critics point out that they are still cast in narrow and stereotypical terms, or in comical roles (such as in *Modern Family*). Cultural visibility for any group is important because it validates people and can influence the public's acceptance of and generate support for equal rights protection (Gamson 1998).

Television is not the only form of popular culture that influences public consciousness, class, gender, and race. Music, film, books, and other industries play a significant role in molding public consciousness. What images do these cultural forms produce? You can look for yourself. Try to buy a birthday card that contains neither an age nor gender stereotype. Alternatively, watch TV or a movie and see how different gender and race groups are portrayed. You will likely find that women are depicted as trying to get the attention of men; African Americans are more likely than Whites to be seen singing and dancing.

Do these images matter? Studies find that exposure to traditional sexualized imagery in music videos has a negative effect on college students' attitudes, for example, holding more adversarial attitudes about sexual relationships (Kalof 1999). Other studies find that even when viewers see media images as unrealistic, they think that others find the images important and will evaluate them accordingly. Although people do not just passively internalize media images, such images form cultural ideals that have a huge impact on people's behavior, values, and self-image.

Theoretical Perspectives on Culture and the Media

Sociologists study culture and the media in a variety of ways, asking a variety of questions about the relationship of culture to other social institutions and the role of culture in modern life (see ◆ Table 2.2). One important question for sociologists studying the mass media is: Do the media create popular values or reflect them?

The **reflection hypothesis** contends that the mass media reflect the values of the general population (Tuchman 1979). The media try to appeal to the most broad-based audience, so they aim for the middle ground in depicting images and ideas. Maximizing popular appeal is central to television program development; media organizations spend huge amounts on market research to uncover what people think and believe and what they will like. Characters are then created with whom people will identify. Interestingly, the images in the media with which we identify are distorted versions of reality. Real people seldom live like the characters on television, although part of the appeal of these shows is how they build upon, but then mystify, the actual experiences of people.

The reflection hypothesis assumes that images and values portrayed in the media reflect the values existing in the public, but the reverse can also be true—that is, the ideals portrayed in the media also influence the attitudes and values of those who see them. This has been illustrated in research on music videos. In a controlled experiment, the researchers exposed college men and women to hip-hop videos with high sexual content. Following their viewing, men in the sample expressed greater sexual objectification of women, more sexual permissiveness, stereotypical gender attitudes, and acceptance of rape myths; the findings did not hold for women in the sample (Kistler and Lee 2010). Although there is not a simple and direct relationship between the content of mass media images and what people think, clearly these mass-produced images can have a significant impact on who we are and what we think.

Culture and Group Solidarity

Many sociologists have studied particular forms of culture and have provided detailed analyses of the content of cultural artifacts, such as images in certain television programs or genres of popular music. Other

◆ **Table 2.2 Theoretical Perspectives on Culture**

According to:				
Functionalism	**Conflict Theory**	**Symbolic Interaction**	**New Cultural Studies**	**Feminist Theory**
Culture . . .				
Integrates people into groups	Serves the interests of powerful groups	Creates group identity from diverse cultural meanings	Is ephemeral, unpredictable, and constantly changing	Reflects the interests and perspectives of powerful men
Provides coherence and stability in society	Can be a source of political resistance	Changes as people produce new cultural meanings	Is a material manifestation of a consumer-oriented society	Is anchored in the inequality of women
Creates norms and values that integrate people in society	Is increasingly controlled by economic monopolies	Is socially constructed through the activities of social groups	Is best understood by analyzing its artifacts—books, films, and television images	Creates images and values that reproduce sexist and racist images

sociologists take a broader view by analyzing the relationship of culture to other forms of social organization. Beginning with some of the classical sociological theorists (see Chapter 1), sociologists have studied the relationship of culture to other social institutions. Max Weber looked at the impact of culture on the formation of social and economic institutions. In his classic analysis of the Protestant work ethic and capitalism, Weber argued that the Protestant faith rested on cultural beliefs that were highly compatible with the development of modern capitalism. By promoting a strong work ethic and a need to display material success as a sign of religious salvation, the Protestant work ethic indirectly but effectively promoted the interests of an emerging capitalist economy. (We revisit this issue in Chapter 13.) In other words, culture influences other social institutions.

Many sociologists have also examined how culture integrates members into society and social groups. Functionalist theorists, for example, believe that norms and values create social bonds that attach people to society. Culture therefore provides coherence and stability in society. Participation in a common culture is an important social bond—one that unites society (Etzioni et al. 2001).

Classical theoretical analyses of culture have placed special emphasis on nonmaterial culture—the values, norms, and belief systems of society. Sociologists who use this perspective emphasize the integrative function of culture, that is, its ability to give people a sense of belonging in an otherwise complex social system (Smelser 1992). In the broadest sense, they see culture as a major integrative force in society, providing societies with a sense of collective identity and commonly shared worldviews.

Culture, Power, and Social Conflict

Whereas the emphasis on shared values and group solidarity drives one sociological analysis of culture, conflict and power drive another. Conflict theorists (see Chapter 1) analyze culture as a source of power in society. You can find numerous examples throughout human history where conflict between different cultures has actually shaped the course of world affairs. One such example comes from the Middle East and the situation for the Kurdish people. The Kurds are an ethnic group (see Chapter 10) who speak their own language and inhabit an area in the Middle East that includes parts of Iraq, Iran, Turkey, and Syria, although they mostly live in northern Iraq. Most are Sunni Muslims, and they have experienced years of political and economic repression. Numerous examples throughout history show how intense group hatred and powerful forms of domination can drive cultural conflict.

Conflict theorists see contemporary culture as produced within institutions that are based on inequality and capitalist principles. The cultural values and products that are produced and sold promote the economic and political interests of the few—those who own or benefit from these cultural industries. As we have seen, this is especially evident in the study of the mass media and popular culture marketed to the masses by entities with a vast economic stake in distributing their products. Conflict theorists conclude that the cultural products most likely to be produced are those consistent with the values, needs, and interests of the most powerful groups in society. The evening news, for example, is typically sponsored by major financial institutions and oil companies. Conflict theorists then ask how this commercial sponsorship influences the content of the news. If the news were sponsored by labor unions, would conflicts between management and workers always be defined as "labor troubles," or might newscasters refer instead to "capitalist troubles"?

Conflict theorists see culture as increasingly controlled by economic monopolies. Whether it is books, music, films, news, or other cultural forms, monopolies in the communications industry (where culture is increasingly located) have a strong interest in protecting the status quo. As media conglomerates swallow up smaller companies and drive out smaller, less-efficient competitors, the control that economic monopolies have over the production and distribution of culture becomes enormous. Mega-communications companies then influence everything—from the movies and television shows you see to the books you read in school.

Culture can also be a source of political resistance and social change. Reclaiming an indigenous culture that had been denied or repressed is one way that groups mobilize to assert their independence. An example from within the United States is the *repatriation movement* among American Indians who have argued for the return of both cultural artifacts and human remains held in museum collections. Many American Indians believe that, despite the public good that is derived from studying such remains and objects, cultural independence and spiritual respect outweigh such scientific arguments (Thornton 2001). Other social movements, such as the gay and lesbian movement, have also used cultural performance as a means of political and social protest. Cross-dressing, drag shows, and other forms of "gender play" can be seen as cultural performances that challenge homophobia and traditional sexual and gender roles (Rupp and Taylor 2003).

A final point of focus for sociologists studying culture from a conflict perspective lies in the concept of cultural capital. **Cultural capital** refers to the cultural resources that are deemed worthy (such as knowledge of elite culture) and that give advantages to groups possessing such capital. This idea has been most developed by the French sociologist Pierre Bourdieu (1984), who sees the appropriation of culture as one way that groups maintain their social status.

what would a sociologist say?

Classical Theorists on Hip-Hop!

Perhaps you are a fan of hip-hop. You love the beat, the style, and it might even influence how you dress. Fans of different forms of popular culture typically just "like" it—but sociology also provides a way to think about popular culture—where it originates, who and what it influences, and how it is organized in social institutions. This gives you a different way of thinking about popular culture. Suppose some of the classical theorists of sociology were asked to comment on the popularity of hip-hop. What might they say? Here is an imagined conversation among them.

Emile Durkheim: I notice that young people can name hip-hop musicians that others in the society do not recognize. This commonly happens because different generations tend to grow up within a shared music culture. Whether it's hip-hop, country, or pop, music cultures bind groups together by creating a sense of shared and collective identity. For young people, this makes them feel like part of a generation instead of being completely alienated from an otherwise adult-dominated culture.

Karl Marx: It is interesting that White youth are now the major consumers of hip-hop. Hip-hop originated from young, Black youth who are disadvantaged by the economic system of society. Now capitalism has appropriated this creative work and turned it into a highly profitable commodity that benefits dominant groups who control the music industry. As this has happened, the critical perspective originated by young, Black urban men has been supplanted by race and gender stereotypes that support the interests of the powerful.

Max Weber: Emile and Karl, you just see it one way. It's not that you are wrong, but you have to take a multidimensional view. Yes, hip-hop is an economic and a cultural phenomenon, but it is also linked to power in society. Haven't you noticed how political candidates try to use popular music to appeal to different political constituencies? Don't be surprised to find hip-hop artists performing at political conventions! That's what I find so intriguing: Hip-hop is an economic, cultural, *and* political phenomenon.

W. E. B. DuBois: I've said that Black people have a "double consciousness"—one where they always have to see themselves through the eyes of a world that devalues them—American and "Black" at the same time. But, concurrently, the "two-ness" that Black people experience generates wonderful cultural forms such as hip-hop that reflect the unique spirit of African Americans. I once wrote that "there is no true American music but the wild sweet melodies of the Negro slave" (DuBois 1903: 14), but I wish I had lived to see this new spirited and soulful form of musical expression!

Bourdieu argues that members of the dominant class have distinctive lifestyles that mark their status in society. Their ability to display this cultural lifestyle signals their importance to others; that is, they possess cultural capital. From this point of view, culture has a role in reproducing inequality among groups. Those with cultural capital use it to improve their social and economic position in society. Sociologists have found a significant relationship, for example, between cultural capital and grades in school. Those from the more well-to-do classes (those with more cultural capital) are able to parlay their knowledge into higher grades, thereby reproducing their social position by being more competitive in school admissions and, eventually, in the labor market (Hill 2001; Treiman 2001).

Symbolic Interaction and the Study of Culture

Especially productive when applied to the study of culture has been *symbolic interaction theory*—a perspective that analyzes behavior in terms of the meaning people give it (see Chapter 1). The concept of culture is central to this orientation. Symbolic interaction emphasizes the interpretive basis of social behavior, and culture provides the interpretive framework through which behavior is understood.

Symbolic interaction also emphasizes that culture, like all other forms of social behavior, is socially constructed. That is, culture is produced through social relationships and in social groups, such as the media organizations that produce and distribute culture. People do not just passively submit to cultural norms. People actively make, interpret, and respond to the culture around them. Culture is not one-dimensional; it contains diverse elements and provides people with a wide range of choices from which to select how they will behave (Swidler 1986). Culture, in fact, represents the creative dimension of human life.

In recent years, a new interdisciplinary field known as *cultural studies* has emerged that builds on the insights of the symbolic interaction perspective in sociology. Sociologists who work in cultural studies are often critical of classical sociological approaches to studying culture, arguing that the classical approach has overemphasized nonmaterial culture, that is, ideas, beliefs, values, and norms. The new scholars of cultural studies find that material culture has increasing importance in modern society (Walters 1999; Crane 1994). This includes cultural forms that are recorded through print, film, artifacts, or the electronic media. Postmodernist theory has greatly influenced new cultural studies (see Chapter 1). *Postmodernism* is based on the idea that society is not an objective thing; rather, it is found in the words and images that people use to represent behavior and ideas. Given this orientation, postmodernism often analyzes common images and cultural products found in everyday life.

Classical theorists have tended to study the unifying features of culture; cultural studies researchers tend to see culture as more fragmented and unpredictable. To them, culture is a series of images that can be interpreted in multiple ways, depending on the viewpoint of observers. From the perspective of new cultural studies theorists, the ephemeral and rapidly changing quality of contemporary cultural forms is reflective of the highly technological and consumer-based culture on which the modern economy rests. Modern culture, for example, is increasingly dominated by the ever-changing, but ever-present, images that the media bombard us with in everyday life. The fascination that cultural studies theorists have for these images is partially founded in illusions that such a dynamic and rapidly changing culture produces.

Feminist Theory and Culture

Feminist theory also adds to our understanding of culture. Feminist theory analyzes the power that men have in controlling cultural institutions. In addition, feminist theory analyses the gendered stereotypes that are culturally reproduced. It is also critical of women's exclusion from important leadership roles within cultural institutions. We explore gender and cultural stereotypes in more detail in Chapter 11, but here it is important to understand how culture reflects and reinforces gendered images that maintain gender inequality.

Feminists also note that changes are appearing in the dominant culture as women assume more significant roles in the production of culture. By and large, however, even with such changes, men still dominate both popular and elite culture. Women remain a minority on the boards of most elite cultural institutions. In film, television, and the Internet, women's bodies are routinely sexualized. When women assume positions of leadership (such as in roles as news anchors), much of the commentary about them focuses on their looks or their roles as mothers—attributes not so frequently expressed on commentaries about men.

Violence against women is also routinized in the media—in video games, on crime dramas, and even in comedy. On any given night, if you only listen to television in the background, you might be amazed at how frequently you will hear women screaming.

Feminist theory analyzes cultural imagery and also criticizes the taken-for-granted nature of gender stereotyping and beliefs in the culture. On the positive side, feminist theory also encourages the production and distribution of images that counter and challenge sexism. Thus, rediscovering artistic works by and for women is one way that feminism can alter the cultural landscape. As with conflict theory, feminist theory emphasizes that the transformation of imagery is an important part of social movements for human liberation.

Cultural Change

In one sense, culture is a conservative force in society. Culture tends to be based on tradition and is passed on through generations, conserving and regenerating the values and beliefs of society. Culture is also increasingly based on institutions that have an economic interest in maintaining the status quo. People are also often resistant to cultural change because familiar ways and established patterns of doing things are hard to give up. But in other ways, culture is completely taken for granted, and it may be hard to imagine a society different from what is familiar.

Imagine, for example, the United States without fast food. Can you do so? Probably not. Fast food is so much a part of contemporary culture that it is hard to imagine life without it. Consider these facts about fast-food culture:

- The average person in the United States consumes three hamburgers and four orders of French fries per week.
- People in the United States spend more money on fast food than on movies, books, magazines, newspapers, videos, music, computers, and higher education combined.
- Ninety-six percent of American schoolchildren can identify Ronald McDonald—only exceeded by the number who can identify Santa Claus (Schlosser 2001).

Eric Schlosser, who has written about the permeation of society by fast-food culture, writes that "a nation's diet can be more revealing than its art or literature" (2001: 3). He relates the growth of the fast-food industry to other fundamental changes in American society, including the vast numbers of women entering the paid labor market, the development of an automobile culture, the increased reliance on low-wage service jobs, the decline of family farming, and the growth of agribusiness. One result is a cultural emphasis on uniformity, not to mention increased fat and calories in people's diets.

This example shows how cultures can change over time, sometimes in ways that are hardly visible to us unless we take a longer-range view or, as sociologists would do, question that which surrounds us. Culture is

Key Terms

concept **62**
content analysis **68**
controlled experiment **67**
correlation **68**
covert participant observation **67**
cross-tabulation **68**
data **61**
data analysis **65**
debriefing **72**
deductive reasoning **58**
dependent variable **61**
evaluation research **71**
generalization **65**
Hawthorne effect **64**
hypothesis **61**
independent variable **61**
indicator **62**
inductive reasoning **59**
informant **67**
informed consent **74**
mean **68**
median **68**
mode **68**
overt participant observation **67**
participant observation **58**
percentage **68**
population **65**
qualitative research **61**
quantitative research **61**
random sample **65**
rate **68**
reliability **63**
replication study **60**
research design **60**
sample **65**
scientific method **58**
serendipity **65**
spurious correlation **68**
validity **63**
variable **61**

4

Ariel Skelley/Blend Images/Jupiter Images

The Media

As we saw in Chapter 2, the mass media increasingly are important agents of socialization. Television alone has a huge impact on what we are socialized to believe and become. Add to that films, music, video games, radio, Facebook, Twitter, Instagram, and other social media outlets, and you begin to see the enormous influence the media have on the values we form, our images of society, our desires for ourselves, and our relationships with others. These images are powerful throughout our lifetimes, but many worry that their effect during childhood may be particularly deleterious.

The high degree of violence in the media resulted in a rating system for televised programming, movies, music lyrics, and video games. There is no doubt that violence is extensive in the media. Analysts estimate that by age 18, the average child will have witnessed at least 18,000 simulated murders on television (Wilson et al. 2002). Research continues to examine the relationship between exposure to media violence and different types of aggressive behaviors (Gentile, Mathieson, and Crick 2011).

Media violence also tends to desensitize children to the effects of violence, including engendering less sympathy for victims of violence (Baumeister and Bushman 2008; Huesmann et al. 2003). Many also think that violent video games (another form of media) may contribute to school shootings, where an armed individual—often a student at a particular school—randomly shoots and wounds or kills one or more individuals, usually other students but also teachers (Newman et al. 2006). Media portrayals of violence desensitize viewers regarding the danger of weapons. Guns such as the ones used in the school shooting in Newtown, Connecticut, have deadly consequences; yet, media images of them pervade society.

Perhaps there is some link here, but it is too simplistic to see a direct causal connection between viewing violence and actually engaging in it. For one thing, such an argument ignores the broader social context of violent behavior, including such things as the availability of guns, family characteristics, parental control, youth alienation from school, to name a few (Newman et al. 2006).

Violence in the media is not solely to blame for violent behavior in society. Children do not watch television in a vacuum. Children live in families where they learn different values and attitudes about violent behavior. They observe the society around them, not just the images they see in fictional representations. Children are influenced not only by the images of televised and filmed violence but also by the social context in which they live. The images of violence in the media in some ways only reflect the violence in society. The sociological question is whether or not media reflect societal reality, or if reality is influenced by the images presented in the media.

The media expose us to numerous images that shape our definitions of ourselves and the world around us. What we think of as beautiful, sexy, politically acceptable, and materially necessary is strongly influenced by the media. If every week, as you read a newsmagazine, someone shows you the new car that will give you status and distinction, the message is clear and we begin to think that our self-worth can be measured by the car we drive. If every weekend, as we watch televised sports, someone tells us that to have fun we should drink the right beer, we come to believe that parties are perceived as better when everyone is drinking. The values represented in the media, whether they are about violence, racist and sexist stereotypes, or any number of other social images, have a great effect on what we think and who we come to be.

Peers

Peers are those with whom you interact on equal terms, such as friends, fellow students, and coworkers. Among peers, there are no formally defined superior and subordinate roles, although status distinctions commonly arise in peer group interactions. Without peer approval, most people find it hard to feel socially accepted.

Peers are important agents of socialization. Young girls and boys learn society's images of what they are supposed to be through the socialization process, and peers are enormously important in that process. Peer cultures for young people often take the form of *cliques*—friendship circles where members identify with each other and hold a sense of common identity.

PhotoEuphoria/iStockphoto.com

Peers are important agents of socialization. Young girls and boys learn society's images of what they are supposed to be through the socialization process.

what would a sociologist say?

Interaction in Cyberspace

As a student at a college or university, you likely do not remember a world without the Internet. Use of the Internet started out small but has grown to include an online version of almost all activities that can be done face-to-face. Everything from watching the news, researching the latest statistics, staying in touch with friends, and participating in a class discussion are now possible in front of a computer screen instead of in front of another person.

Socialization involves the ongoing process of learning how to interact with others and what the social norms are for communicating. George Herbert Mead and Charles Horton Cooley, sociologists you will read about a bit later in the chapter, explain how a sense of self develops through the expectations and judgments of others in a social environment. Cooley outlined the importance of the *looking-glass self*. Identity is developed by balancing how we think we appear to others and how those others judge us. Mead emphasized our different roles that we take on as a result of our relationship to others. Both Cooley and Mead provided the groundwork for symbolic interaction theory to explain how socialization happens within a social environment.

When communication happens in an online community, how are norms for interaction altered? In what ways have you been socialized into the expected behavior of online communication?

Research details how participants in an online discussion group navigate self-presentation (Lee 2006). Private information is often concealed with techniques such as using a false name or a "user" name that is unidentifiable. Public information, however, is carefully revealed through personal narratives that others in the discussion group can use to learn about the person. Cooley's concept that we develop identity by how others perceive us takes on new meaning for online communication. If the user limits how much private information is made public, then the user has greater control over the perception by others. Consider your online interactions. What do your privacy settings on social networking sites say about what you are trying to conceal? How much can you influence the way you are perceived online?

Additionally, the roles assumed by participants in an online discussion group are different than in-class course discussions. Mead argues that there is distinction between the part of our personality that is self-defining (the "I") and the part of our personality that is conforming to what others expect of us (the "me"). If online discussions allow participants to be unseen, will this change what contribution will be made to the discussion? Are you more comfortable saying something contrary in a discussion in person or online?

Symbolic-interactionist theory provides a good starting point for how to think about socialization in online communities. Social interaction is still crucial to understanding the development of self. Online communities, through e-courses, social networking sites, or digital chat rooms, provide a new forum for social interaction. Sociologists utilize core theoretical ideas like the looking-glass self and role-taking to explain this type of interaction.

You probably had cliques in your high school and may even be able to name them. Did your school have "jocks," "goths," "tech geeks," "freaks," "stoners," and so forth? Sociologists studying cliques have found that they are formed based on a sense of exclusive membership, like in-groups and out-groups. Cliques are cohesive but also have an internal hierarchy, with certain group leaders having more power and status than other members. Interaction techniques, like inside jokes and high fives, produce group boundaries, defining who's in and who's out. The influence of peers is strong in childhood and adolescence, but it also persists into adulthood.

A phenomenon of concern on high school grounds and on college campuses is bullying—the systematic, consistent long-time beating or verbally berating of a single student, who is chosen to be the victim by a clique. School bullying is serious business and nothing to be ignored, because it often has dire consequences. There are instances in which bullying has resulted in the suicide of a victim. In 2010, a gay student at Rutgers University in New Jersey took his own life after his college roommate posted a video linking him romantically to another male student. The incident received national media attention, highlighting the bullying of gay students. The roommate who filmed the victim was convicted on fifteen criminal counts, including invasion of privacy and bias intimidation.

As agents of socialization, peers are important sources of social approval, disapproval, and support. This is one reason groups without peers of similar status are often at a disadvantage in various settings, such as women in male-dominated professions or minority students on predominantly White campuses. Being a "token" or an "only," as it has come to be called, places unique stresses on those in settings with relatively few peers from whom to draw support (Thoits 2009). This is one reason those who are minorities in a dominant group context often form same-sex or same-race groups for support, social activities, and the sharing of information about how to succeed in their environment.

Religion

Religion is another powerful agent of socialization, and religious instruction contributes greatly to the identities children construct for themselves. Children tend to develop the same religious beliefs as their parents. Even those who renounce the religion of their youth are deeply affected by the attitudes, images, and beliefs instilled by early religious training. Very often, those who disavow religion return to their original faith at some point in their life, especially if they have strong ties to their family of origin and if they form families of their own (Wuthnow 2010).

Religious socialization influences a large number of beliefs that guide adults in how they organize their lives, including beliefs about moral development and behavior, the roles of men and women, and sexuality, to name a few. Higher religiosity is connected to sexist views, especially among men (Maltby et al. 2010). Religious socialization also influences beliefs about sexuality, including the likelihood of tolerance for gay and lesbian sexuality (Whitehead and Baker 2012). Religion can even influence child-rearing practices, including the use of physical nurturing and strict discipline.

Sports

Most people perhaps think of sports as something that is just for fun and relaxation—or perhaps to provide opportunities for college scholarships and athletic careers—but sports are also an agent of socialization. Through sports, men and women learn concepts of self that stay with them in their later lives.

Sports are also where many ideas about gender differences are formed and reinforced (Eitzen 2012; Messner 2009). For men, success or failure as an athlete can be a major part of a man's identity. Even for men who have not been athletes, knowing about and participating in sports is an important source of men's gender socialization. Men learn that being competitive in sports is considered a part of manhood. Indeed, the attitude that "sports builds character" runs deep in American culture. Sports are supposed to pass on values such as competitiveness, the work ethic, fair play, and a winning attitude. Sports are considered to be where one learns to be a man.

DEBUNKING Society's Myths

Myth: Youth sports are simply games children play for fun.

Sociological Perspective: Although sports are a form of entertainment, playing sports, much like playing with dolls, is also a source for socialization into roles, such as gender roles (Messner 2002).

Michael Messner's research on men and sports reveals the extent to which sports shape masculine identity. His research shows that, for most men, playing or watching sports is often the context for developing relationships with fathers, even when the father is absent or emotionally distant in other areas of life. Through sports relationships with male peers, more than anyone else, however, the men's identity was shaped. As boys, the men could form "safe" bonds with other men (Messner 2002).

Part of the socialization of masculine identity in sports is learning homophobic attitudes (that is, fear and hatred of homosexuals, discussed in Chapter 11). "Gay" and "athlete" were rarely words used together. The socialization of athletes includes the expectation that men are heterosexual. In April 2013, Jason Collins, a professional basketball player in the NBA, shocked the sports world when he announced he was gay. While retired athletes in the past have come out as homosexual, Collins was the first active player to go public with his sexuality. Other professional athletes have done the same, forcing dialogue within the sports media community about sexuality, gender, and sports.

Still, athletic prowess, highly esteemed in men, is not tied to cultural images of womanliness. Quite the contrary, women who excel at sports are sometimes stereotyped as lesbians, or "butches," and may be ridiculed for not being womanly enough. These stereotypes reinforce traditional gender roles for women, as do media images of women athletes that emphasize family images and the personality of women athletes (Eitzen 2012; Cavalier 2003). Research in the sociology of sports shows how activities as ordinary as shooting baskets on a city lot, playing on the soccer team for one's high school, or playing touch football on a Saturday afternoon can convey powerful cultural messages about our identity and our place in the world. Sports are a good example of the power of socialization in our everyday lives.

Schools

Once young people enter kindergarten (or, even earlier, day care), another process of socialization begins. At home, parents are the overwhelmingly dominant source of socialization cues. In school, teachers and other students are the source of expectations that encourage children to think and behave in particular ways. The expectations encountered in schools vary for different groups of students. These differences are shaped by a number of factors, including teachers' expectations for different groups and the resources that different parents can bring to bear on the educational process. The parents of children attending elite, private schools, for example, often have more influence on school policies and classroom activities than do parents in low-income communities. In any context, studying socialization in the schools is an excellent way to see the influence of gender, class, and race in shaping the socialization process.

DEBUNKING Society's Myths

Myth: Schools are primarily places where young people learn skills and other knowledge.
Sociological Perspective: There is a *hidden curriculum* in schools where students learn expectations associated with race, class, and gender relations in society as influenced by the socialization process (Henson 1995).

For example, research finds that teachers respond differently to boys and girls in school. Boys receive more attention from teachers than do girls. Even when teachers respond negatively to boys who are misbehaving, they are paying more attention to the boys (American Association of University Women 2010). Social class stereotypes also affect teachers' interactions with students. Teachers are likely to perceive working-class children and poor children as less bright and less motivated than middle-class children; teachers are also more likely to define working-class students as troublemakers (Dunne and Gazely 2008; Oakes et al. 2000). These negative appraisals are *self-fulfilling prophecies*, meaning that the expectations they create often become the cause of actual behavior in the children; thus they affect the odds of success for children. (We will return to a discussion of self-fulfilling prophecies in Chapter 14.)

Boys also receive more attention in the curriculum than girls. The characters in texts are more frequently boys; the accomplishments of boys are more likely portrayed in classroom materials; and boys and men are more typically depicted as active players in history, society, and culture (Sadker and Zittleman 2009; Loewen 2007). This is called the *hidden curriculum* in the schools—the informal and often subtle messages about social roles that are conveyed through classroom interaction and classroom materials—roles that are clearly linked to gender, race, and class.

Socialization in schools influences students in everything from classroom behavior to subjects they choose to study. This can differ by gender. Recent research focused specifically on how children are socialized to choose science courses and consider careers in science. Socialization into science fields differs for students when they are not surrounded by members of the opposite sex. Findings suggest that, although girls generally study the life sciences and boys study the physical sciences, choices are different when students attend same-sex schools. Boys, for example, when not surrounded by girls, are much more likely to choose the life sciences. The socialization process differs for boys in an all-boys school (Sikora 2014).

While in school, young people acquire identities and learn patterns of behavior that are congruent with the needs of other social institutions. Sociologists using conflict theory to understand schools would say that U.S. schools reflect the needs of a capitalist society. School is typically the place where children are first exposed to a hierarchical, bureaucratic environment. Not only do schools teach them the skills of reading, writing, and other subject areas, but they also train children to respect authority, be punctual, and follow rules—thereby preparing them for their future lives as workers in organizations that value these traits.

SEE for **YOURSELF**

Visit a local day-care center, preschool, or elementary school and observe children at play. Record the activities they are involved in, and note what both girls and boys are doing. Do you observe any differences between boys' and girls' play? What do your observations tell you about *socialization* patterns for boys and girls?

Theories of Socialization

Knowing that people become socialized does not explain how it happens. People tend to think of socialization solely in psychological terms. The influence of **Sigmund Freud** (1856–1939), for example, permeates our culture. Perhaps Freud's greatest contribution was the idea that the unconscious mind shapes human behavior. Freud is also known for developing the technique of *psychoanalysis* to help discover the causes of psychological problems in the recesses of troubled patients' minds. Freud's approach depicts the human psyche in three parts: the id is about impulses; the superego is about the standards of society and morality; and the ego is about reason and common sense. The psychoanalytic perspective interprets human identity as relatively fixed at an early age in a process greatly influenced by one's family.

Psychological theories of socialization hold much in common with sociology, but increasingly rely on studies of the brain to understand how people operate. Within sociology, socialization is explained using social learning theory, functionalism, conflict theory, and symbolic interaction theory. Each sociological perspective focuses on interactions with others and within social institutions to explain socialization and its effect on the development of the self (see ◆ Table 4.1).

a nursing home, and so forth), and ask them to describe this new experience. Ask questions such as what others expect of them in this new role, how these expectations are communicated to them, what changes they see in their own behavior, and what expectations they have of their new situation. What do your observations tell you about *adult socialization*?

Resocialization also occurs in many settings such as college sport teams, fraternities, and the military. During initiation rituals, new members may be given menial and humiliating tasks and be expected to act in a subservient manner. Such behaviors reinforce one's new identity. Although the participants may not think of this as resocialization, that is precisely what is happening.

The Process of Conversion

Resocialization also occurs during what people popularly think of as conversion. A conversion is a far-reaching transformation of identity, often related to religious or political beliefs. People usually think of conversion in the context of cults, but it happens in other settings as well.

John Walker Lindh was a U.S. citizen when the United States entered the Iraq war in 2000. He joined the Taliban in Afghanistan and was later charged with conspiring to kill Americans abroad and supporting terrorist organizations. Lindh is an example of an *extreme conversion*. He was raised Catholic in an affluent family, but he converted to Islam as a teenager, changing not just his ideas, but also his dress. Neighbors described him as being transformed from "a boy who wore blue jeans and T-shirts to an imposing figure in flowing Muslim garb" (Robertson and Burke 2001). Lindh's case can now be compared to the numerous American converts to radical Islam who have been arrested trying to travel to Syria to join ISIS and other extremist groups.

Jorge Blanco/Alamy

Hazings are good examples of rites of passage that often accompany induction into a group.

As when people join religious cults, these are extreme conversion, but conversion happens in less extreme situations, too. People may convert to a different religion, thereby undergoing resocialization by changing beliefs and religious practices. Or someone may become strongly influenced by the beliefs of a *social movement*, such as the tea party political movement, and abruptly or gradually change beliefs—even identity—as a result.

The Brainwashing Debate

Extreme examples of resocialization are seen as "brainwashing." In the popular view of brainwashing, converts have their previous identities totally stripped. The transformation is seen as so complete that only deprogramming can restore the former self. Potential candidates of brainwashing include people who enter religious cults, prisoners of war, and hostages. Sociologists have examined so-called brainwashing to illustrate the process of resocialization, but they note that even with extreme conversions, converts do not necessarily drop their former identity. Resocialization in its most extreme form can be seen in the radicalization of youth into terrorist organizations. Terrorist groups across the globe recruit children and young adults to join extremist movements, engage in violent acts, and possibly even detonate "suicide bombs" in the name of the radical group. Both political and religious doctrines can be used to resocialize people to radicalized belief systems (Nawaz 2013).

Forcible confinement and physical torture can be instruments of extreme resocialization. Under severe captivity and deprivation, a captured person may come to identify with the captor; this is known as the **Stockholm syndrome**. In traditional psychology, this same phenomenon was called "identification with the aggressor." In such instances, the captured person has become *dependent* on the captor. On release, the captive frequently needs debriefing, or deprogramming. Prisoners of war and hostages may not lose free will altogether, but they do lose freedom of movement and association, which makes prisoners intensely dependent on their captors and therefore vulnerable to the captor's influence.

The Stockholm syndrome can help explain why some battered women do not leave their abusing spouses or boyfriends. Dependent on their abuser both financially and emotionally, battered women often develop identities that keep them attached to men who abuse them, a clear example of identification with an aggressor. In these cases, outsiders often think the women should leave instantly, whereas the women themselves may find leaving difficult, even in the most abusive situations.

The socialization process begins at birth and continues throughout life. How we are socialized as children defines much of who we are and how we interact with society. The various agents of socialization influence the roles we take on as children, young adults, and into old age.

Chapter Summary

What is socialization, and why is it significant for society?

Socialization is the process by which human beings learn the social expectations of society. Socialization creates the expectations that are the basis for people's attitudes and behaviors. Through socialization, people conform to social expectations, although people still express themselves as individuals.

What are the agents of socialization?

Socialization agents are those who pass on social expectations. They include the family, the media, peers, sports, religious institutions, and schools, among others. The family is usually the first source of socialization. The media also influence people's values and behaviors. *Peers* are an important source of individual identity; without peer approval, most people find it hard to be socially accepted. Schools also pass on expectations that are influenced by gender, race, and other social characteristics of people and groups.

What theoretical perspectives do sociologists use to explain socialization?

Psychoanalytic theory sees the self as driven by unconscious drives and forces that interact with the expectations of society. *Social learning theory* sees identity as a learned response to social stimuli such as reward-punishment and role models. *Functionalism* interprets socialization as key to social stability because socialization establishes shared roles and values. *Conflict theory* interprets socialization in the context of inequality and power relations. *Symbolic interaction theory* sees people as "constructing" the self as they interact with the environment and give meaning to their experience. Charles Horton Cooley described this process as the *looking-glass self.* Another sociologist, George Herbert Mead, described childhood socialization as occurring in three stages: imitation, play, and games.

Does socialization mean that everyone grows up the same?

Socialization is not a uniform process. Growing up in different environments and in such a diverse society means that different people and different groups are exposed to different expectations. Factors such as family structure, social class, regional differences, and many others influence how one is socialized.

Does socialization end during childhood?

Socialization continues through a lifetime, although childhood is an especially significant time for the formation of *identity*. Adolescence is also a period when peer cultures have an enormous influence on the formation of people's self-concepts. *Adult socialization* involves the learning of specific expectations associated with new roles.

What are the social dimensions of the aging process?

Although aging is a physiological process, its significance stems from social meanings attached to aging. *Age prejudice* and *age discrimination* result in the devaluation of older people. *Age stratification*—referring to the inequality that occurs among different age groups—is the result.

What does resocialization mean?

Resocialization is the process by which existing social roles are radically altered or replaced. It can take place in an organization that maintains strict social control and demands that the individual conform to the needs of the group or organization. Examples are religious conversion, excessive influence via social interaction ("brainwashing"), and the Stockholm syndrome.

Key Terms

age cohort **95**
age discrimination **94**
age prejudice **94**
age stereotype **94**
age stratification **95**
ageism **95**
anticipatory socialization **94**
disengagement theory **96**
game stage **89**
generalized other **89**
identity **78**
imitation stage **89**
internalization **78**
life course **90**
looking-glass self **88**
peers **83**
personality **78**
play stage **89**
psychoanalytic theory **87**
resocialization **98**
rite of passage **97**
roles **78**
self **88**
self-concept **80**
significant others **89**
social control **80**
social learning theory **87**
socialization **78**
socialization agents **81**
Stockholm syndrome **99**
taking the role of the other **88**

John Henley/Blend Images/Getty Images

organic solidarity, and each is cohesive because of the differentiation within each. Roles are no longer necessarily similar, but they are necessarily interlinked—the performance of multiple roles is necessary for the execution of society's complex and integrated functions.

Durkheim described this state as the **division of labor**, defined as the relatedness of *different* tasks that develop in complex societies. The labor force within the contemporary U.S. economy, for example, is divided according to the kinds of work people do. Within any division of labor, tasks become distinct from one another, but they are still woven into a whole.

The division of labor is a central concept in sociology because it represents how the different pieces of society fit together. The division of labor in most contemporary societies is often marked by distinctions such as age, gender, race, and social class. In other words, if you look at who does what in society, you will see that women and men tend to do different things; this is the gender division of labor. Similarly, old and young to some extent do different things; this is a division of labor by age. This is crosscut by the racial division of labor, the pattern whereby those in different racial-ethnic groups tend to do different work—or are often forced to do different work—in society. At the same time, the division of labor is also marked by class distinctions, with some groups providing work that is highly valued and rewarded and others doing work that is devalued and poorly rewarded. As you will see throughout this book, gender, race, and class intersect and overlap in the division of labor in society.

Gemeinschaft and Gesellschaft

Different societies are held together by different forms of solidarity. Some societies are characterized by what the German sociologist Ferdinand Tönnies called **gemeinschaft**, a German word that means "community"; other societies are characterized as **gesellschaft**, which literally means "society" (Tönnies 1963/1887). Each involves a type of solidarity or cohesiveness. Those societies that are *gemeinschafts* (communities) are characterized by a sense of "we" feeling, a very moderate division of labor, strong personal ties, strong family relationships, and a sense of personal loyalty. The sense of solidarity between members of the gemeinschaft society arises from personal ties; small, relatively simple social institutions; and, a collective sense of loyalty to the whole society. People tend to be well integrated into the whole, and social cohesion comes from deeply shared values and beliefs (often, sacred values). Social control need not be imposed externally because control comes from the internal sense of belonging that members share. You might think of a small community church as an example.

In contrast, in societies marked by *gesellschaft*, importance is placed on the secondary relationships people have—that is, less intimate and more instrumental relationships such as work roles instead of family or community roles. Gesellschaft is characterized by less prominence of personal ties, a somewhat diminished role of the nuclear family, and a lessened sense of personal loyalty to the total society. The solidarity and cohesion remain, and it can be very cohesive, but the cohesion comes from an elaborated *division of labor* (thus, *organic* solidarity), greater flexibility in social roles, and the instrumental ties that people have to one another.

Social solidarity under gesellschaft is weaker than in the gemeinschaft society. Although class conflict is still present in gemeinschaft, it is less prominent, making gesellschaft societies more at risk for class conflict. Racial-ethnic conflict is also more likely within gesellschaft societies because the gemeinschaft tends to be ethnically and racially very homogeneous, meaning it is often characterized by only one racial or ethnic group.

In sum, complexity and differentiation are what make the gesellschaft cohesive, whereas similarity and unity bond the gemeinschaft society. In a single society, such as the United States, you can conceptualize the whole society as gesellschaft, with some internal groups marked by gemeinschaft.

Types of Societies

In addition to comparing how different societies are bound together, sociologists are interested in how social organization evolves in different societies. Simple things such as the size of a society can also shape its social organization, as do the different roles that men and women engage in as they produce goods, care for the old and young, and pass on societal traditions. Societies also differ according to their resource base—whether they are predominantly agricultural or industrial, for example, and whether they are sparsely or densely populated.

Thousands of years ago, societies were small, sparsely populated, and technologically limited. In the competition for scarce resources, larger and more technologically advanced societies dominated smaller ones. Today, we have arrived at a global society with highly evolved degrees of social differentiation and inequality, notably along class, gender, racial, and ethnic lines (Nolan and Lenski 2014).

Sociologists distinguish six types of societies based on the complexity of their social structure, the amount of overall cultural accumulation, and the level of their technology. They are *foraging, pastoral, horticultural, agricultural* (these four are called *preindustrial* societies), and then *industrial* and *postindustrial* societies (see ◆ Table 5.1). Each type of society can still be found on Earth, although all but the most isolated societies are rapidly moving toward the industrial and postindustrial stages of development.

Table 5.1 Types of Societies

		Economic Base	Social Organization	Examples
Preindustrial Societies	*Foraging societies*	Economic sustenance dependent on hunting and foraging	Gender is important basis for social organization, although division of labor is not rigid; little accumulation of wealth	Pygmies of central Africa
	Pastoral societies	Nomadic societies, with substantial dependence on domesticated animals for economic production	Complex social system with an elite upper class and greater gender role differentiation than in foraging societies	Bedouins of Africa and Middle East
	Horticultural societies	Society marked by relatively permanent settlement and production of domesticated crops	Accumulation of wealth and elaboration of the division of labor, with different occupational roles (farmers, traders, craftspeople, and so on)	Ancient Aztecs of Mexico; Inca Empire of Peru
	Agricultural societies	Livelihood dependent on elaborate and large-scale patterns of agriculture and increased use of technology in agricultural production	Caste system develops that differentiates the elite and agricultural laborers; may include system of slavery	American South, pre-Civil War
Industrial Societies		Economic system based on the development of elaborate machinery and a factory system; economy based on cash and wages	Highly differentiated labor force with a complex division of labor and large formal organizations	Nineteenth and most of twentieth-century United States and western Europe
Postindustrial Societies		Information-based societies in which technology plays a vital role in social organization	Education increasingly important to the division of labor	Contemporary United States, Japan, and others

These different societies vary in the basis for their organization and the complexity of their division of labor. Some, such as foraging societies, are subsistence economies, where men and women hunt and gather food but accumulate very little. Others, such as pastoral societies and horticultural societies, develop a more elaborate division of labor as the social roles that are needed for raising livestock and farming become more numerous. With the development of agricultural societies, production becomes more large-scale, and strong patterns of social differentiation develop, sometimes taking the form of a caste system or even slavery.

The key driving force that distinguishes these different societies from each other is the development of technology. All societies use technology to help fill human needs, and the form of technology differs for the different types of society.

Preindustrial Societies

A **preindustrial society** is one that directly uses, modifies, and/or tills the land as a major means of survival. There are four kinds of preindustrial societies, listed here by degree of technological development: foraging

(or hunting-gathering) societies, pastoral societies, horticultural societies, and agricultural societies (see Table 5.1).

In *foraging (hunting-gathering) societies*, the technology enables the hunting of animals and gathering of vegetation. The technology does not permit the refrigeration or processing of food, hence these individuals must search continuously for plants and game. Because hunting and gathering are activities that require large amounts of land, most foraging societies are nomadic, constantly traveling as they deplete the plant supply or follow the migrations of animals. The central institution is the family, which serves as the means of distributing food, training children, and protecting its members. There is usually role differentiation on the basis of gender, although the specific form of the gender division of labor varies in different societies. The pygmies of central Africa are an example of a foraging society.

In *pastoral societies*, technology is based on the domestication of animals. Such societies tend to develop in desert areas that are too arid to provide rich vegetation. The pastoral society is nomadic, necessitated by the endless search for fresh grazing grounds for the herds of their domesticated animals. The animals are used as sources of hard work that enable the creation of a material surplus. Unlike a foraging society, this surplus frees some individuals from the tasks of hunting and gathering and allows them to create crafts, make pottery, cut hair, build tents, and apply tattoos. The surplus generates a more complex and differentiated social system with an elite class or an upper class and more role differentiation on the basis of gender. The nomadic Bedouins of Africa and the Middle East are pastoral societies.

In *horticultural societies*, hand tools are used to cultivate the land, such as the hoe and the digging stick. The individuals in horticultural societies practice ancestor worship and conceive of a deity or deities (God or gods) as a creator. Horticultural societies recultivate the land each year and tend to establish relatively permanent settlements and villages. Role differentiation is extensive, resulting in different and interdependent occupational roles such as farmer, trader, and craftsperson. The ancient Aztecs of Mexico and the Incas of Peru represent examples of horticultural societies.

The *agricultural society* is exemplified by the pre-Civil War American South, a society of slavery. Such societies have a large and complex economic system that is based on large-scale farming. Such societies rely on technologies such as use of the wheel and metals. Farms tend to be considerably larger than the cultivated land in horticultural societies. Large and permanent settlements characterize agricultural societies, which also create dramatic social inequalities. A rigid caste system develops, separating the peasants, or slaves, from the controlling elite caste, which is then freed from manual work, allowing time for art, literature, and philosophy, activities of which they can then claim the lower castes are incapable. The American pre-Civil War South and its system of slavery is a good example of an agricultural society. In fact, some argue that the system of sharecropping in the American South and Southwest was a slave-like agricultural society.

AP Images/Sergei Grits

Andersen Ross/Stockbyte/Jupiter Images

Different types of societies produce different kinds of social relationships. Some may involve more direct and personal relationships (called gemeinschafts), whereas others produce more fragmented and impersonal relationships (called gesellschafts).

Industrial Societies

An *industrial society* is one that uses machines and other advanced technologies to produce and distribute goods and services. The Industrial Revolution began over 250 years ago when the steam engine was invented in England, delivering previously unattainable amounts of

mechanical power for the performance of work. Steam engines powered locomotives, factories, and dynamos and transformed societies as the Industrial Revolution spread. The growth of science led to advances in farming techniques such as crop rotation, harvesting, and ginning cotton, as well as industrial-scale projects such as dams for generating hydroelectric power. Joining these advances were developments in medicine, new techniques to prolong and improve life, and the emergence of birth control to limit population growth.

Unlike agricultural societies, industrial societies rely on a highly differentiated labor force and the intensive use of capital and technology. Large formal organizations are common. The task of holding society together falls more on the institutions that have a high division of labor, such as the economy and work, government, politics, and large bureaucracies.

Within industrial societies, the forms of gender inequality that we see in contemporary U.S. society tend to develop. With the advent of industrialization, societies move to a cash-based economy, with labor performed in factories and mills paid on a wage basis and household labor remaining unpaid. This introduced what is known as the family wage economy, in which families become dependent on wages to support themselves, but work within the family (housework, child care, and other forms of household work) is unpaid and therefore increasingly devalued. In addition, even though women (and young children) worked in factories and mills from the first inception of industrialization, the family wage economy is based on the idea that men are the primary breadwinners. A system of inequality in men's and women's wages was introduced—an economic system that even today continues to produce a wage gap between men and women.

Industrial societies tend to be highly productive economically, with a large working class of industrial laborers. People become increasingly urbanized as they move from farmlands to urban centers or other areas where factories are located. Immigration is common in industrial societies, particularly because industries are forming where there is a high demand for more, cheap labor.

Industrialization has brought many benefits to U.S. society—a highly productive and efficient economic system, expansion of international markets, extraordinary availability of consumer products, and for many, a good working wage. Industrialization has, at the same time, also produced some of the most serious social problems that our nation faces: industrial pollution, an overdependence on consumer goods, wage inequality and job dislocation for millions, and problems of crime and crowding in urban areas.

Postindustrial Societies

In the contemporary era, a new type of society is emerging. Whereas most twentieth-century societies can be characterized in terms of their making of material goods, **postindustrial society** depends economically on the production and distribution of services, information, and knowledge. Postindustrial societies are information-based societies in which technology plays a vital role in the social organization. The United States is fast becoming a postindustrial society, and Japan may be even further along. Many of the workers provide services such as administration, education, legal services, scientific research, and banking, or they engage in the development, management, and distribution of information, particularly in the areas of computer use and design. Central to the economy of the postindustrial society are the highly advanced technologies of computers, robotics, and genetic engineering. Multinational corporations globally link the economies of postindustrial societies.

The transition to a postindustrial society has a strong influence on the character of social institutions. Educational institutions become extremely important in the postindustrial society, and science takes an especially prominent place. For some, the transition to a postindustrial society means more discretionary income for leisure activities like tourism and entertainment. Companies that specialize in relaxation and health (spas, massage centers, and exercise) become more prominent, at least for people in the upper classes. As with the United States in the last recession, the transition to postindustrialism has meant permanent joblessness for many. For others, it has meant the need to hold down more than one job simply to make ends meet.

Social Interaction and Society

You can see by now that society is an entity that exists above and beyond individuals. Also, different societies are marked by different forms of *social organization.* Although societies differ, emerge, and change, they are also highly predictable. Your society shapes virtually every aspect of your life from the structure of its social institutions to the more immediate ways that you interact with people. This is the micro level of society.

Groups

At the micro level, society is made up of many different social groups. At any given moment, each of us is a member of many groups simultaneously, and we are subject to their influence: family, friendship groups, athletic teams, work groups, racial and ethnic groups, and so on. Groups impinge on every aspect of our lives and are a major determinant of our attitudes and values regarding everything from personal issues, such

as sexual attitudes and family values, to major social issues, such as the death penalty and physician-assisted suicide.

To sociologists, a **group** is a collection of individuals who

- interact and communicate with each other;
- share goals and norms; and,
- have a subjective awareness of themselves as "we," that is, as a distinct social unit.

To be a group, the social unit in question must possess all three of these characteristics. We will examine the nature and behavior of groups in greater detail in Chapter 6.

In sociological terms, not all collections of people are groups. People may be lumped together into *social categories* based on one or more shared characteristics, such as teenagers (an age category) or truck drivers (an occupational category).

Social categories can become social groups, depending on the amount of "we" feeling the group has. Only when there is this sense of common identity, as defined in the previous characteristics of groups, is a collection of people an actual group. For example, all people nationwide watching television programs at 8 o'clock Wednesday evening form a distinct social unit, an *audience*, but they are not a group because they do not interact with one another, nor do they possess an awareness of themselves as "we." However, if many viewers were to come together for a convention where they could interact and develop a "we" feeling, such as do fans of comic books who attend Comic-Con, then they would constitute a group.

We now know that people do not need to be face-to-face to constitute a group. Online communities, for example, are people who interact with each other regularly, share a common identity, and think of themselves as being a distinct social unit. On the Internet community Facebook, for example, you may have a group of "friends," some of whom you know personally and others whom you only know online. These *friends*, as they are known on Facebook, make up a social group that might interact on a regular, indeed, daily basis—possibly even across great distances.

Groups also need not be small or "close up" and personal. *Formal organizations* are highly structured social groupings that form to pursue a set of goals. Bureaucracies such as business corporations or municipal governments or associations such as the National Parent Teacher Association (PTA) are examples of formal organizations.

Status

Within groups, people occupy different statuses. **Status** is an established position in a social structure that carries with it a degree of social rank or value. A status is a rank in society. For example, the position "vice president of the United States" is a status, one that carries relatively high prestige. "High school teacher" is another status; it carries less prestige than "vice president of the United States," but more prestige than, say, "cabdriver." Statuses occur within institutions and also within groups. "High school teacher" is a status within the education institution. Other statuses in the same institution are "student," "principal," and "school superintendent." Within a given group, people may occupy different statuses that can be dependent on a variety of factors, such as age or seniority within the group.

Typically, a person occupies many statuses simultaneously. The combination of statuses composes a **status set**, which is the complete set of statuses occupied by a person at a given time. (*Status set* is a term originally introduced by sociological theorist Robert Merton [1968].) A person may occupy different statuses in different institutions. Simultaneously, a person may be a bank president (in the economic institution), voter (in the political institution), church member (in the religious institution), and treasurer of the PTA (in the education institution). Each status may be associated with a different level of prestige.

Sometimes the multiple statuses of an individual conflict with one another. **Status inconsistency** exists where the statuses occupied by a person bring with them significantly different amounts of prestige and thus differing expectations. For example, someone trained as a lawyer, but working as a cabdriver, experiences status inconsistency. Some recent immigrants from Vietnam and Korea have experienced status inconsistency. Many refugees who had been in high status occupations in their home country, such as teachers, doctors, and lawyers, could find work in the United States only as grocers or technicians—jobs of relatively lower status than the jobs they left behind. A relatively large body of research in sociology has demonstrated that status inconsistency—in addition to low status itself—can lead to stress and depression (Taylor et al. 2006; Thoits 2009).

Achieved statuses are those attained by virtue of individual effort. Most occupational statuses—police officer, pharmacist, or boat builder—are achieved statuses. In contrast, **ascribed statuses** are those occupied from the moment a person is born. Your biological sex is an ascribed status. Yet, even ascribed statuses are not exempt from the process of social construction. For most individuals, race is an ascribed status fixed at birth. But African American individuals with light skin may appear to be White and be treated as White people throughout their lifetime. Finally, ascribed statuses can arise long after birth, through means beyond an individual's control, such as severe disability or chronic illness.

Some seemingly ascribed statuses, such as gender, can become achieved statuses. Gender, typically thought

of as fixed at birth, is a social construct. You can be born female or male (this is your sex), but becoming a woman or a man is the result of social behaviors associated with your ascribed status. In other words, gender is also achieved. People who cross-dress, have a sex change, or develop some characteristics associated with the other sex are good examples of how gender is achieved, separate and apart from your ascribed sex status. All people "do" gender in everyday life. They put on appearances and behaviors that are associated with their presumed gender (Andersen 2015; West and Fenstermaker 1995). If you doubt this, ask yourself what you did today to "achieve" your gender status. Did you dress a certain way? Wear "manly" cologne or deodorant? Splash on a "feminine" fragrance? These behaviors—all performed at the micro level—reflect the macro level of your gender status.

DEBUNKING Society's Myths

Myth: Gender is an *ascribed status* where one's gender identity is established at birth.

Sociological Perspective: Although one's biological sex identity is an ascribed status, gender is a social construct and thus is also an *achieved status*—that is, accomplished through routine, everyday behavior, including patterns of dress, speech, touch, and other social behaviors. Sex is not the same as gender (Andersen 2015).

The line between achieved and ascribed status can be hard to draw. Social class, for example, is determined by occupation, education, and annual income—all of which are achieved statuses—yet one's job, education, and income are known to correlate strongly with the social class of one's parents. Hence, one's social class status is at least partly—though not perfectly—determined at birth. It is an achieved status that includes an inseparable component of ascribed status as well.

Although people occupy many statuses at one time, it is usually the case that one status is dominant, called the **master status**, overriding all other features of the person's identity. The master status may be imposed by others, or a person may define his or her own master status. A woman judge, for example, may carry the master status "woman" in the eyes of others. She is seen not just as a judge, but as a woman judge, thus making gender a master status. A master status can completely supplant all other statuses in someone's status set. Being in a wheelchair is another example of a master status. Consider, for example, the case of a person in a wheelchair who is at the same time a medical doctor, an author, and a painter. People will see the wheelchair, at least at first, as the most important, or salient, part of identity, ignoring other statuses that define someone as a person. For a time, that person will be stereotyped as "that wheelchair guy that paints" or "that wheelchair doctor."

THINKING Sociologically

Make a list of terms that describe who you are. Which of these are *ascribed statuses* and which are *achieved statuses*? What do you think your *master status* is in the eyes of others? Does one's *master status* depend on who is defining you? What does this tell you about the significance of social judgments in determining who you are?

Roles

A **role** is the behavior others expect from a person associated with a particular status. Statuses are occupied; roles are acted or "played." The status of police officer carries with it many expectations; these expected behaviors comprise the role of police officer. Police officers are expected to enforce the law, pursue suspected criminals, assist victims of crime, complete forms for reports, and obey laws themselves. Usually, people behave in their roles as others expect them to, but not always. When a police officer commits a crime, such as physically brutalizing someone, he or she has violated the role expectations. Role expectations may vary according to the role of the observer—whether the person observing the police officer is a member of a minority group, for example.

As we saw in Chapter 4, social learning theory predicts that we learn attitudes and behaviors in response to the positive reinforcement and encouragement received from those around us. This is important in the formation of our own identity in society. We embrace certain statuses, and the roles associated with them, based on our interactions with others. The "Thinking Sociologically" feature suggests you consider your own status. What is your identity? Are you a college

elenaleonova/Getty Images

In role modeling, a person imitates the behavior of an admired other or attempts to conform to a group.

student first? Are there particular roles you feel identify you because others see you that way? These identities are often obtained through **role modeling**, a process by which we imitate the behavior of another person we admire who is in a particular role. A college freshman might admire a senior student in his dorm. The student's self-identity is influenced by his attempts to imitate the senior.

A person may occupy several statuses and roles at one time. A person's **role set** includes all the roles occupied by the person at a given time. Thus a person may be not only a student, but a store cashier, a roommate, and an admissions tour guide. Roles can also clash with each other, a situation called **role conflict**, wherein two or more roles are associated with contradictory expectations. Notice that in ▲ Figure 5.1 some of the roles diagrammed for this college student may conflict with others. Can you speculate about which might and which might not? Can you draw your own role set?

In U.S. society, some of the most common forms of role conflict arise from the dual responsibilities of job and family. The parental role demands extensive time and commitment, and so does the role of worker. Time given to one role is time taken away from the other. Although the norms pertaining to working women and working men have changed over time, it is still true that women are more often expected to uphold traditional role expectations associated with their gender role and are more likely responsible for tending to family issues even when job and family conflict. The sociologist Arlie Hochschild captured the predicament of today's women when she described the "second shift." An employed mother spends time and energy all day on the job, only to come home to the "second shift" of family and home responsibilities (Hochschild 2003, 1997; Hochschild and Machung 1989).

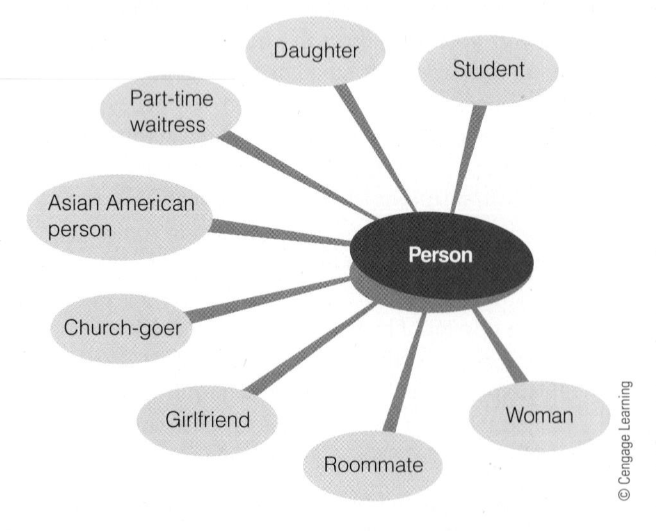

▲ **Figure 5.1** Roles in a College Student's Role Set
Identify the different roles that you occupy and draw a similar diagram of your own role set. Then identify which roles are consistent with each other and which might produce *role conflict* and *role strain*.

Hochschild's studies point to the conflict between two social roles: family roles and work roles. This conflict also highlights the sociological concept of **role strain**, a condition wherein a single role brings conflicting expectations. Different from role conflict, which involves tensions *between* two roles, role strain involves conflicts within a single role. When considering work-family balance for women, the work role has the expectations traditionally associated with work but also the expectation that she "love" her work and be as devoted to it as to her family. The same is expected of men. The result is role strain. The role of a high school student also often involves role strain. For example, students are expected to be focused on academics and performing their best, yet students also feel pressure to be involved in sports, music, community service, or other extracurricular activities. The tension between these two competing expectations is an example of role strain.

Everyday Social Interaction

You can see the influence of society in everyday behavior, including such basics as how you talk, patterns of touch, and who you are attracted to. Although you might think these things just come "naturally," they are deeply patterned by society. The cultural context of social interaction really matters in our understanding of what given behaviors mean. An action that is positive in one culture can be negative in another. For example, shaking the right hand in greeting is a positive action in the United States, but the same action in East India or certain Arab countries might be an insult. Social and cultural contexts matter. A kiss on the lips is a positive act in most cultures, yet if a stranger kissed you on the lips, you would probably consider it a negative act, perhaps even a crime.

Verbal and Nonverbal Communication. Patterns of social interaction are embedded in the language we use, and language is deeply influenced by culture and society. Furthermore, communication is not just what you say, but also how you say it and to whom. You can see the influence of society on *how* people speak, especially in different contexts. The gender of the speaker is also part of that cultural context—there are masculine and feminine styles of conversation. Japanese women, for example, are more polite and supportive when speaking to Japanese men. In conversations with English-speaking men, women are more self-assured and express their own opinions (Itakura 2014). Americans may mistakenly believe Japanese women are submissive, not realizing their conversation style changes with the context, depending on who they are talking to.

Nonverbal communication is also a form of social interaction and can be seen in various social patterns. A surprisingly large portion of our everyday communication with others is nonverbal, although we are generally only conscious of a small fraction of the nonverbal "conversations" in which we take part. Consider all the nonverbal signals exchanged in a casual chat: body position, head nods, eye contact, facial expressions, touching, and so on. Studies of nonverbal communication, like those of verbal communication, show how it is influenced by social forces, including the relationships between diverse groups of people. The meanings of nonverbal communications depend heavily on race, ethnicity, social class, and gender.

For example, patterns of touch (called *tactile communication*) are strongly influenced by gender. Parents vary their touching behavior depending on whether the child is a boy or a girl. Boys tend to be touched more roughly; girls, more tenderly and protectively. Such patterns continue into adulthood, where women touch each other more often in everyday conversation than do men. Women are on the average more likely to touch and hug as an expression of emotional support, whereas men touch and hug more often to assert power or to express sexual interest (Baumeister and Bushman 2008). Clearly, there are also instances where women touch to express sexual interest and/or dominance, but in general, touch is a supportive activity for women and an expression of sexual interest for men. In the context of sports, however, men hug and pat other men as a show of support.

In observing patterns of touch, you can see where social status influences the meaning of nonverbal behaviors. Professors, male or female, may pat a man or woman student on the back as a gesture of approval; students will rarely do this to a professor. Male professors touch students more often than do female professors, showing the additional effect of gender. Because such patterns of touching reflect power relationships between women and men, they can also be offensive and may even involve sexual harassment.

You can also see the social meaning of interaction by observing how people use personal space. *Proxemic communication* refers to the amount of space between interacting individuals. Although people are generally unaware of how they use personal space, usually the more friendly people feel toward each other, the closer they will stand. In casual conversation, friends stand closer to each other than do strangers. People who are sexually attracted to each other stand especially close, whether the sexual attraction is gay, lesbian, or heterosexual. According to anthropologist E. T. Hall (1966), we all carry around us a *proxemic bubble* that represents our personal, three-dimensional space. When people we do not know enter our proxemic bubble, we feel threatened and may take evasive action. Friends stand close; enemies tend to avoid interaction and keep far apart. According to Hall's theory, we attempt to exclude from our private space those whom we do not know or do not like, even though we may not be fully aware that we are doing so.

Patterns of touch reflect different types of social relationships. In this photo, a young man is helping an older man with the computer.

SEE for YOURSELF

Riding in Elevators

1. Try a simple experiment. Ride in an elevator and closely observe the behavior of everyone in the elevator with you. Write down in a notebook such things as how far away people stand from each other. Note the differences carefully, even in estimated inches. What do they look at? Do they tend to stand in the corners? Do they converse with strangers or the people they are with? If so, what do they talk about?
2. Now return to the same elevator and do something that breaks the usual norms of elevator behavior, such as standing too close to someone. (You will have to get up a lot of nerve to do this!) How did people react? What did they do? How did you feel? How does this experiment show how social norms are maintained through informal norms of social control?

The proxemic bubbles of different ethnic groups on average have different sizes. Hispanic people tend to stand much closer to each other than do White, middle-class Americans; their proxemic bubble is, on average, smaller.

In a society as diverse as the United States, understanding how diversity shapes social interaction is an essential part of understanding human behavior.

Ignorance of the meanings that gestures have in a society can get you in trouble. For example, some Mexicans and Mexican Americans may display the right hand held up, palm inward, all fingers extended, as an obscene gesture directed at someone in anger. This provocative gesture has no meaning at all in Anglo (White) society. Instead, extending the middle finger up, as an aggressive form of communication, is understood in many societies.

Likewise, people who grow up in urban environments learn to avoid eye contact on the streets. Staring at someone for only two or three seconds can be interpreted as a hostile act, if done man to man (Anderson 1999). If a woman maintains mutual eye contact with a male stranger for more than two or three seconds, she may be assumed by the man to be sexually interested in him. In contrast, during sustained conversation with acquaintances, women maintain mutual eye contact longer than do men (Romain 1999).

Nina Leen/The Life Picture Collection/Getty Images

Konrad Lorenz, the animal behaviorist, shows that adult greylag geese that have *imprinted* on him the moment they were hatched will follow him anywhere, as though he were their mother goose (from Tweed Roosevelt, personal communication)!

Interpersonal Attraction. We have already asked, "What holds society together?" This was asked at the macroanalysis level—that is, the level of society. But what holds relationships together—or, for that matter, makes them fall apart? You will not be surprised to learn that formation of relationships has a strong social structural component—that is, it is patterned by social forces.

Humans have a powerful desire to be with other human beings; in other words, they have a strong need for *affiliation*. We tend to spend about 75 percent of our time with other people when doing all sorts of activities—eating, watching television, studying, doing hobbies, working, and so on (Cassidy and Shaver 2008). People who lack all forms of human contact are very rare in the general population, and their isolation is usually rooted in psychotic or schizophrenic disorders. Extreme social isolation at an early age causes severe disruption of mental, emotional, and language development.

The affiliation tendency has been likened to *imprinting*, a phenomenon seen in newborn or newly hatched animals who attach themselves to the first living creature they encounter, even if it is of another species (Lorenz 1966). Studies of geese and squirrels show that once the young animal attaches itself to a human experimenter, the process is irreversible. The young animal prefers the company of the human to the company of its own species! A degree of imprinting may be discernible in human infant attachment, but researchers note that the process is more complex, more changeable, and more influenced by social factors in infants.

Somewhat similar to affiliation is *interpersonal attraction*, a nonspecific positive response toward another person. Attraction occurs in ordinary day-to-day interaction and varies from mild attraction (such as thinking your grocer is a "nice person") all the way to deep feelings of love. According to one view, attractions fall on a continuum ranging from hate to strong dislike to mild dislike to mild liking to strong liking to love. Another view is that attraction and love are two different feelings, able to exist separately. In this view, you can actually like someone a whole lot, but not be in love. Conversely, you can feel passionate love for someone, including strong sexual feelings and intense emotion, yet not really "like" the person.

Can attraction be scientifically predicted? Can you identify with whom you are most likely to fall in love? The surprising answer to these questions is "yes," with some qualifiers. Most of us have been raised to believe that love is impossible to measure and certainly impossible to predict scientifically. We think of love, especially romantic love, as quick and mysterious—a lightning bolt. Couples report falling in love at first sight, thinking that they were "meant for each other." Countless novels and stories support this view, but extensive research in sociology and social psychology suggests otherwise. Love can be predicted beyond the level of pure chance.

A strong determinant of your attraction to others is simply whether you live near them, work next to them, or have frequent contact with them. (This is a *proxemic* determinant.) You are more likely to form friendships with people from your own city than with people a thousand miles away. One classic study even showed that you are more likely to be attracted to someone on your floor, your residence hall, or your apartment building than to someone even two floors down or two

Stephen Coburn/Shutterstock.com

Romantic love is idealized in this society as something that "just happens." Despite research that interpersonal attraction follows predictable patterns, there is an increase in interracial couples.

streets over (Festinger et al. 1950). Subsequent studies continue to show this effect (Baumeister and Bushman 2008). Such is the effect of proximity in the formation of human friendships.

Now, though the general principle still holds, many people form relationships without being in close proximity, such as in online dating. In earlier societies, people would only date, fall in love, and marry people they knew from their communities. Now with social media and the ease in which we interact with one another across long distances, there is much greater likelihood to form romantic relationships with people far away. Studies of Internet dating show that people can form love relationships with people they hardly know (Rosenfeld and Thomas 2012).

We hear that "beauty is only skin deep." Apparently, that is deep enough. To a surprisingly large degree, the attractions we feel toward people of either gender are based on our perception of their physical attractiveness. Assumptions about gender differences were that men wanted beautiful women but that women cared less about attractiveness in their mate. The evidence suggests, however, that both men and women highly value attractiveness when pursuing romantic or sexual relationships (McClintock 2011). Although there are societal standards for attractiveness, there are individual preferences. "Beauty is in the eye of the beholder." Men and women see their romantic partners as attractive, even when others may not (Solomon and Vazire 2014). The point is that romantic relationships are more likely to develop between people who feel physically attracted to one another.

Of course, standards of attractiveness vary between cultures and between subcultures within the same society. What is highly attractive in one culture may be repulsive in another. In the United States, there is a maxim that you can never be too thin—a major cause of eating disorders such as *anorexia* and *bulimia*, especially among White women (Hesse-Biber 2007). The maxim is oppressive for women in U.S. society, yet it is clearly highly culturally relative, even within U.S. culture. What is considered "overweight" or "fat" is indeed a social construction (Atkins 2011). Among many African Americans, the standard of thinness is different, and larger body sizes are more ideal. Similar cultural norms often apply in certain U.S. Hispanic populations. The skinny woman is not necessarily considered attractive. Nonetheless, studies show that anorexia and bulimia are now increasing among women of color, showing how cultural norms can change (Atkins 2011; Warren et al. 2010).

Perceived physical attractiveness may predict who is attracted to whom initially, but other variables are better predictors of how long a relationship will last. So, do "opposites attract"? Not according to the research. We have all heard that people are attracted to their "opposite" in personality, social status, background, and other characteristics. Many of us grow up believing this to be true. However, if the research tells us one thing about interpersonal attraction, it is that with only a few exceptions we are attracted to people who are *similar* or *even identical* to us in socioeconomic status, race, ethnicity, religion, perceived personality traits, and general attitudes and opinions (Taylor et al. 2006). Couples tend to have similar opinions about political issues of great importance to them, such as attitudes about abortion, crime, animal rights, gun violence, and whom to vote for as president. Overall, couples tend to exhibit strong cultural or subcultural similarity, not difference.

There are exceptions, of course. We sometimes fall in love with the *exotic*—the culturally or socially different. Novels and movies return endlessly to the story of the rich young woman who falls in love with a rough and ready biker, but such a pairing is by far the exception and not the rule. That rich young woman is far more likely to fall in love with a rich young man. When it comes to long-term relationships, including both friends and lovers (whether heterosexual, lesbian, gay, or bisexual), humans vastly prefer a great degree of similarity, even though, if asked, they might deny it. In fact, the less similar a heterosexual relationship is with respect to race, social class, age, and educational aspirations (how far in school the person wants to go), then the quicker the relationship is likely to break up (Silverthorne and Quinsey 2000).

DEBUNKING Society's Myths

Myth: Love is purely an emotional experience that you cannot predict or control.

Sociological Perspective: Whom you fall in love with can be predicted beyond chance by such factors as proximity, how often you see the person (frequency, or mere exposure

effect), how physically attractive you perceive the person to be, and whether you are similar (not different) to her or him in social class, race/ethnicity, religion, age, educational aspirations, and general attitudes, including political attitudes and beliefs (Taylor et al. 2006).

Theories about Analyzing Social Interaction

Groups, statuses, and roles form a web of social interaction. Sociologists have developed different ways of conceptualizing and understanding social interaction. Functionalist theory offers one such concept. Here we detail four others: the social construction of reality, ethnomethodology, impression management, and social exchange theory (refer to ◆ Table 5.2). The first three theories come directly from the symbolic interaction perspective.

The Social Construction of Reality

What holds society together? This is a basic question for sociologists, one that, as we have seen, has long guided sociological thinking. Sociologists note that society cannot hold together without something that is shared—a shared social reality.

Some sociological theorists have argued convincingly that there is little actual reality beyond that produced by the process of social interaction itself. This is the principle of the *social construction of reality*, the idea that our perception of what is real is determined by the subjective meaning that we attribute to an experience. This is a principle central to symbolic interaction theory (Blumer 1969; Berger and Luckmann 1967). Hence, there is no objective "reality" in itself. Things do not have their own intrinsic meaning. We subjectively *impose* meaning on things.

A simple example of the social construction of reality is to consider a desk and chair in a classroom. We assign meaning to these objects based on the social context within which we use them. Students sit in the chair with notebook or computer on the desk. This is a desk. Now consider these same objects, but with a tablecloth on the desk and a plate, fork, knife, and glass set up there. Now this is not a desk and chair, but a dining table and chair. The meaning assigned to these things is influenced by the interaction we have to them. Let's take the same desk and chair and put them upside down or balancing on the corners, bolted to a cement base, with an up-light shining on them. We can paint the surface with bright colors or add a mosaic of tiles. Now the same objects are a work of art. The social context and the social interaction people have with the object give those objects meaning.

Ethnomethodology

Our interactions are guided by rules that we follow. Sometimes these rules are nonobvious and subtle. These rules are the *norms* of social interaction. Again, what holds society together? Society cannot hold together without norms, but what rules do we follow? How do we know what these rules or norms are? An approach in sociology called *ethnomethodology* is a clever technique for finding out.

Ethnomethodology (Garfinkel 1967), after *ethno* for "people" and *methodology* for "mode of study," is a technique for studying human interaction by deliberately disrupting social norms and observing how individuals attempt to restore normalcy. The idea is that to study such norms, one must first break them, because the subsequent behavior of the people involved will reveal just what the norms were in the first place. In the "See for Yourself" elevator example you were asked to perform previously, an application of ethnomethodology would be standing too close to someone on the elevator (this is the norm violation) and observing what

◆ **Table 5.2** Theories of Social Interaction

	The Social Construction of Reality	Ethnomethodology	Dramaturgy	Social Exchange Theory
Interprets society as:	Organized around the subjective meaning that people give to social behavior	Held together through the consensus that people share around social norms; you can discover these norms by violating them	A stage on which actors play their social roles and give impression to those in their "audience"	A series of interactions that are based on estimates of rewards and punishments
Analyzes social interaction as:	Based on the meaning people give to, or attribute to, actions in society	A series of encounters in which people manage their impressions in front of others	Enactment of social roles played before a social audience	A rational balancing act involving perceived costs and benefits of a given behavior

doing sociological research

Vegetarians versus Omnivores: A Case Study of Impression Management

Research Question: Author Jessica Greenebaum is a vegan. She noticed tension between herself and her meat-eating family and friends, possibly because of stereotypes about vegetarians and vegans. She did research to ask: How do vegetarians interact with omnivores to avoid negative impressions of vegetarians? What tactics do they use in their *presentation of self*?

Research Method: Greenebaum interviewed 19 vegans and 7 vegetarians, finding her research subjects through a website for educated, upper-middle-class adults who identify as vegetarian activists. She conducted face-to-face interviews and telephone interviews, averaging about one hour per interview.

Research Results: Many of the people she interviewed spoke about avoiding confrontation. One woman explained that she used to be "in your face" with meat eaters, but changed her approach to be more gentle. Another man explained that the "activist" approach makes conversations difficult.

Another key theme in her research findings was the timing of when to engage in discussions about vegetarianism or veganism. Most respondents indicated they did not want to be the first to bring it up in conversation. Vegetarians used "face-saving" tactics to make interactions more pleasant. Respondents did not try to recruit omnivores to become vegetarian, but instead emphasized the health benefits of not eating meat. By focusing on how healthy they are and how much better they feel, they encountered fewer negative impressions of vegetarians and vegans. In conversations with omnivores, the vegetarians and vegans turned away from the topic of animal rights and, instead, highlighted how well they felt. This meant talking about how they were strong, healthy, capable people by no longer eating meat. Face-saving also occurred by presenting a no-meat diet as easy to do and joyful. Greenebaum asserts that "If vegans are perceived as wheat grass–drinking hippies, people are less likely to keep an open mind about veganism" (Greenebaum 2012: 321).

Conclusions and Implications: Greenebaum concludes that interactions between two groups of people who have opposing views about diet require *impression management.* Vegetarians and vegans used particular tactics to prepare themselves for conversations with omnivores, presenting themselves in a more positive way.

Questions to Consider

The next time you are talking with someone about food, diet, and overall health, observe the social interaction with particular attention to similar and differing opinions. Seek out people with different diets from your own.

1. What do you do to manage others' impressions of you and your food choices?
2. With so much media attention on the dangers of the American diet, do you worry about the impression other people get based on what you choose to eat?

Source: Greenebaum, Jessica B. 2012. "Managing Impressions: 'Face-Saving' Strategies of Vegetarians and Vegans." *Humanity & Society* 36(4): 309–325.

that person does as a result (which would be the norm restoration behavior).

Ethnomethodology is based on the premise that human interaction takes place within a consensus, and interaction is not possible without this consensus. The consensus is part of what holds society together. According to Garfinkel, this consensus will be revealed by people's *background expectancies,* namely, the norms for behavior that they carry with them into situations of interaction. The presumption is that these expectancies are to a great degree shared, and thus studying norms by deliberately violating them will reveal the norms that most people bring with them into interaction. The ethnomethodologist argues that you cannot simply walk up to someone and ask what norms the person has and uses, because most people will not be able to articulate them. We are not wholly conscious of what norms we use even though they are shared. Ethnomethodology is designed to "uncover" those norms.

The recently aired television programs called "What Would You Do?" employs what is in effect ethnomethodology, though in a nonsystematic and relatively uncontrolled way. For example, in one episode, a father is seen in a restaurant very loudly scolding his own small child for accidentally dropping a few crumbs on the floor. The extremely loud scolding represents a norm violation in this context. The father is in alliance with the television producers. The point is to see what the observing people in the restaurant do, namely, engage in what the ethnomethodologist would call norm restoration behavior. They found that many people looked but did not intervene. A few did intervene, such as by asking the father why he was so loud, saying that his punishment was too severe.

Sociological studies that use the ethnomethodology approach point to the importance of the context within which the interaction takes place as well as the result of that interaction. Research on police interrogations

6

D. Trozzo/Alamy

not to the in-group member's inherent disposition or personality. For example, a White person may see another White person carrying a knife and conclude, without further information, that the weapon must be carried for protection in a dangerous area.
3. If an out-group member is seen to perform in some laudable way, the behavior is often attributed to a variety of special circumstances, and the out-group member is seen as "the exception."
4. An in-group member who performs in the same laudable way is given credit for a worthy personality disposition.

Typical attribution errors include misperceptions between racial groups and between men and women (Taylor et al. 2013; Kluegel and Bobo 1993). In the case of race, recent events in the news tend to bear this out: In the case of a Black youth shot and killed by a White policeman, and *without further information*, a White person will tend to conclude that the White policeman was probably justified in the shooting. All else being equal, a Black person, without further information, will conclude that the shooting was probably unjustified and may well have been murder. These conclusions accurately describe—on the average—the reactions of Blacks and Whites to the well-publicized shootings of Black men by White police, such as the shootings of Trayvon Martin in Sanford, Florida; Michael Brown in Ferguson, Missouri; and, tragically, several others. Even in the aftermath of these police shootings, surveys find a large gap in how much White and Black Americans have confidence in the police (Drake 2014).

A related phenomenon has been seen in men's perceptions of women coworkers. Meticulous behavior in a man is perceived positively and is seen by other men as "thorough"; in a woman, the *exact same* behavior is perceived negatively and is considered "picky." Behavior applauded in a man as "aggressive" is condemned in a woman exhibiting the same behavior as "pushy" or "bitchy" (Uleman et al. 1996).

Social Networks

As already noted, no individual is a member of only one group. Social life is far richer than that. A **social network** is a set of links between individuals, between groups, or between other social units, such as bureaucratic organizations or even entire nations (Salganik 2015; Aldrich and Ruef 2006; Hargittai and Centeno 2001; Mizruchi 1992). One could say that any given person belongs simultaneously to several networks (Wasserman and Faust 1994). With the development of *social media* (see Chapter 2), networks that may have once been face-to-face have now developed through electronic media, such as on Facebook and Twitter. The development of social media brings a new dimension to the study and analysis of networks because you may be in a network with people you do not even know. Nonetheless, your group of friends, or all the people on an electronic mailing list to which you subscribe, or all of your Facebook subscribers are social networks, some human, some electronic.

Let us do a bit of network analysis (including group size effects) right now. Assume first that a group consisting of only two people has by definition one two-way relationship (each knows the other personally). A group of three people will thus have three possible two-way relationships; and a group of four people will have six possible two-way relationships. Extending this simple counting of the number of pairs (i.e., two-way relationships) shows that a group of five people has ten possible two-way relationships; a group of six people fifteen possible two-way relationships; and so on. With even as few as thirty "friends" on Facebook, there are actually 435 possible two-way (i.e., mutual) relationships! That is a *lot* of mutual relationships! As early as the 1950s, Robert Bales (1951) estimated that thirty is the limit for the number of people that can be in an intimate "small group" such that each person can have or recall a perception of each other member as an individual person. How "close" can your Facebook friends really be?

Networks can be critical to your success in life. Numerous research studies indicate that people get jobs via their personal networks more often than through formal job listings, want ads, or placement agencies (Ruef et al. 2003; Petersen et al. 2000; Granovetter 1995, 1974). Getting a job is more often a matter of whom you

Jim West/Alamy

Research finds that social networks are critical to finding jobs, even when those networks might be relatively weak.

what would a sociologist say?

Finding a Job: The Invisible Hand

Is getting a job simply a matter of getting the right credentials and training? Hardly, according to sociologists. Even in a good job market, one needs the help of social networks to find a job. This has been clearly demonstrated by sociologist Deirdre Royster, who compared the experiences of two groups of men: One group was White, the other Black. Both had graduated from vocational school. The Black and White men had comparable educational credentials, the same values, and the same work ethic; yet, the White men were far more likely to gain employment than were the Black men. Why? Royster's research revealed that the most significant difference between the two groups was access to job networks, just as Granovetter's work on job networks would predict (Royster 2003; Granovetter 1995).

know than what you know. Who you know, and whom they know in turn, is a social network that may have a marked effect on your life and career.

Networks form with all the spontaneity of other forms of human interaction (Wasserman and Faust 1994). Networks evolve, such as social ties within neighborhoods, professional contacts, and associations formed in fraternal, religious, occupational, and volunteer groups. Networks to which you are only *weakly* tied (you may know only one person in your neighborhood) provide you with access to that entire network, hence the sociological paradox that there is "strength in weak ties" (Granovetter 1973).

Networks based on race, class, and gender form with particular readiness. This has been especially true of job networks. The person who leads you to a job is likely to have a similar social background. Research indicates that the "old boy network"—any network of White, male corporate executives—is less important than it used to be, although it is certainly not by any means gone. The diminished importance of the old boy network is because of the increasing prominence of women and minorities in business organizations. In fact, among African American and Latino individuals, one's family can provide network contacts that can lead to jobs and upward mobility (Dominguez and Watkins 2003). Still, as we will see in various places later in this book, women and minorities are considerably underrepresented in corporate life, especially in high-status jobs, and since 2004, the presence of women and racial minorities on corporate boards has actually *declined* (Alliance for Board Diversity 2013). Some recent research shows that Blacks and Latinos relative to Whites are still disproportionately harmed by a lack of network contacts (Smith 2007).

The recent research of Gile et al. (2015) on network size has shown that network size is a very important force in human existence. His network research has been used to identify populations and subpopulations that are at risk of being struck with the human immunodeficiency virus (HIV) syndrome, the basis of the venereal disease AIDS. Using network sampling of *pairs* of individuals (network pairs as referred to previously), the researchers found that the estimated number of HIV-infected individuals in Curitiba, Brazil, turned out to be *five to ten times higher* than estimates that used standard survey sampling (that is, sampling individuals, not pairs). Thus, sampling using a network approach revealed a surprisingly large number of infected individuals.

Social Networks as "Small Worlds"

Networks can reach around the world, but how big is the world? How many of us, when we discover someone we just met is a friend of a friend, have remarked, "My, it's a small world, isn't it?"? Research into what has come to be known as the small world problem has shown that networks make the world a lot smaller than you might otherwise think.

Original small world researchers Travers and Milgram wanted to test whether a document could be routed via the U.S. postal system to a complete stranger more than 1000 miles away using only a chain of acquaintances (Watts 1999; Watts and Strogatz 1998; Kochen 1989; Lin 1989; Travers and Milgram 1969). If so, how many steps would be required? The researchers organized an experiment back in 1969 in which approximately 300 senders were all charged with getting a document to one receiver, a complete stranger. (Remember that all this was well before the advent of the desktop computer in the 1980s.) The receiver was a male Boston stockbroker. The senders were one group of Nebraskans and one group of Bostonians chosen completely at random. Every sender in the study was given the receiver's name, address, occupation, alma mater, year of graduation, wife's maiden name, and hometown. They were asked to send the document directly to the stockbroker only if they knew him on a first-name basis. Otherwise, they were asked to send the folder

We all like to think we stand on our own two feet, immune to a phenomenon as superficial as group pressure. The conviction that one is impervious to social influence results in what social psychologist Philip Zimbardo calls the *not me syndrome:* When confronted with a description of group behavior that is disappointingly conforming and not individualistic, most individuals counter that some people may conform to social pressure, "but not me"; or "some people yield quickly to styles of dress, but not me"; or "some people yield to autocratic authority figures, but not me" (Taylor et al. 2013; Zimbardo et al. 1977). Sociological experiments often reveal a dramatic gulf between what people *think* they will do and what they *actually do.* The original conformity study by Solomon Asch discussed next is a case in point.

The Asch Conformity Experiment

We learned in the previous sections that social influences are evidently quite strong. Are they strong enough to make us disbelieve our own senses? Are they strong enough to make us misperceive what is obviously objective, actual fact? In a classic piece of work known as the Asch conformity experiment, researcher Solomon Asch showed that even simple objective facts cannot withstand the distorting pressure of group influence (Asch 1955, 1951).

Examine the two illustrations in ▲ Figure 6.1. Which line on the right is more nearly equal in length to the line on the left (Line S)? Line B, obviously. Could anyone fail to answer correctly?

In fact, Solomon Asch discovered that social pressure of a rather gentle sort was sufficient to cause an astonishing rise in the number of wrong answers. Asch lined up five students at a table and asked which line in the illustration on the right is the same length as the line on the left. Unknown to the fifth student, the first four were *confederates*—collaborators with the experimenter who only pretended to be participants. For several rounds, with similar photos, the confederates gave correct answers to Asch's tests. The fifth student also answered correctly, suspecting nothing. Then on subsequent trials the first student gave a wrong answer. The second student gave the same wrong answer. Third, wrong. Fourth, wrong. Then came the fifth student's turn. If you were the fifth student, what would you have done?

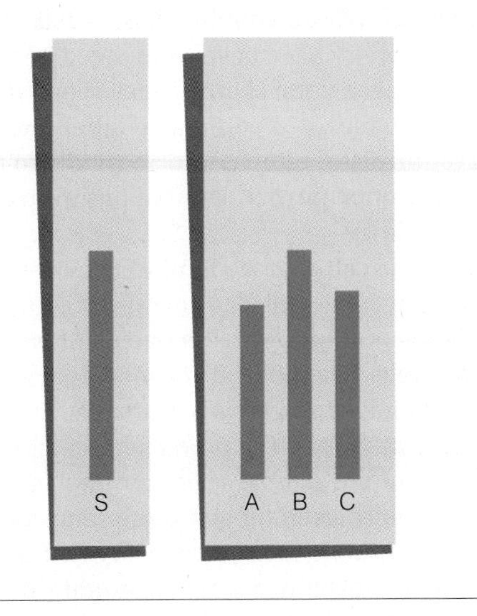

▲ **Figure 6.1** Lines from Asch Experiment

Source: Asch, Solomon. 1956. "Opinion and Social Pressure." *Scientific American* 19 (July): 31–36.

In Asch's experiment, fully *one-third* of all students in the fifth position gave the same wrong answer as the confederates at least half the time. Forty percent gave "some" wrong answers. Only one-fourth of the students consistently gave correct answers in defiance of the invisible pressure to conform.

Line length is not a vague or ambiguous stimulus. It is clear and objective, yet one-third of all subjects, a very high proportion, gave wrong answers. The subjects fidgeted and stammered while doing it, but they did it nonetheless. Those who did not yield to group pressure showed even more stress and discomfort than those who yielded to the (apparent) opinion of the group.

Would you have gone along with the group? Perhaps, perhaps not. Sociological insight grows when we acknowledge the fact that fully a third of all participants will yield to the group. The Asch experiment has been repeated many times over the years, with students and nonstudents, old and young, in groups of different sizes, and in different settings (Baumeister and Bushman 2008; Worchel et al. 2000). The results remain essentially the same! A third to a half of the participants make a judgment contrary to fact, yet in conformity with the group. Finally, the Asch findings have consistently revealed a *group size effect:* The greater the number of individuals (confederates) giving an incorrect answer (from five up to fifteen confederates), then the greater the number of subjects per group giving an incorrect answer.

The Milgram Obedience Studies

What are the limits of social group pressure? In terms of moral and psychological issues, judging the length of a line is a small matter. What happens if an authority figure demands obedience—a type of conformity—even if the task is something the test subject (the person) finds morally wrong and reprehensible? A chilling answer emerged from the now famous Milgram obedience studies done from 1960 through 1973 by Stanley Milgram (Milgram 1974).

In this study, a naive research subject entered a laboratory-like room and was told that an experiment on learning was to be conducted. The subject was to act as a "teacher," presenting a series of test questions to another person, the "learner." Whenever the learner gave a wrong answer, the teacher would administer an electric shock.

to a friend, relative, or acquaintance known on a first-name basis who might be more likely than the sender to know the stockbroker.

How many intermediaries do you think it took, on average, for the document to get through? (Most people estimate from twenty to hundreds.) The average number of intermediate contacts was only 6.2! However, only about one-third of the documents actually arrived at the target. This was still quite impressive, considering that the senders did not know the target person—hence, the current expression that any given person in the country is on average only about "six degrees of separation" from any other person. In this sense, the world is indeed "small."

This original small world research has recently been criticized on two grounds: First, only one-third of the documents actually reached the target person. The 6.2 average intermediaries applied only to these completed chains. Thus, two-thirds of the initial documents never reached the target person. For these people, the world was certainly not "small." Second, the sending chains tended to closely follow occupational, social class, and ethnic lines, just as general network theory would predict (Kleinfeld 1999; Wasserman and Faust 1994). Thus, the world may indeed be "small," but only for people in your immediate social network (Ruef et al. 2003; Watts 1999).

A study of Black national leaders by Taylor and associates (Jackson et al. 1995; Taylor 1992) shows that Black leaders form a very closely knit network, one considerably more closely knit than longer-established White leadership networks (Domhoff 2002; Jackson 2000; Jackson et al. 1995, 1994; Alba and Moore 1982; Moore 1979; Kadushin 1974). The world is indeed quite "small" for America's Black leadership. Included in the study were Black members of Congress, mayors, business executives, military officers (generals and full colonels), religious leaders, civil rights leaders, media personalities, entertainment and sports figures, and others. The study found that when considering only direct personal acquaintances—not indirect links involving intermediaries—one-fifth of the entire national Black leadership network know each other directly as a friend or close acquaintance. The Black leadership network is considerably more closely connected than White leadership networks. The Black network has greater *density*. Add only one intermediary, the friend of a friend, and the study estimated that almost *three-quarters* of the entire Black leadership network are included. Therefore, any given Black leader can generally get in touch with three-quarters of all other Black leaders in the country either by knowing them personally (a "friend") or via only one common acquaintance (a "friend of a friend"). That's pretty amazing when one realizes that the study is considering the population of Black leaders in the entire country.

Rene Meier/Reuters /Landov

Streaking, or running nude in a public place—relatively popular among college students in the 1970s and early 1980s and still popular on some campuses—is more common as a group activity than as a strictly individual one. This illustrates how the group can provide the people in it with deindividuation, or diffusion of responsibility among group members—a type of merging of self with group. This allows individuals to feel less responsibility or blame for their actions, thus convincing them that the group must share the blame.

Social Influence in Groups

The groups in which we participate exert tremendous influence on us. We often fail to appreciate how powerful these influences are. For example, who decides what you should wear? Do you decide for yourself each morning, or is the decision already made for you by fashion designers, role models, and your peers? Consider how closely your hair length, hair styling, and choice of jewelry have been influenced by your peers. Did you invent your skinny jeans, your dreadlocks, or your blue blazer? People who label themselves as "nonconformists" often conform rigidly to the dress-code and other norms of their in-group (a type of small network).

A group such as one's family even influences your adult life long after children have grown up and formed households of their own. The choice of political party among adults (Republican, Democratic, or Independent) correlates strongly with the party of one's parents, again demonstrating the power of the primary group. Seven out of ten people vote with the political party of their parents, even though these same people insist that they think for themselves when voting (Worchel et al. 2000). Furthermore, most people share the religious affiliation of their parents, although they will insist that they chose their own religion, free of any influence by either parent.

the shuttle to lift off given these conditions and their prior knowledge? The managers had all the information about the O-rings before the launch. Furthermore, engineers had warned them against the danger. In a detailed analysis of the decision to launch, sociologist Diane Vaughan (1996) (yes—a sociologist!) uncovered both risky shift and organizational ritualism within the organization. The NASA insiders, confronted with signals of danger, proceeded as if nothing was wrong when they were repeatedly faced with the evidence that something was indeed *very* wrong. They, in effect, *normalized* their own behavior so that their actions became acceptable to them, representing nothing out of the ordinary. This is an example of organizational ritualism, as well as what Vaughan calls the "normalization of deviance."

Unfortunately, history repeated itself on February 1, 2003, when the space shuttle *Columbia*, upon its return from space, broke up in a fiery descent into the atmosphere above Texas, killing all who were aboard. The evidence shows that a piece of hard insulating foam separated from an external fuel tank during launch and struck the shuttle's left wing, damaging it and dislodging its heat-resistant tiles that are necessary for reentry. The absence of these tiles caused a burn-up upon reentry into the atmosphere. With eerie similarity to the earlier 1986 *Challenger* accident, subsequent analysis concluded that a "flawed institutional culture" and—citing sociologist Diane Vaughan—a normalization of deviance accompanying a gradual erosion of safety margins were among the causes of the *Columbia* accident (Schwartz and Wald 2003).

No single individual was at fault in either accident. The story is not one of evil but rather of the ritualism of organizational life in one of the most powerful bureaucracies in the United States. It is a story of rigid group conformity within an organizational setting and of how deviant behavior is redefined, that is, socially constructed, just as also happened in the Penn State Sandusky scandal, already discussed. Organizational culture overshadows individual good judgment, creating a decrease in safety and increased risk. This is one of the hazards of organizational behavior.

Alienation. The stresses on rules and procedures within bureaucracies can result in a decrease in the overall cohesion of the organization. This often psychologically separates a person from the organization and its goals. This state of *alienation* results in increased turnover, tardiness, absenteeism, and overall dissatisfaction with an organization.

Alienation can be widespread in organizations where workers have little control over what they do, or where workers themselves are treated like machines employed on an assembly line, doing the same repetitive action for an entire work shift. Alienation is not restricted to manual labor, however. In organizations where workers are isolated from others, where they are expected only to implement rules, or where they think they have little chance of advancement, alienation can be common. As we will see, some organizations have developed new patterns of work to try to minimize worker alienation and thus enhance their productivity.

The McDonaldization of Society

Sometimes the problems and peculiarities of bureaucracy can have effects on the total society. This has been the case with what George Ritzer (2010) has called **McDonaldization**, a term coined from the well-known fast-food chain. In fact, 90 percent of U.S. children between ages 3 and 9 visit McDonald's each month! Ritzer noticed that the principles that characterize fast-food organizations are increasingly dominating more aspects of U.S. society, indeed, of societies around the world. McDonaldization refers to the increasing and ubiquitous presence of the fast-food model in most organizations that shape daily life. Work, travel, leisure,

George Frey/Bloomberg/Getty Images

Evidence of the "McDonaldization of society" can be seen everywhere, perhaps including on your own campus. Shopping malls, food courts, sports stadiums, even cruise ships reflect this trend toward standardization.

shopping, health care, politics, and even education have all become subject to McDonaldization. Each industry is based on a principle of high and efficient productivity, which translates into a highly rational social organization, with workers employed at low pay but with customers experiencing ease, convenience, and familiarity.

Ritzer argues that McDonald's has been such a successful model of business organization that other industries have adopted the same organizational characteristics, so much so that their nicknames associate them with the McDonald's chain: McPaper for *USA Today*, McChild for child-care chains like KinderCare, and McDoctor for the drive-in clinics that deal quickly and efficiently with minor health and dental problems.

Based in part upon Max Weber's concept of the ideal bureaucracy mentioned earlier, Ritzer identifies four dimensions of the McDonaldization process—efficiency, calculability, predictability, and control:

1. *Efficiency* means that things move from start to finish in a streamlined path. Steps in the production of a hamburger are regulated so that each hamburger is made exactly the same way—hardly characteristic of a home-cooked meal. Business can be even more efficient if the customer does the work once done by an employee. In fast-food restaurants, the claim that you can "have it your way" really means that you assemble your own sandwich or salad.
2. *Calculability* means there is an emphasis on the quantitative aspects of products sold: size, cost, and the time it takes to get the product. At McDonald's, branch managers must account for the number of cubic inches of ketchup used per day; likewise, ice cream scoopers in chain stores measure out predetermined and exact amounts of ice cream.
3. *Predictability* is the assurance that products will be exactly the same, no matter when or where they are purchased. Eat an Egg McMuffin in New York, and it will likely taste just the same as an Egg McMuffin in Los Angeles or Paris!
4. *Control* is the primary organizational principle that lies behind McDonaldization. Behavior of the customers and workers is reduced to a series of machinelike actions. Ultimately, efficient technologies replace much of the work that humans once performed.

McDonaldization clearly brings many benefits. There is a greater availability of goods and services to a wide proportion of the population; instantaneous service and convenience to a public with less free time; predictability and familiarity in the goods bought and sold; and standardization of pricing and uniform quality of goods sold, to name a few benefits. However, this increasingly rational system of goods and services also spawns irrationalities. For example, the majority of workers at McDonald's lack full-time employment, have no worker benefits, have no control over their workplace, have no pension, and quit on average after only four or five months.

Diversity in Organizations

The hierarchical structuring of positions within organizations results in the concentration of power and influence with a few individuals at the top. Because organizations tend to reflect patterns within the broader society, this hierarchy, like that of society, is marked by inequality in race, gender, and class relations. Although the concentration of power in organizations is incompatible with the principles of a democratic society, organizations are structured by hierarchies and discrimination is still quite pervasive, especially among the power elite, White men still predominate, and studies find that, even though the presence of women and people of color is growing in positions of organizational leadership, they tend to take on the same values as the dominant group, evidence once again of group conformity (Zweigenhaft and Domhoff 2006).

DEBUNKING Society's Myths

Myth: Putting more women and people of color into positions of power will transform institutions.
Sociological Perspective: Not necessarily. Although having diverse people involved in decision making tends to produce more innovation, there is also a tendency for new members of a group to conform to the values and orientations of the dominant group (Zweigenhaft and Domhoff 2006).

A classic study by Rosabeth Moss Kanter (1977) shows how the structure of organizations leads to obstacles in the advancement of groups who are underrepresented in the organization. People who are underrepresented in the organization become tokens; they feel put "out front" and under the all-too-watchful eyes of their superiors as well as peers. As a result—as research since Kanter's has shown—they often suffer severe stress (Smith 2007; Jackson 2000). They may be assumed to be incompetent, getting their position simply because they are women, minorities, or both—even in instances where the person has had superior admissions qualifications. This is stressful for a person and shows that tokenism can have very negative consequences (Guttierez y Muhs et al. 2012).

Social class, in addition to race and gender, plays a part in determining people's place within formal organizations. Employees of middle- and upper-class origins in organizations make higher salaries and wages and are more likely to get promoted than are people of

Avava/iStockphoto.com

Few organizational boards and executive committees contain minorities and women, unlike what is pictured here.

lower social class origins, even for individuals who are of the same race or ethnicity. Few organizational boards and executive committees contain minorities and women: When present, they are often tokens. This even holds for people coming from families of lower social class status who are as well educated as their middle- and upper-class coworkers. Thus their lower salaries and lack of promotion cannot necessarily be attributed to a lack of education. In this respect, their treatment in the bureaucracy only perpetuates rather than lessens the negative effects of the social class system in the United States.

The social class stratification system in the United States produces major differences in the opportunities and life chances of individuals, and the bureaucracy simply carries these differences forward. Class stereotypes also influence hiring practices in organizations. Personnel officers look for people with "certain demeanors," a code phrase for those who convey middle-class or upper-middle-class standards of dress, language, manners, and so on, which some people may be unable to afford or may not possess.

Even as the structure of organizations reproduces the race, class, and gender inequalities that permeate society, ample research now finds that diversity within organizations has numerous benefits. Diverse groups—that is, diverse people—bring different experiences and perspectives to organizations and to organizational decision making. Of course, the problem is that there is still pressure on such people to conform to the dominant culture of the organization. That pressure to conform can silence dissent, as groupthink would suggest, especially if those who bring diversity to the organization, such as women, gays, lesbians, bisexuals, transgender people, and people of color, are treated as tokens or silenced because they are different. When people are tokens in organizations, they are pressured not to stand out or, when they speak out, they may be ignored—or, worse, others take credit for their ideas.

Nonetheless, new research on diversity is consistently demonstrating the benefit of diversity for all kinds of organizations. In schools, all students learn more when in classrooms where there are people from different backgrounds (Gurin et al. 2002). In business organizations, racial diversity is associated with increased sales revenues, more customers, a stronger market share, and higher profits (Herring 2009). There is ample evidence that companies are now much more aware of the fact that innovation is more likely to occur in diverse work organizations (Page 2007). On college campuses, more cross-race interaction produces a more positive campus climate (Valentine et al. 2012).

DEBUNKING Society's Myths

Myth: Diversity is a real problem for organizations.
Sociological Perspective: Despite the challenges posed by trying to create more diverse work organizations, research shows that more diverse organizations have greater profits and are more innovative (Bell 2011; Herring 2009; Page 2007).

Functionalism, Conflict Theory, and Symbolic Interaction: Theoretical Perspectives

All three major sociological perspectives—functionalism, conflict theory, and symbolic interaction—are exhibited in the analysis of formal organizations and bureaucracies (see ◆ Table 6.1). The functional perspective, based in this case on the early writing of Max Weber, argues that certain functions, called *eufunctions* (that is, positive functions), characterize bureaucracies and contribute to their overall unity. The bureaucracy exists to accomplish these eufunctions, such as efficiency, control, impersonal relations, and a chance for individuals to develop a career within the organization. As we have seen, however, bureaucracies develop the "other face" (informal interaction and culture, as opposed to formal or bureaucratic interaction and culture), as well as problems of ritualism and alienation of people from the organization. These latter problems are called *dysfunctions* (negative

◆ Table 6.1 Theoretical Perspectives on Organizations

	Functionalist Theory	Conflict Theory	Symbolic Interaction Theory
Central Focus	Positive functions (such as efficiency) contribute to unity and stability of the organization.	Hierarchical nature of bureaucracy encourages conflict between superiors and subordinates, men and women, and people of different racial or class backgrounds. Tokenism often results.	This theory stresses the role of self in the bureaucracy and how the self develops and changes.
Relationship of Individual to the Organization	Individuals, like parts of a machine, are only partly relevant to the operation of the organization.	Individuals are subordinated to systems of power and experience stress and alienation as a result.	Interaction between superiors and subordinates forms the structure of the organization.
Criticism	Hierarchy can result in dysfunctions such as ritualism and alienation.	This theory de-emphasizes the positive ways that organizations work.	This theory tends to downplay overall social organization.

functions), which have the consequence of contributing to disunity, lack of harmony, and less efficiency in the bureaucracy. Finally, with increasing diversity in organizations, tokenism, an organizational dysfunction, may result.

The conflict perspective argues that the hierarchical or stratified nature of the bureaucracy in effect encourages rather than inhibits conflict among individuals within it. These conflicts are between superiors and subordinates, as well as between racial and ethnic groups, men and women, and people of different social class backgrounds, hampering smooth and efficient running of the bureaucracy. Furthermore, conflict theory helps us understand the power structures that exist in organizations—both the formal ones that come from the organizational hierarchy and the less formal ways that power is exercised between people and among groups within the organization.

Symbolic interaction theory stresses the role of the self in any group and especially how the self develops as a product of social interaction. Within organizations, people may feel that their "self" becomes subordinated to the larger structure of the organization. This is especially true in bureaucratic organizations where individuals often feel overwhelmed by the sheer complexity of working through bureaucratic structures. But symbolic interaction also emphasizes the creativity of human beings as social actors and thus would be a good perspective to use if analyzing how people change organizational structures and cultures.

Chapter Summary

What are the types of groups?

Groups are a fact of human existence and permeate virtually every facet of our lives. Group size is important, and determines quite a bit, as does the otherwise simple distinction between dyads and triads. *Primary groups* form the basic building blocks of social interaction in society. *Reference groups* play a major role in forming our attitudes and life goals, as do our relationships with in-groups and out-groups. *Social networks* partly determine things such as who we know and the kinds of jobs we get. Networks based on race or ethnicity, social class, and other social factors are extremely closely connected and are very dense. Network research has shown that network sampling (for example, sampling pairs of individuals) predicts who is at risk for AIDS/HIV better than does traditional survey sampling.

How strong is social influence?

The social influence groups exert on us is tremendous, as seen by the Asch conformity experiments. The Milgram experiments demonstrated that the interpersonal influence of an authority figure can cause an individual to act against his or her deep convictions. The torture and abuse of Iraqi prisoners of war by American soldiers/prison guards serves as testimony to the powerful effects of both social influence and authority structures. The Iraqi tortures were in effect experimentally predicted by a simulated prison study done in the United States over thirty years earlier.

What is the importance of groupthink and risky shift?

Groupthink can be so pervasive that it adversely affects group decision making and often results in group decisions that by any measure are simply stupid. *Risky shift* (and *polarization shift*) similarly often compel individuals to reach decisions that are at odds with their better judgment.

What are the types of formal organizations and bureaucracies, and what are some of their problems?

There are several types of *formal organizations,* such as *normative, coercive,* or *utilitarian.* Weber typified *bureaucracies* as organizations with an efficient division of labor, an authority hierarchy, rules, impersonal relationships, and career ladders. Bureaucratic rigidities often result in organizational problems such as ritualism and resulting "normalization of deviance." The *McDonaldization* of society has resulted in greater efficiency, calculability, and control in many industries, probably at the expense of some individual creativity. Formal organizations perpetuate society's inequalities on the basis of race or ethnicity, gender, and social class. Current research finds, however, that innovation in organizations is more likely if there is greater diversity—and thus a variety of perspectives—within the organization.

What do functional, conflict, and symbolic interaction theories say about organizations?

Functional, conflict, and symbolic interaction theories highlight and clarify the analysis of organizations by specifying both organizational functions and dysfunctions (functional theory); by analyzing the consequences of hierarchical, gender, race, and social class conflict in organizations (conflict theory); and, finally, by studying the importance of social interaction and integration of the self into the organization (symbolic interaction theory).

Key Terms

7

Deviance and Crime

In the early 1970s, an airplane carrying forty members of an amateur rugby team crashed in the Andes Mountains in South America. The twenty-seven survivors were stranded at 12,000 feet in freezing weather and deep snow. There was no food except for a small amount of chocolate and some wine. A few days after the crash, the group heard on a small transistor radio that the search for them had been called off.

Scattered in the snow were the frozen bodies of dead passengers. Preserved by the freezing weather, these bodies became, after a time, sources of food. At first, the survivors were repulsed by the idea of eating human flesh, but as the days wore on, they agonized over the decision about whether to eat the dead crash victims, eventually concluding that they had to eat if they were to live.

In the beginning, only a few ate the human meat, but soon the others began to eat too. The group experimented with preparations as they tried different parts of the body. They developed elaborate rules (social norms) about how, what, and whom they would eat.

After two months, the group sent out an expedition of three survivors to find help. The group was rescued, and the world learned of their ordeal. Their cannibalism (the eating of other human beings) generally came to be accepted as something they had to do to survive. Although people might have been repulsed by the story, the survivors' behavior was understood as a necessary adaptation to their life-threatening circumstances. The survivors also maintained a sense of themselves as good people even though what they did profoundly violated ordinary standards of socially acceptable behavior in most cultures in the world (Henslin 1993; Read 1974).

Was the behavior of the Andes crash survivors socially deviant? Were the people made crazy by their experience, or was this a normal response to extreme circumstances?

Compare the Andes crash to another case of human cannibalism. In 1991, in Milwaukee, Wisconsin, Jeffrey Dahmer pled guilty to charges of murdering at least fifteen men in his home. Dahmer lured the men to his apartment, where he murdered and dismembered them, then cooked and ate some of their body parts. For those he considered most handsome, he boiled the flesh from their heads so that he could save and

in this chapter, you will learn to:

- Present a sociological definition of deviance as a social construction
- Compare and contrast theoretical approaches to understanding deviant behavior
- Explain the importance of labels in determining deviant behavior
- Relate the social structure associated with deviant identities and deviant communities
- Examine the race, class, and gender disparities within the criminal justice system

admire their skulls. Dahmer was seen as a total social deviant, someone who violated every principle of human decency. Even hardened criminals were disgusted by Dahmer. In fact, he was killed by another inmate in prison in 1994.

Why was Dahmer's behavior considered so deviant when that of the Andes survivors was not? The answer can be found by looking at the situation in which these behaviors occurred. For the Andes survivors, eating human flesh was essential for survival. For Dahmer, however, it was murder. From a sociological perspective, the deviance of cannibalism resides not just in the act itself but also in the social context in which it occurs. The exact same behavior—eating other human beings—is considered reprehensible in one context and acceptable in another. That is the essence of the sociological explanation. The nature of deviance is not simply about the deviant act itself, nor is deviance just about the individual who engages in the behavior. Instead, social deviance is socially constructed and a product of social structure.

Defining Deviance

Sociologists define **deviance** as behavior that is recognized as violating expected rules and norms. Deviance is more than simple nonconformity—most of us may "break the rules" now and again. Behavior that departs significantly from social expectations is deviant. In the sociological perspective on deviance, there are four main identifying characteristics:

- Deviance emerges in a social context, not just in the behavior of individuals; sociologists see deviance in terms of group processes and judgments. Deviance, therefore, can change overtime, or from one setting to another.
- Groups judge behaviors differently. What is deviant to one group may be normative to another.
- Established rules and norms are socially created, not just morally decided or individually imposed.
- Deviance lies not just in behavior itself but also in the social responses to behavior and people engaged in the behavior.

Sociological Perspectives on Deviance

Strange, unconventional, or nonconformist behavior is often understandable in its sociological context. Consider suicide. Are all people who commit suicide suffering with mental illness? Might their behavior instead be explained by social factors? Think about it. There are conditions under which suicide may well be acceptable behavior—for example, someone who voluntarily receives a lethal dose of medicine to end her life in the face of a terminal illness, compared to a despondent person who jumps from a window.

Sociologists distinguish two types of deviance: formal and informal. *Formal deviance* is behavior that breaks laws or official rules. Crime is an example. There are formal sanctions against formal deviance, such as imprisonment and fines. *Informal deviance* is behavior that violates customary norms. Although such deviance may not be specified in law, it is judged to be deviant by those who uphold the society's norms.

SEE for YOURSELF

Perform an experiment by doing something mildly deviant for a period, such as carrying around a teddy bear doll and treating it as a live baby, or standing in the street and looking into the air, as though you are looking at something up there. Make a record of how others respond to you, and then ask yourself how labeling theory is important to the study of deviance.

How might reactions to you differ had you been of another race or gender? You might want to structure this question into your experiment by teaming up with a classmate of another race or gender. You could then compare each of your responses to the same behavior. A note of caution: Do not do anything illegal or dangerous. Even the most seemingly harmless acts of deviance can generate strong reactions, so be careful in planning your sociological exercise!

The study of deviance includes both the study of why people violate laws or norms and the study of how society reacts. *Labeling theory* is discussed in detail later, but it recognizes that deviance is not just in the breaking of norms or rules but it includes how people react to those behaviors. Sociologists consider the social context of the behavior, the social construction of that behavior as deviant, and the response to the behavior (Becker 1963).

DEBUNKING Society's Myths

Myth: Deviance is bad for society because it disrupts normal life.

Sociological Perspective: Deviance tends to stabilize society. By defining some forms of behavior as deviant,

what would a sociologist say?

Drugs as Deviance or Crime

Many people consider all drug use to be criminal or deviant behavior, but "drugs" includes a broad spectrum of substances. Alcohol, tobacco, caffeine, and prescription drugs are all legal drugs. Let's consider prescription drugs. Most people would agree that using a medication that has been prescribed to you by a medical doctor for a legitimate illness is neither deviant nor criminal. The same drug used by someone who does not have an illness or a prescription to use the medication would be considered deviance. Sociological research shows that the nonmedical use of prescription drugs is increasingly a problem on college campuses and among high school adolescents.

So-called "study drugs" are used to help students maintain focus on their academics. The drugs, like Adderall or Ritalin, are intended to aid those with diagnosed ADHD, but other students are using them to help improve academic performance. There are health risks and criminal consequences, yet most students do not consider using these study drugs as deviant (Nargiso et al. 2015).

people are affirming the social norms of groups. In this sense, society actually *creates* deviance to some extent (Durkheim 1951/1897).

The Social Context of Deviance. Even the most unconventional behavior can be understood if we know the context in which it occurs. Behavior that is deviant in one circumstance may be normal in another, or behavior may be ruled deviant only when performed by certain people. For example, people who break gender stereotypes may be judged as deviant even though their behavior is considered normal for the other sex. Women who have painted fingernails, shaved legs, and wear eye makeup are feminine and "normal." Except for those who are on stage or on camera, men who wear nail polish and makeup are usually regarded as "deviant."

Another example regards the consumption of alcohol, a legal drug. Whether someone who drinks is judged to be an alcoholic depends in large part on the social context in which one drinks, not solely on the amount of alcohol consumed. Drinking wine from a bottle in a brown bag on the street corner is considered highly deviant. Having martinis in a posh bar is perfectly acceptable among adults. The act of drinking alcohol is not intrinsically deviant. The societal reaction to it determines deviance.

The definition of deviance can also change over time. Acquaintance rape (also called "date rape"), for example, was not considered social deviance until fairly recently. Women have been presumed to mean yes when they said no, and men were expected to "seduce" women through aggressive sexual behavior. Even now, women who are raped by someone they know may not think of it as rape. If they do, they may find that prosecuting the offender is difficult because others do not think of it as rape, especially under certain circumstances, such as the woman being drunk.

Understanding the context in which deviance occurs and the context in which it is punished reveals much about the norms of society. The sociologist Emile Durkheim argued that one reason acts of deviance are publicly punished is that the social order is threatened by deviance. Judging those behaviors as deviant and punishing them confirms general social standards. Therein lies the value of widely publicized trials and public executions. The punishment affirms the collective beliefs of the society, reinforces social order, and inhibits future deviant behavior, especially as defined by those with the power to judge others.

The Influence of Social Movements. The perception of deviance may also be influenced by *social movements*, which are networks of groups that organize to support or resist changes in society (see Chapter 16). With a change in the social climate, formerly acceptable behaviors may be newly defined as deviant. Cigarette smoking, for instance, was once considered glamorous, sexy, and "cool." The social climate toward smoking, however, has changed. In 1987, only 17 percent of people thought that smoking should be banned in public places. Recent estimates are that over half (56 percent) supported a ban on smoking cigarettes in public spaces (Riffkin 2014). The increase in public disapproval of smoking results as much from social and political movements as it does from the known health risks.

The antismoking movement was successful in articulating to the public that smoking is dangerous. The ability of people to mobilize against something, in this case cigarette smoking, is just as important to creating the context for deviance as is any evidence of risk of harm from cigarettes. In other words, there has to be a social response for deviance to be defined as such; scientific evidence of harm in and of itself is not enough.

The Social Construction of Deviance

Perhaps because it violates social conventions or because it sometimes involves unusual behavior, deviance captures the public imagination. Commonly, however, the public understands deviance as the result of individualistic or personality factors. Many people see deviants as crazy, threatening, "sick," or in some other way inferior, but sociologists see deviance as influenced by society—the same social processes and institutions that shape all social behavior.

Deviance, for example, is not necessarily irrational or "sick" and may be a positive and rational adaptation to a situation. Think of the Andes survivors discussed in this chapter's opener. Was their action (eating human flesh) irrational, or was it an inventive and rational response to a dreadful situation? To use another example, is marijuana use part of a deviant subculture, or are some people using marijuana as a rational response to personal circumstances, such as illness?

Marijuana use, especially by smoking a "joint" or inhaling through a "bong," although legal in some states, is considered deviant behavior. Most Americans do not use it and consider marijuana use undesirable behavior. Using medical marijuana, though, to help with the pain and nausea from cancer treatments is a rational choice and blurs the lines of deviant and "normative" behavior. In fact, if marijuana is distributed to patients in pill form, consistent with other types of medication, most people think marijuana use is acceptable. The actual *use* of marijuana is only seen as deviant in the context of how it is administered. Taking a marijuana pill is not considered as deviant as smoking a joint or inhaling marijuana smoke through a bong (Rudski 2014). The social context in which marijuana use takes place includes both how it is used and why.

High-risk drinking of alcohol is acceptable behavior among students on many college campuses. This same behavior among adults outside of the college culture is considered problematic and possibly alcoholism.

Smoking for recreational purposes and using pipes or bongs is considered more deviant than using marijuana for medicinal purposes.

Also, in some subcultures or situations, deviant behavior is encouraged and praised. Have you ever been egged on by friends to do something that you thought was deviant, or have you done something you knew was wrong? Many argue that the reason so many college students drink excessively is that the student subculture encourages them to do so—even though students know it is harmful. High-risk drinking is characterized by drinking to the point of vomiting, blacking out, and possibly even dying from alcohol poisoning. Still, a college campus with a culture of drinking will encourage people to drink heavily, despite the known risks.

The Medicalization of Deviance

Commonly, people will say that someone who commits a very deviant act is "sick." This common explanation is what sociologists call the **medicalization of deviance** (Conrad and Schneider 1992). Medicalizing deviance attributes deviant behavior to a "sick" state of mind, where the solution is to "cure" the deviance through therapy or other psychological treatment.

As an example, some evidence indicates that alcoholism may have a genetic basis, and certainly alcoholism must be understood at least in part in medical terms, but viewing alcoholism *solely* from a medical perspective ignores the social causes that influence the development and persistence of this behavior. Practitioners know that medical treatment alone does not solve the problem. The social relationships, social

that the economic organization of capitalist societies produces deviance and crime. The high rates of crime among the poorest groups, especially economic crimes such as theft, robbery, prostitution, and drug selling, are a result of the economic status of these groups. Rather than emphasizing values and conformity as a source of deviance as do functional analyses, conflict theorists see crime in terms of power relationships and economic inequality (Grant and Martínez 1997).

The upper classes, conflict theorists point out, can also better hide crimes they commit because affluent groups have the resources to mask their deviance and crime. As a result, a working-class man who beats his wife is more likely to be arrested and prosecuted than an upper-class man who engages in the same behavior. In addition, those with greater resources can afford to buy their way out of trouble by paying bail, hiring expensive attorneys, or even resorting to bribes.

Corporate crime is crime committed within the legitimate context of doing business. Conflict theorists expand our view of crime and deviance by revealing the significance of such crimes. They argue that appropriating profit based on exploitation of the poor and working class is inherent in the structure of capitalist society. **Elite deviance** refers to the wrongdoing of wealthy and powerful individuals and organizations (Simon 2011). Elite deviance includes what early conflict theorists called *white-collar crime* (Sutherland and Cressey 1978; Sutherland 1940). Elite deviance includes tax evasion; illegal campaign contributions; illegal investment schemes that steal money from innocent investors; corporate scandals, such as fraudulent accounting practices that endanger or deceive the public but profit the corporation or individuals within it; and even government actions that abuse the public trust.

The ruling groups in society develop numerous mechanisms to protect their interests according to conflict theorists who argue that law, for example, is created by elites to protect the interests of the dominant class. Thus law, supposedly neutral and fair in its form and implementation, works in the interest of the most well-to-do (Weisburd et al. 2001, 1991; Spitzer 1975).

Conflict theory emphasizes the significance of social control in managing deviance and crime. **Social control** is the process by which groups and individuals within those groups are brought into conformity with dominant social expectations. Social control can take place simply through socialization, but dominant groups can also control the behavior of others through marking them as deviant. An example is the historic persecution of witches during the Middle Ages in Europe and during the early colonial period in America (Ben-Yehuda 1986; Erikson 1966). Witches often were women who were healers and midwives—those whose views were at odds with the authority of the exclusively patriarchal hierarchy of the church, then the ruling institution.

One implication of conflict theory, especially when linked with labeling theory, is that the power to define deviance confers an important degree of social control. **Social control agents** are those who regulate and administer the response to deviance, such as the police and mental health workers. Members of powerless groups may be defined as deviant for even the slightest infraction against social norms, whereas others may be free to behave in deviant ways without consequence. Oppressed groups may actually engage in more deviant behavior, but it is also true that they have a greater likelihood of being labeled deviant and incarcerated or institutionalized, whether or not they have actually committed an offense. This is evidence of the power wielded by social control agents.

When powerful groups hold stereotypes about other groups, the less powerful people are frequently assigned deviant labels. As a consequence, the least powerful groups in society are subject most often to social control. You can see this in the patterns of arrest data. All else being equal, poor people are more likely to be considered criminals and therefore are more likely to be arrested, convicted, and imprisoned than middle- and upper-class people. The same is true of Latinos, Native Americans, and African Americans. Sociologists point out that this does not necessarily mean that these groups are somehow more criminally prone; rather, they take it as partial evidence of the differential treatment of these groups by the criminal justice system.

Conflict Theory: Strengths and Weaknesses. The strength of conflict theory is its insight into the significance of power relationships in the definition, identification, and handling of deviance. It links the commission, perception, and treatment of crime to inequality in society and offers a powerful analysis of how the injustices of society produce crime and result in different systems of justice for disadvantaged and privileged groups. This theory is not without its weaknesses, however, and critics point out that laws protect most people, not just the affluent, as conflict theorists argue.

In addition, although conflict theory offers a powerful analysis of the origins of crime, it is less effective in explaining other forms of deviance. For example, how would conflict theorists explain the routine deviance of middle-class adolescents? They might point out that consumer marketing drives much of middle-class deviance. Profits are made from the accoutrements of deviance—rings in pierced eyebrows, "gangsta" rap music, and so on—but economic interests alone cannot explain all the deviance observed in society. As Durkheim argued, deviance is functional for the whole of society, not just those with a major stake in the economic system.

Symbolic Interaction Theories of Deviance

Whereas functionalist and conflict theories are *macrosociological* theories, certain *microsociological* theories of deviance look directly at the interactions people have with one another as the origin of social deviance. *Symbolic interaction theory* holds that people behave as they do because of the meanings people attribute to situations. This perspective emphasizes the meanings surrounding deviance, as well as how people respond to those meanings. Symbolic interaction emphasizes that deviance originates in the interaction between different groups and is defined by society's reaction to certain behaviors.

Symbolic interactionist theories of deviance originated in the perspective of the Chicago School of Sociology. **W. I. Thomas** (1863–1947), one of the early sociologists from the University of Chicago, was among the first to develop a sociological perspective on social deviance. Thomas explained deviance as a normal response to the social conditions in which people find themselves. Thomas was one of the first to argue that delinquency was caused by the social disorganization brought on by slum life and urban industrialism. He saw deviance as a problem of social conditions, less so of individual character or individual personality.

Differential Association Theory. Thomas's work laid the foundation for a classic theory of deviance: differential association theory. **Differential association theory**, a type of symbolic interaction theory, interprets deviance, including criminal behavior, as behavior one learns through interaction with others (Sutherland and Cressey 1978; Sutherland 1940). Edwin Sutherland argued that becoming a criminal or a juvenile delinquent is a matter of learning criminal ways within the primary groups to which one belongs. To Sutherland, people become criminals when they are more strongly socialized to break the law than to obey it. Differential association theory emphasizes the interaction people have with their peers and others in their environment. Those who "differentially associate" with delinquents, deviants, or criminals learn to value deviance. The greater the frequency, duration, and intensity of their immersion in deviant environments, the more likely it is that they will become deviant.

Consider the case of cheating on college tests and assignments. Students learn from others about the culture of cheating, namely that because everyone does it, cheating is okay. Students also share the best ways to cheat without getting caught. Students, who would ordinarily not engage in criminal or unethical behavior, are socialized to become cheaters themselves. Sociologists found that students who were told by another student in the room how to cheat on a word memorization experiment were much more likely to do it (Paternoster et al. 2013). Differential association theory offers a compelling explanation for how deviance is culturally transmitted—that is, people pass on deviant expectations through the social groups in which they interact.

Critics of differential association theory have argued that this perspective tends to blame deviance on the values of particular groups. Differential association has been used, for instance, to explain the higher rate of crime among the poor and working class, arguing that this higher rate of crime occurs because they do not share the values of the middle class. Such an explanation, critics say, is class biased, because it overlooks the deviance that occurs in the middle-class culture and among elites. Disadvantaged groups may share the values of the middle class but cannot necessarily achieve them through legitimate means.

Deviance: The Importance of Labels. **Labeling theory** is a branch of symbolic interaction theory that interprets the responses of others as the most significant factor in understanding how deviant behavior is both created and sustained (Becker 1963). The work of labeling theorists such as Becker stems from the work of W. I. Thomas, who wrote, "If men define situations as real, they are real in their consequences" (Thomas and Thomas 1928: 572). A *label* is the assignment or attachment of a deviant identity to a person by others, including by agents of social institutions. People's reactions, not the action itself, produce deviance as a result of the labeling process.

Linked with conflict theory, labeling theory shows how those with the power to label an act or a person deviant and to impose sanctions—such as police, court officials, school authorities, experts, teachers, and official agents of social institutions—wield great power in determining societal understandings of deviance. Furthermore, because deviants are handled through bureaucratic organizations, the workers within these bureaucracies "process" people according to rules and procedures, seldom questioning the basis for those rules.

Once the label is applied, it sticks, and it is difficult for a person labeled deviant to shed the label—namely, to recover a nondeviant identity. To give an example, once a social worker or psychiatrist labels clients as mentally ill, those people will be treated as mentally ill, regardless of their actual mental state. In a kind of "catch-22," when people labeled as mentally ill plead that they are indeed mentally sound, this is taken as evidence that they are, in fact, mentally ill!

A person need not have actually engaged in deviant behavior to be labeled deviant and for that label to stick. Labeling theory helps explain why convicts released

doing sociological research

The Rich Get Richer and the Poor Get Prison

Research Question: Jeffrey Reiman and Paul Leighton (2012) have studied U.S. prisons by asking: (1) What happens in prisons? and (2) What are the perceptions of prisons held by those in society?

Research Method: Reiman and Leighton used field research in prisons to answer these questions.

Research Results: The researchers found that the prison system in the United States, instead of serving as a way to rehabilitate criminals, is in effect designed to train and socialize inmates into a career of crime. It is also designed in such a way as to assure the public that crime is a threat primarily from the poor and that it originates at the lower rungs of society. Prisons contain elements that seem designed to accomplish this view.

Conclusions and Implications: One can "construct" a prison that ends up looking like a U.S. prison. First, continue to label as criminal those who engage in crimes that have no unwilling victim, such as prostitution or gambling. Second, give prosecutors and judges broad discretion to arrest, convict, and sentence based on appearance, dress, race, and apparent social class. Third, treat prisoners in a painful and demeaning manner, as one might treat children. Fourth, make certain that prisoners are not trained in a marketable skill that would be useful upon their release. And, finally, assure that prisoners will forever be labeled and stigmatized as different from "decent citizens," even after they have paid their debt to society. Once an ex-con, always an ex-con. One has thus just socially constructed a U.S. prison, an institution that will continue to generate the very thing that it claims to eliminate.

Questions to Consider

1. In your own opinion, how accurate is this "construction" of the U.S. prison? Do you know anyone who is currently in or recently in prison? Interview them and get their opinion.
2. How persistent in the coming years do you think this vision of the U.S. prison system will be?

Sources: Reiman, Jeffrey H., and Paul Leighton. 2012. *The Rich Get Richer and the Poor Get Prison*, 10th ed. Upper Saddle River, NJ: Pearson.

from prison have such high rates of *recidivism* (return to criminal activities). Convicted criminals are formally and publicly labeled wrongdoers. They are treated with suspicion ever afterward and have great difficulty finding legitimate employment: The label "ex-con" defines their future options.

Former inmates struggle to find employment after release from prison, especially if the person is male and Black or Hispanic. Sociologist Devah Pager has shown this clearly through her research. Pager (2007) had pretrained role-players pose as ex-cons looking for a job. These role-players went into the job market and were interviewed for various jobs; all of them used the same preset script during the interview. The idea of the study was to see how many of them would be invited back for another interview. The results were staggering: Blacks who were *not* ex-cons were *less* likely to be invited back for a job interview than were Whites who *were* ex-cons, even though White ex-cons were not invited back in large numbers. All ex-cons had trouble being invited back, but even more so for Black and Hispanic ex-cons. So the effect of race alone exceeded the effect of incarceration alone. These upsetting differences could not be attributed to differences in interaction displayed during the interview, because everyone used the exact same prepared script.

The prison system in the United States also shows the power of labeling theory. Prisons, in effect, *train* and *socialize* prisoners into a career of secondary deviance. (See the box, "Doing Sociological Research: The Rich Get Richer and the Poor Get Prison.") Reiman (2012) argues that the goal of the prison system is not to reduce crime but to impress upon the public that crime is inevitable, originating only from the lower classes. Prisons accomplish this, even if unintentionally, by demeaning prisoners and stigmatizing them as different from "decent citizens," not training them in marketable skills. As a consequence, these people will never be able to pay their debt to society, and the prison system has created the very behavior it intended to eliminate.

Labeling theory suggests that deviance refers not just to something one does but to something one becomes. **Deviant identity** is the definition a person has of himself or herself as a deviant. Most often, deviant identities emerge over time (Simon 2011; Lemert 1972). A person addicted to drugs, for example, may not think of herself as a junkie until she realizes she no longer has nonusing friends. The formation of a deviant identity, like other identities, involves a process of social transformation in which a new self-image and new public definition of a person emerge. This is a process that involves how people view deviants and how deviants view themselves.

A social **stigma** is an attribute that is socially devalued and discredited. Some stigmas result in people being labeled deviant. The experiences of people who are disabled, disfigured, or in some other way stigmatized are studied in much the same way as other forms

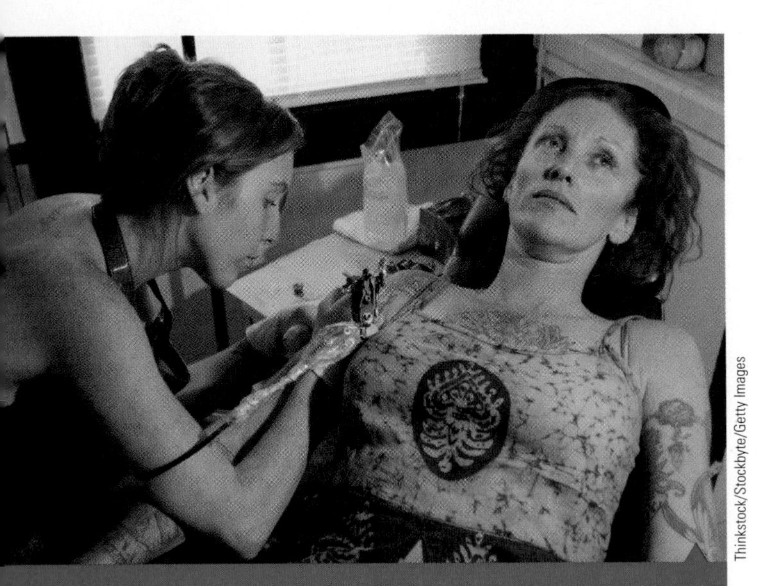

Extensive tattooing is regarded by many as deviant, although it may seem perfectly ordinary in the context of some peer groups.

of social deviance. Like other deviants, people with stigmas are stereotyped and defined only in terms of their presumed deviance.

Think, for example, of how people in a wheelchair are treated in society. Their disability can become a **master status** (see Chapter 5), a characteristic of a person that overrides all other features of the person's identity (Goffman 1963). Physical disability can become a master status when other people see the disability as the defining feature of the person. People with a particular stigma are often all seen to be alike. This may explain why stigmatized individuals of high visibility are often expected to represent the whole group.

People who suddenly become disabled often have the alarming experience of their new master status rapidly erasing their former identity. People they know may treat and see them differently. A master status may also prevent people from seeing other parts of a person. A person with a disability may be assumed to have no meaningful sex life, for example, even if the disability is unrelated to sexual ability or desire. Sociologists have argued that the negative judgments about people with stigmas tend to confirm the "usualness" of others (Goffman 1963). For example, when welfare recipients are stigmatized as lazy and undeserving of social support, others are indirectly promoted as industrious and deserving. Stigmatized individuals are thus measured against a presumed norm and may be labeled, stereotyped, and discriminated against.

Sometimes, people with stigmas bond with others, perhaps even strangers. This can involve an acknowledgment of kinship or affiliation that can be as subtle as an understanding look, a greeting that makes a connection between two people, or a favor extended to a stranger who the person sees as sharing the presumed stigma. Public exchanges are common between various groups that share certain forms of disadvantage, such as people with disabilities, lesbians and gays, or members of other minority groups.

The strength of labeling theory is its recognition that the judgments people make about presumably deviant behavior have powerful social effects. Labeling theory does not, however, explain why deviance occurs in the first place. It may illuminate the consequences of a young man's violent behavior, but it does not explain the actual origins of the behavior. Labeling theory helps us understand how some *are considered* deviant while others are not, but it does not explain why some people initially engage in deviant behaviors and others do not.

Deviant Careers and Communities. In the ordinary context of work, a career is the sequence of movements a person makes through different positions in an occupational system (Becker 1963). A **deviant career**—a direct outgrowth of the labeling process—is the sequence of movements people make through a particular subculture of deviance. Deviant careers can be studied sociologically, like any other career. Within deviant careers, people are socialized into new "occupational" roles, and are encouraged, both materially and psychologically, to engage in deviant behavior. The concept of a deviant career emphasizes that there is a progression through deviance: Deviants are recruited, given or denied rewards, and promoted or demoted. As with legitimate careers, deviant careers involve an evolution in the person's identity, values, and commitment over time. Deviants, like other careerists, may have to demonstrate their commitment to the career to their superiors, perhaps by passing certain tests, such as when a gang expects new members to commit a crime, perhaps even shoot someone.

Within deviant careers, rites of passage may bring increased social status among peers. Punishments administered by the authorities may even become badges of honor within a deviant community. Similarly, labeling a teenager as a "bad kid" for poor behavior in school may actually encourage the behavior to continue because the juvenile may take this as a sign of success as a deviant.

The preceding discussion continues to indicate an important sociological point: Deviant behavior is not just the behavior of maladjusted individuals; it often takes place within a group context and involves group response. Some groups are actually organized around particular forms of social deviance; these are called **deviant communities** (Mizruchi 1983; Blumer 1969; Erikson 1966; Becker 1963).

Jim West/Alamy

Some deviance develops in deviant communities, such as the neo-Nazis/"skinheads" shown marching here. Such right-wing extremist groups have increased significantly in recent years, as monitored by the Southern Poverty Law Center.

Like subcultures and countercultures, deviant communities maintain their own values, norms, and rewards for deviant behavior. Joining a deviant community closes one off from conventional society and tends to solidify deviant careers because the deviant individual receives rewards and status from the in-group. Disapproval from the out-group may only enhance one's status within. Deviant communities also create a worldview that solidifies the deviant identity of their members. They may develop symbolic systems such as emblems, forms of dress, publications, and other symbols that promote their identity as a deviant group. Gangs wear their "colors," and skinheads have their insignia and music. Both are examples of deviant communities. Ironically, subcultural norms and values reinforce the deviant label both inside and outside the deviant group, thereby reinforcing the deviant behavior.

Some deviant communities are organized specifically to provide support to those in presumed deviant categories. Groups such as Alcoholics Anonymous, Weight Watchers, and various twelve-step programs help those identified as deviant overcome their deviant behavior. These groups, which can be quite effective, accomplish their mission by encouraging members to accept their deviant identity as the first step to recovery.

Crime and Criminal Justice

The concept of deviance in sociology is a broad one, encompassing many forms of behavior—legal and illegal, ordinary and unusual. **Crime** is one form of deviance, specifically, behavior that violates particular criminal laws. Not all deviance is crime. Deviance becomes crime when institutions of society designate it as violating a law or laws.

Criminology is the study of crime from a scientific perspective. Criminologists include social scientists such as sociologists who stress the societal causes and treatment of crime. All the theoretical perspectives on deviance that we examined earlier contribute to our understanding of crime (see ◆ Table 7.2). According to the functionalist perspective, crime may be *necessary* to hold society together. By singling out criminals as socially deviant, others are defined as good. The nightly reporting of crime on television is a demonstration of this sociological function of crime. Conflict theory suggests that disadvantaged groups are more likely to become criminal. Conflict theory also sees the well-to-do as better able to hide their crimes and less likely to be punished. Symbolic interaction helps us understand how people learn to become criminals or come to be accused of criminality, even when they may be innocent. Each perspective traces criminal behavior to social conditions rather than only to the intrinsic tendencies or personalities of individuals.

◆ **Table 7.2** Sociological Theories of Crime

Functionalist Theory	Symbolic Interaction Theory	Conflict Theory
Societies require a certain level of crime in order to clarify norms.	Crime is behavior that is learned through social interaction.	The lower the social class, the more the individual is *forced* into criminality.
Crime results from social structural strains (such as class inequality) within society.	Labeling criminals and stigmatizing them tends to reinforce rather than deter crime.	Inequalities in society by race, class, gender, and other forces tend to produce criminal activity.
Crime may be functional to society, thus difficult to eradicate.	Institutions with the power to label, such as prisons, actually produce rather than lessen crime.	Reducing social inequality in society is likely to reduce crime.

Measuring Crime: How Much Is There?

Is crime increasing in the United States? One would certainly think so from watching the media. Images of violent crime abound and give the impression that crime is a constant threat and is on the rise. Data about crime come from the Federal Bureau of Investigation (FBI) based on reports from police departments across the nation. The data are distributed annually in the *Uniform Crime Reports* and are the basis for official reports about the extent of crime and its rise and fall over time. Although media coverage of crime has remained high and about the same, data on crime actually show that violent crime peaked in 1990, but *decreased* through the 1990s and has continued to decline through 2013 (see ▲ Figure 7.2). The officially reported rate of assault and robbery has decreased, although rape and murder have remained roughly the same.

A second major source of crime data is the *National Crime Victimization Surveys* published by the Bureau of Justice Statistics in the U.S. Department of Justice. These data are based on surveys in which national samples of people are periodically asked if they have been the victims of one or more criminal acts. These surveys clearly show that the likelihood of being a victim of crime is influenced by one's race, gender, and social class.

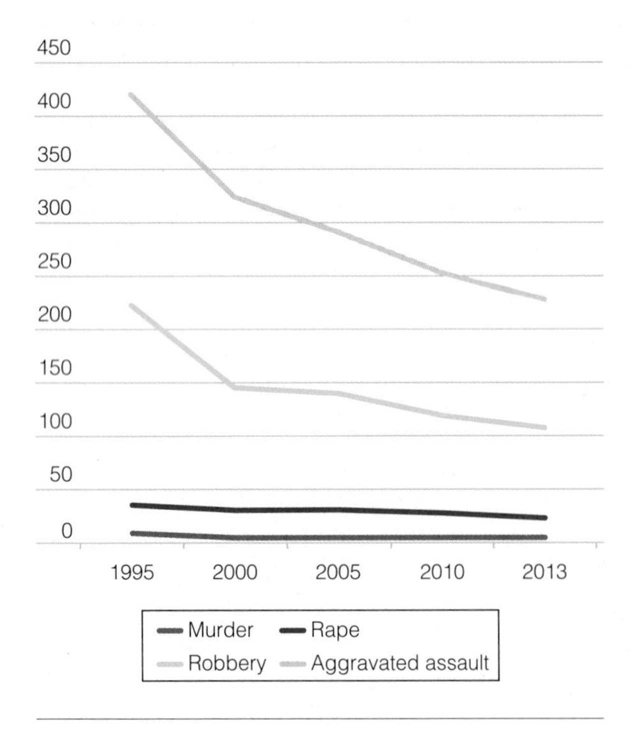

▲ **Figure 7.2** Violent Crime in the United States, 1995–2013 This graph shows that, despite many news stories about violent crime, the rates of violent crimes have gone down since 1995.

Source: Federal Bureau of Investigation. 2013. *Crime in the United States 2013*. Washington, DC: Federal Bureau of Investigation. **www.fbi.gov**

Both of these sources of data—the *Uniform Crime Reports* and the *National Crime Victimization Surveys*—are subject to the problem of underreporting. About half to two-thirds of all crimes may not be reported to police, meaning that much crime never shows up in the official statistics. Rape is particularly known to be vastly underreported. Victims may be reluctant to report for a variety of reasons, including that the police will not take the rape seriously, especially if the assailant was known to the victim. Also, the victim may not want to undergo the continued stress of an investigation and trial.

A Problem with Official Statistics. Official statistics on crime are important for describing the extent of crime and various patterns in the perpetration and victimization by crime. You have to be cautious, however, in relying on these official reports because of the logic of labeling theory. Recall that labeling theorists would see crime statistics as produced by those with the power to assign labels. Reported rates of deviant behavior are, like crime itself, the product of social behavior.

Official statistics are produced by people in various agencies (police, courts, and other bureaucratic organizations). These people define, classify, and record certain behaviors as falling into the category of crime—or not. Labeling theorists think that official rates of crime (and deviance) do not necessarily reflect the actual commission of crime; instead, the official rates reflect social judgments.

In an interesting example, in the aftermath of the terrorist attacks on the World Trade Center in 2001, officials debated whether to count the deaths of thousands as murder or as a separate category of terrorism. The decision would change the official murder rate in New York City that year. In the end, these deaths were not counted in the murder rate. That is a unique example, but an ongoing example is the official reporting of rape. Research finds that the police are less likely to "count" some rapes, such as those in which the victim is a prostitute, was drunk at the time of the assault, or had a previous relationship with the assailant. Rapes resulting in the victim's death are classified as homicides and thus do not appear in the official statistics on rape.

Types of Crime

When people think of crime, they may imagine a stereotypical criminal—someone who is a stranger, someone who randomly assaults you, or someone who commits a quick street crime, like a mugging. Stereotypes about crime, however, hide the many different kinds of crime committed—and the characteristics of those who commit them. The different types of crime reveal various social patterns in the commission of crime and victimization by crime, little of which is random as the stereotype suggests.

a sociological eye on the media

Images of Violent Crime

The media routinely drive home two points to the consumer: Violent crime is always high and may be increasing over time, and there is much random violence constantly around us. The media bombard us with stories of "wilding," in which bands of youths kill random victims. Many of us think road rage is extensive (which it is not) and completely random. The media vividly and routinely report such occurrences as pointless, random, and probably increasing.

The evidence shows that although violent crime in the United States increased during the 1970s and 1980s, it nonetheless began to decrease in 1990 and continues to decrease nationally through the present. For example, both robbery and physical assaults have declined dramatically since 1990. Yet, according to research (Best 2011, 2008, 1999; Glassner 1999), the media have consistently given a picture that violent crime has increased during this same period and, furthermore, that the violence is completely unpatterned and random.

No doubt there are occasions when victims are indeed picked at random. But the statistical rule of randomness could not possibly explain what has come to be called random violence, a vision of patternless chaos that is advanced by the media. If randomness truly ruled, then each of us would have an equal chance of being a victim—and of being a criminal. This is assuredly not the case. The notion of random violence, and the notion that it is increasing, ignores virtually everything that criminologists, psychologists, sociologists, and extensive research studies know about crime: It is highly patterned and significantly predictable, beyond sheer chance, by taking into account the social structure, social class, location, race and ethnicity, gender, labeling, age, whom one's family members are, and other such variables and forces in society that affect both criminals and victims.

The correct central picture, then, is clearly not conveyed in the media. Some have speculated that the picture maintained in the media of increasing crime is simply a tool to increase viewer ratings. Criminal violence is not increasing though, but decreasing, and it is not random, but highly patterned and predictable.

Personal and Property Crimes. The *Uniform Crime Reports* report something called the *crime index*. The crime index includes the violent crimes of murder, manslaughter, rape, robbery, and aggravated assault, plus property crimes of burglary, larceny theft, and motor vehicle theft. The crime index includes both *personal crimes* (violent or nonviolent crimes directed against people, including murder, aggravated assault, forcible rape, and robbery) and *property crimes* (those involving theft of property without threat of bodily harm, such as burglary, larceny, auto theft, and arson). Property crimes are the most frequent criminal infractions.

DEBUNKING Society's Myths

Myth: Crimes, especially very violent ones, are committed by people who are mentally ill.

Sociological Perspective: People are often shocked when someone who may live near them commits a violent crime; "He seemed so normal," they typically say. This reaction shows how much the public believes crime and deviance to be the behavior of poorly adjusted people, but sociologists find clear patterns in the commission of crime. Violent crime, for example, is most likely committed by someone who knows the victim, probably well. Most crime is not random (Best 1999).

Hate Crimes. Hate crime is a relatively new official category of crime, although hate crimes have certainly been committed throughout the nation's history. Lynching, vandalism of synagogues, and the assault of gay people are not new, but the formal reporting of hate crime did not begin until 1980. Now the U.S. Congress, via the FBI, defines **hate crime** as a criminal offense that is motivated in whole or part by bias against a "race, religion, disability, ethnic origin, or sexual orientation" (www.fbi.gov). This form of crime has been increasing in recent years, especially against gays and lesbians, but also because of the ability now to report and then track such heinous acts. The vast majority of hate crimes are committed by White offenders—or, in many cases, unknown offenders. More than half of all reported hate crimes are committed against people because of race or ethnicity; 20 percent are based on sexual orientation of the victim; and another 20 percent are based on religion (Federal Bureau of Investigation 2012).

Human Trafficking. Human trafficking has long played a role in the national and international economy. Slavery, for example, is a pernicious example of human trafficking, but this is a crime that continues in various forms. The FBI defines *human trafficking* as compelling or coercing a person to engage in some form of labor, service, or commercial sex. Sometimes the coercion is overtly physical, but it can also be psychological and subtle, such as a pimp who recruits prostitutes into a network of sex work by initially seeming to be a boyfriend. Undocumented immigrants are particularly prone to trafficking as they are a very vulnerable population. Children are also among some of the most vulnerable, especially when coming from war-torn regions. Estimates of the extent of human trafficking are difficult to come by, but in 2013, the U.S. State Department identified nearly 45,000 victims. One of the problems in getting accurate data is not only the covert nature of this crime, but also lack of uniformity in how nations tabulate known cases (U.S. Department of State 2014).

Gender-Based Violence. **Gender-based violence** is the term used to describe the various forms of violence that are associated with unequal power relationships between men and women. Gender-based violence takes many forms, including, but not limited to rape, domestic violence, sexual abuse and incest, stalking, and more. Although both men and women can be victims of gender-based violence, it far more frequently victimizes women and girls (Bloom 2008).

For all women, victimization by rape is probably the greatest fear. Although rape is the most underreported crime, even with underreporting, the FBI estimates that one rape occurs in the United States every 6.6 minutes (Federal Bureau of Investigation 2013).

Recently, the nation has focused its attention on the widespread phenomenon of campus rape, also referred to as sexual assault. *Acquaintance rape* is that committed by an acquaintance or someone the victim has just met. The extent of acquaintance rape is difficult to measure. The Bureau of Justice Statistics finds that 3 percent of college women experience rape or attempted rape in a given college year, and 13 percent report being stalked (Fisher et al. 2000). Acquaintance rape is linked to men's acceptance of various rape myths, such as believing that a woman's "no" means "yes." Excessive drinking also increases one's chances of being raped during campus parties. Some campus cultures and environments are especially likely to put women at risk of rape, particularly in some all-male groups and organizations, especially those organized around hierarchy, secrecy among "brothers," and loyalty, which create an atmosphere where rape can occur. This can help you understand why different organizations, such as some fraternities, sports teams, churches, and military schools, have high rates of rape (Langton and Sinozich 2014; Martin and Hummer 1989).

Sociologists have argued that the causes of rape lie in women's status in society—that women are treated as sexual objects for men's pleasure. The relationship between women's status and rape is also reflected in data revealing who is most likely to become a rape victim. African American women, Latinas, and poor women have the highest likelihood of being raped, as do women who are single, divorced, or separated. Young women are also more likely to be rape victims than older women (U.S. Bureau of Justice Statistics 2013). Sociologists interpret these patterns to mean that the most powerless women are also most subject to this form of violence.

Identity Theft. A new type of crime has also emerged in the context of the technological revolution that is changing daily habits. *Identity theft* is defined as the use of someone else's personal identifying information, usually for purposes of some kind of fraud (Allison et al. 2005). The cost of such crimes is staggering: Estimates are that financial losses to individual victims total about five billion dollars per year. Corporate losses are even greater—47 billion dollars a year (Holt and Turner 2012; Federal Trade Commission 2003). Not surprisingly, individuals who use the Internet for routine activities, such as banking, email, and instant messaging, are about 50 percent more likely to be victims of identity theft than others. Online shopping increases risk by about 30 percent. Men, older people, and those with higher incomes are most likely to experience victimization from identity theft (Reyns 2013).

Victimless Crimes. Victimless crimes are those that violate laws but where there is no complainant. Victimless crimes include various illicit activities, such as gambling, illegal drug use, and prostitution. Although there is no victim per se, there is clearly some degree of victimization in such crimes: Some researchers see prostitution, in many instances, as containing at least one victim because of the consequences for one's health, safety, and well-being through participation in such activities. Enforcement of these crimes is typically not as rigorous as enforcement of crimes against people or property.

Elite and White-Collar Crime. The term *white-collar* crime refers to criminal activities by people of high social status who commit crime in the context of their occupation (Sutherland and Cressey 1978). White-collar crime includes activities such as embezzlement (stealing funds from one's employer), involvement in illegal stock manipulations (insider trading), and a variety of violations of income tax law, including tax evasion. Until very recently, white-collar crime seldom generated great public concern, far less than the concern about street crime. In terms of total dollars, however, white-collar crime is even more consequential for society. Scandals involving prominent white-collar criminals have come to the public eye more frequently, such as during the recession of 2008, which many say resulted from very risky financial practices and excessive borrowing by the nation's banks and on Wall Street.

Corporate Crime. *Corporate crime* is wrongdoing that occurs within the context of a formal organization or bureaucracy that is actually sanctioned by the norms and operating principles of the bureaucracy (Simon 2011). This can occur within any kind of organization—corporate, educational, governmental, or religious. Sociological studies of corporate crime show that it is embedded in the ongoing and routine activities of organizations (Ermann and Lundman 2001). Individuals within the organization may participate in the behavior with little awareness that their behavior is illegitimate. In fact, their actions are likely to be defined as in the best interests of the organization—business as usual. New members who enter the organization learn to comply with the organizational expectations or leave.

One of the most upsetting recent examples of massive corporate malfeasance involved a world-famous and time-honored American institution: the Johnson and Johnson Co., manufacturer of medical supplies such as bandages, baby oil, and artificial limbs. The company is the manufacturer of the now infamous DePuy artificial hip joint, adopted by thousands since 2005 to replace their own failing hip joints (Meier 2013). It turns out that Johnson and Johnson executives knew *years before* they officially recalled the faulty DePuy artificial hip joint in 2010 that it had a deadly design flaw. In the interest of maintaining high profits, the company deliberately concealed evidence of the design flaw from physicians, patients, and their families. Evidently, the wish to maintain high profits exceeded the wish to make the patients well and to save their lives. Consultants and medical researchers discovered the flaw several years before Johnson and Johnson recalled the DePuy joint, yet company executives totally ignored these research results. The company eventually reached a settlement with the thousands of victims, having to pay out four billion dollars for damages caused by this corporate crime.

Organized Crime. The structure of crime and criminal activity in the United States often takes on an organized, almost institutional character. This is crime in the form of mob activity and racketeering, known as organized crime. *Organized crime* is crime committed by structured groups typically involving the provision of illegal goods and services to others. Organized crime syndicates are typically stereotyped as the Mafia, but the term can refer to any group that exercises control over large illegal enterprises, such as the drug trade, illegal gambling, prostitution, weapons smuggling, or money laundering. These organized crime syndicates are often based on racial, ethnic, or family ties, with different groups dominating and replacing each other in different criminal "industries," at different periods in U.S. history.

A key concept in sociological studies of organized crime is that these industries are organized along the same lines as legitimate businesses; indeed, organized crime has taken on a corporate form. There are likely senior partners who control the profits of the business, workers who manage and provide the labor for the business, and clients who buy the services that organized crime provides. In-depth studies of the organized crime underworld are difficult, owing to its secretive nature and dangers.

Terrorism. The FBI includes *terrorism* in its definition of crime, defining it as "the unlawful use of force or violence against persons or property to intimidate or coerce a government, the civilian population, or any segment thereof, in furtherance of political or social objectives" (Federal Bureau of Investigation 2011). Terrorism crosses national borders, and to understand it requires a global perspective. Terrorism is also linked to other forms of international crime. It is suspected that profits from international drug trade fund the terrorist organization al Qaeda.

One of the most frightening things about terrorism as a crime is that its victims, unlike most other crime, may be somewhat randomly targeted. Suicide bombers or other armed attackers may select particular groups because of their identification with the West or because they are associated with Jewish people. This is what happened in Paris, in 2015, when terrorists who were possibly associated with the terrorist group ISIS (Islamic State in Iraq and Syria) attacked and slaughtered at least seventeen people in a kosher market and in the offices of a satirical magazine.

Race, Class, Gender, and Crime

Arrest data show a very clear pattern of differential arrests along lines of race, gender, and class. A key question is whether this pattern reflects actual differences in the commission of crime by different groups or whether it reflects differential treatment by the criminal justice system. The answer is "both." Prosecution by the criminal justice system is significantly related to patterns of race, gender, and class inequality. We see this in the bias of official arrest statistics, in treatment by the police, in patterns of sentencing, and in studies of imprisonment.

Arrest statistics show a strong correlation between social class and crime, the poor being more likely than others to be arrested for crimes. Does this mean that the poor commit more crimes? To some extent, yes, as unemployment and poverty are related to crime (Reiman and Leighton 2012). And the reason is simple: Those who are economically deprived often see no alternative to crime, as Merton's structural strain theory would predict.

Moreover, law enforcement is concentrated in lower-income and minority areas. People who are better off are further removed from police scrutiny and better able to hide their crimes. When and if white-collar criminals are prosecuted and convicted, they tend to receive somewhat lighter sentences. Middle- and upper-income people may be perceived as being less in need of imprisonment because they likely have a job and high-status people to testify for their good character. White-collar crime is simply perceived as less threatening than crimes by the poor. Class also predicts who most likely will be victimized by crime, with those at the highest ends of the socioeconomic scale least likely to be victims of violent crime (Barak et al. 2015).

Bearing in mind the factors that affect the official rates of arrest and conviction—bias of official statistics, the influence of powerful individuals, discrimination in patterns of arrest, differential policing—there remains evidence that the actual commission of crime varies by race. Why? Sociologists find a compelling explanation in social structural conditions. Racial minority groups are far more likely than Whites to be poor, unemployed, and living in single-parent families. These social facts are all predictors of a higher rate of crime. Note, too, as ▲ Figure 7.3 shows, that African Americans and Hispanics are generally more likely to be victimized by crime.

Recently, women's participation in crime has been increasing, the result of several factors. Women are now more likely to be employed in jobs that present opportunities for crimes, such as property theft, embezzlement, and fraud. Violent crime by women has also increased notably since the early 1980s, possibly because the images that women have of themselves are changing, making new behaviors possible. Most significant, crime by women is related to their continuing disadvantaged status in society. Just as crime is linked to socioeconomic status for men, so is it for women (Belknap 2001).

Women are somewhat less likely than men to be victimized by crime, with the exception of gender-based crimes, although this varies significantly by race and age. Black women are more likely than White women to be victims of assault; young Black women are especially vulnerable. Divorced, separated, and single women are more likely than married women to be crime victims.

The Criminal Justice System: Police, Courts, and the Law

Whether in the police station, the courts, or prison, the factors of race, class, and gender are highly influential in the administration of justice in this society. Those in the most disadvantaged groups are more likely to be defined and identified as deviant independently of their behavior and, having encountered these systems of authority, are more likely to be detained and arrested, found guilty, and punished.

DEBUNKING Society's Myths

Myth: The criminal justice system treats all people according to the neutral principles of law.
Sociological Perspective: Race, class, and gender continue to have an influential role in the administration

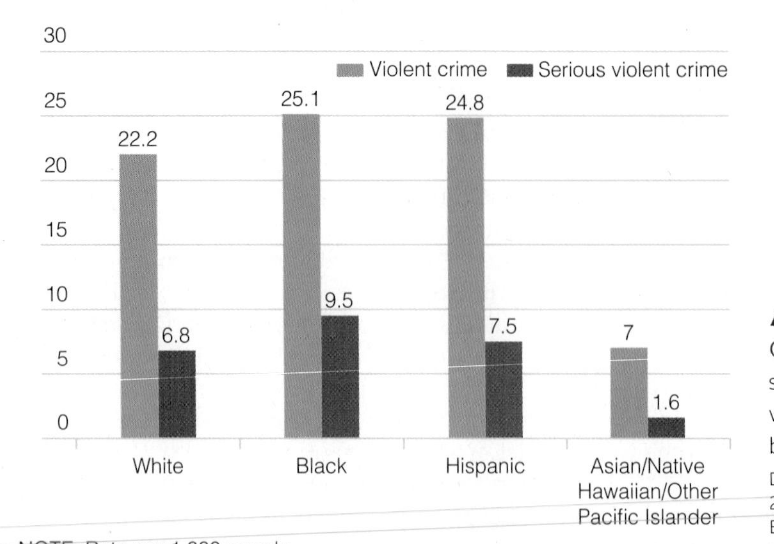

▲ **Figure 7.3** Victimization by Violent Crime: Inequalities by Race This chart shows that the likelihood of victimization by violent crime and serious violent crime varies by racial–ethnic group.

Data Source: Truman, Jennifer L., and Lynn Langton. 2014. *Criminal Victimization, 2013*. Washington, DC: Bureau of Justice Statistics, U.S. Department of Justice. **www.bjs.gov**

Social Class and Social Stratification

One afternoon in a major U.S. city, two women go shopping. They are friends—wealthy, suburban women who shop for leisure. They meet in a gourmet restaurant and eat imported foods while discussing their children's private schools. After lunch, they spend the afternoon in exquisite stores—some of them large, elegant department stores; others, intimate boutiques where the staff knows them by name. When one of the women stops to use the bathroom in one store, she enters a beautifully furnished room with an upholstered chair, a marble sink with brass faucets, fresh flowers on a wooden pedestal, shining mirrors, an ample supply of hand towels, and jars of lotion and soaps. The toilet is in a private stall with solid doors. In the stall, there is soft toilet paper and another small vase of flowers.

The same day, in a different part of town, another woman goes shopping. She lives on a marginal income earned as a stitcher in a textiles factory. Her daughter badly needs a new pair of shoes because she has outgrown last year's pair. The woman goes to a nearby discount store where she hopes to find a pair of shoes for under $15, but she dreads the experience. She knows her daughter would like other new things—a bathing suit for the summer, a pair of jeans, and a blouse. But this summer, the daughter will have to wear hand-me-downs because bills over the winter have depleted the little money left after food and rent. For the mother, shopping is not recreation but a bitter chore reminding her of the things she is unable to get for her daughter.

While this woman is shopping, she, too, stops to use the bathroom. She enters a vast space with sinks and mirrors lined up on one side of the room and several stalls on the other. The tile floor is gritty and gray. The locks on the stall doors are missing or broken. Some of the overhead lights are burned out, so the room has dark shadows. In the stall, the toilet paper is coarse. When the woman washes her hands, she discovers there is no soap in the metal dispensers. The mirror before her is cracked. She exits quickly, feeling as though she is being watched.

in this chapter, you will learn to:

- Explain how class is a social structure
- Describe the class structure of the United States
- Identify the different components of class inequality
- Analyze the extent of social mobility in the United States
- Compare and contrast theoretical models of class inequality
- Investigate the causes and consequences of U.S. poverty

Two scenarios, one society. The difference is the mark of a society built upon class inequality. The signs are all around you. Think about the clothing you wear. Are some labels worth more than others? Do others in your group see the same marks of distinction and status in clothing labels? Do some people you know never seem to wear the "right" labels? Whether it is clothing, bathrooms, schools, homes, or access to health care, the effect of class inequality is enormous, giving privileges and resources to some and leaving others struggling to get by.

Great inequality divides society. Nevertheless, most people think that equal opportunity exists for all in the United States. The tendency is to blame individuals for their own failure or attribute success to individual achievement. Many think the poor are lazy and do not value work. At the same time, the rich are admired for their supposed initiative, drive, and motivation. Neither is an accurate portrayal. There are many hardworking individuals who are poor, but they seldom get credit for their effort. At the same time, many of the richest people have inherited their wealth or have had access to resources (such as the best schools or access to elite networks) that others can barely imagine.

Observing and analyzing class inequality is fundamental to sociological study. What features of society cause different groups to have different opportunities? Why is there such an unequal allocation of society's resources? Sociologists respect individual achievements but have found that the greatest cause for disparities in material success is the organization of society. Instead of understanding inequality as the result of individual effort, sociologists study the social structural origins of inequality.

Social Differentiation and Social Stratification

All social groups and societies exhibit social differentiation. *Status*, as we have seen earlier, is a socially defined position in a group or society. Different statuses develop in any group, organization, or society. Think of a sports organization. The players, the owners, the managers, the fans, the cheerleaders, and the sponsors all have a different status within the organization. Together, they constitute a whole social system, one that is marked by social differentiation.

SEE for YOURSELF

Take a shopping trip to different stores and observe the appearance of stores serving different economic groups. What kinds of bathrooms are there in stores catering to middle-class clients? The rich? The working class? The poor? Which ones allow the most privacy or provide the nicest amenities? What fixtures are in the display areas? Are they simply utilitarian with minimal ornamentation, or are they opulent displays of consumption? Take detailed notes of your observations, and write an analysis of what this tells you about social class in the United States.

Alex Segre/Alamy

AP Images/Don Ryan

Social class differences make it seem as if some people are living in two different societies.

what would a sociologist say?

Social Class and Sports

Sports are a huge part of American culture. Whether you are an athlete, a fan, or just an observer, sports are a window into how social class shapes some of our most popular activities.

Start with the ideas of functionalism and the work of classical theorist Emile Durkheim. Durkheim was interested in the cultural symbols and events that bind people together. Think of how many sports symbols, such as jerseys, hats, and bumper stickers, are common sights in everyday life. These symbols project an identity to others that make a claim about being part of a collective group, but sometimes, they reflect social class locations, too. Rich people, for example, are not likely to be wearing NASCAR caps, but may well have yacht club logos on their polo shirts and ties.

As Max Weber would point out, class, power, and prestige are all tangled up in sports. There are significant class differences associated with different sports, some having more prestige than others. Prestige is also interwoven with power, as you can see during political elections, where you see politicians all over the place—at tailgate parties and hanging out in the expensive box seats. Beyond the connection between power, politics, and prestige, sports is big business.

Corporate profits and sponsorship are very apparent, even in college sports. Look at the advertisements on most college scoreboards. Various plays in a football game might be featured as an "AT&T All-America Play of the Week!" Think of how much money is spent on commercials.

Social class in the world of sports is everywhere, even though the workers who help put on events are often invisible. Some of the athletes are very highly paid, but working-class people serve the food in stadiums, clean up after the fans leave, and take out all the trash. Sports are an amazing example of a class-based social system.

Status differences can become organized into a hierarchical social system. **Social stratification** is a relatively fixed, hierarchical arrangement in society by which groups have different access to resources, power, and perceived social worth. Social stratification is a system of structured social inequality. Again using sports as an example, you can see that many of the players earn extremely high salaries, although most do not. Those who do are among the elite in this system of inequality, but the owners control the resources of the teams and hold the most power in this system. Sponsors (including major corporations and media networks) are the economic engines on which this system of stratification rests. Fans are merely observers who pay to watch the teams play, but the revenue they generate is essential for keeping this system intact. Altogether, sports are systems of stratification because the groups that constitute the organization are arranged in a hierarchy where some have more resources and power than others. Some provide resources; others take them. Even within the field of sports, there are huge differences in which teams—and which sports—are among the elite.

All societies seem to have a system of social stratification, although they vary in the degree and complexity of stratification. Some societies stratify only along a single dimension, such as age, keeping the stratification system relatively simple. Most contemporary societies are more complex, with many factors interacting to create different social strata. In the United States, social stratification is strongly influenced by class, which is in turn influenced by matters such as one's occupation, income, and education, along with race, gender, and other influences such as age, region of residence, ethnicity, and national origin (see ◆ Table 8.1).

◆ Table 8.1 Inequality in the United States

- One in five children (19.5 percent) in the United States lives in poverty, including 38 percent of African American children, 30 percent of Hispanic children, 11 percent of White (not Hispanic) children, and 9.8 percent of Asian American children (DeNavas-Walt and Proctor 2014).
- The rate of poverty among people in the United States has been increasing since 2000 (DeNavas-Walt and Proctor 2014).
- Among women heading their own households, one-third live below the poverty line (DeNavas-Walt and Proctor 2014).
- One percent of the U.S. population controls 38 percent of the total wealth in the nation; the bottom half hold none or are in debt (Mishel et al. 2012).
- Most American families have seen their net worth decline, largely because of declines in the value of housing. Households at the bottom of the wealth distribution lost the largest share of their wealth; those at the top, the least (Pfeffer et al. 2014).
- The average CEO of a major company has a salary of $13 million per year; workers earning the minimum wage make $15,080 per year if they work 40 hours a week for 52 weeks and hold only one job (www.aflcio.org).

Estate, Caste, and Class

Stratification systems can be broadly categorized into three types: estate systems, caste systems, and class systems. In an **estate system** of stratification, the ownership of property and the exercise of power are monopolized by an elite class who have total control over societal resources. Historically, such societies were feudal systems where classes were differentiated into three basic groups—the nobles, the priesthood, and the commoners. Commoners included peasants (usually the largest class group), small merchants, artisans, domestic workers, and traders. The nobles controlled the land and the resources used to cultivate the land, as well as all the resources resulting from peasant labor.

Estate systems of stratification are most common in agricultural societies. Although such societies have been largely supplanted by industrialization, some societies still have a small but powerful landholding class ruling over a population that works mainly in agricultural production. Unlike the feudal societies of the European Middle Ages, however, contemporary estate systems of stratification display the influence of international capitalism. The "noble class" comprises not knights who conquered lands in war, but international capitalists or local elites who control the labor of a vast and impoverished group of people, such as in some South American societies where landholding elites maintain a dictatorship over peasants who labor in agricultural fields.

In a **caste system**, one's place in the stratification system is an *ascribed status* (see Chapter 5), meaning it is a quality given to an individual by circumstances of birth. The hierarchy of classes is rigid in caste systems and is often preserved through formal law and cultural practices that prevent free association and movement between classes. The system of apartheid in South Africa was a stark example of a caste system. Under apartheid, the travel, employment, associations, and place of residence of Black South Africans were severely restricted. Segregation was enforced using a pass system in which Black South Africans could not be in White areas unless for purposes of employment. Those found without passes were arrested, often sent to prison without ever seeing their families again. Interracial marriage was illegal. Black South Africans were prohibited from voting; the system was one of total social control where anyone who protested was imprisoned. The apartheid system was overthrown in 1994 when Nelson Mandela, held prisoner for twenty-seven years of his life, was elected president of the new nation of South Africa. A new national constitution guaranteeing equal rights to all was ratified in 1996.

In **class systems**, stratification exists, but a person's placement in the class system can change according to personal achievements. That is, class depends to some degree on *achieved status*, defined as status that is earned by the acquisition of resources and power, regardless of one's origins. Class systems are more open than caste systems because position does not depend strictly on birth. Classes are less rigidly defined than castes because class divisions are blurred when there is movement from one class to another.

Despite the potential for movement from one class to another, in the class system found in the United States, class placement still depends heavily on one's social background. Although *ascription* (the designation of ascribed status according to birth) is not the basis for social stratification in the United States, the class a person is born into has major consequences for that person's life. Patterns of inheritance; access to exclusive educational resources; the financial, political, and social influence of one's family; and similar factors all shape one's likelihood of achievement. Although there are not formal obstacles to movement through the class system, individual achievement is very much shaped by one's class of origin.

In common terms, *class* refers to style or sophistication. In sociological use, **social class** (or *class*) is the social structural position that groups hold relative to the economic, social, political, and cultural resources of society. Class determines the access different people have to these resources and puts groups in different positions of privilege and disadvantage. Each class has members with similar opportunities who tend to share a common way of life. Class also includes a cultural component in that class shapes language, dress, mannerisms, taste, and other preferences. Class is not just an attribute of individuals; it is a feature of society.

The social theorist Max Weber described the consequences of stratification in terms of **life chances**, meaning the opportunities that people have in common by virtue of belonging to a particular class. Life chances include the opportunity for possessing goods, having an income, and having access to particular jobs. Life chances are also reflected in the quality of everyday life. Whether you dress in the latest style or wear another person's discarded clothes, have a vacation in an exclusive resort, take your family to the beach for a week, or have no vacation at all, these life chances are the result of being in a particular class.

Class is a structural phenomenon; it cannot be directly observed. Nonetheless, you can "see" class through various displays that people project, often unintentionally, about their class status. Do some objects worn project higher-class status than others? How about cars? What class status is displayed through the car you drive or, for that matter, whether you even have a car or use a bus to get to work? In myriad ways, class is projected to others as a symbol of presumed worth in society.

Social class can be observed in the everyday habits and presentations of self that people project. Common objects, such as clothing and cars, can be ranked not only in terms of their economic value but also in terms of the status that various brands and labels carry.

The interesting thing about social class is that a particular object may be quite ordinary, but with the right "label," it becomes a *status symbol* and thus becomes valuable. Take the example of Vera Bradley bags. These paisley bags are made of ordinary cotton with batting. Not long ago, such cloth was cheap and commonplace, associated with rural, working-class women. If such a bag were sewn and carried by a poor person living on a farm, the bag (and perhaps the person!) would be seen as ordinary, almost worthless. Transformed by the right label (and some good marketing), Vera Bradley bags have become status symbols, selling for a high price (often a few hundred dollars—a price one would never pay for a simple cotton purse). Presumably, having such a bag denotes the status of the person carrying it. (See also the box "See for Yourself: Status Symbols in Everyday Life.")

The early sociologist Thorstein Veblen described the class habits of Americans as **conspicuous consumption**, meaning the ostentatious display of goods to define one's social status. Writing in 1899, Veblen said, "conspicuous consumption of valuable goods is a means of respectability to the gentleman of leisure" (Veblen 1953/1899: 42). Although Veblen identified this behavior as characteristic of the well-to-do (the "leisure class," he called them), conspicuous consumption today marks the lifestyle of many. Indeed, mass consumerism is a hallmark of both the rich and the middle class, and even of many working-class people's lifestyles. What examples of this do you see among your associates?

SEE for YOURSELF

Status Symbols in Everyday Life

You can observe the everyday reality of social class by noting the status that different ordinary objects have within the context of a class system. Make a list of every car brand you can think of—or, if you prefer, every clothing label. Then rank your list with the highest status brand (or label) at the top of the list, going down to the lowest status. Then answer the following questions:

1. Where does the presumed value of this object come from? Does the value come from the actual cost of producing the object or something more subjective?
2. Do people make judgments about people wearing or driving the different brands you have noted? What judgments do they make? Why?
3. What consequences do you see (positive and negative) of the ranking you have observed? Who benefits from the ranking and who does not?

What does this exercise reveal about the influence of status symbols in society?

Because sociologists cannot isolate and measure social class directly, they use other *indicators* to serve as measures of class. A prominent indicator of class is income. Other common indicators are education, occupation, and place of residence. These indicators alone do not define class, but they are often accurate measures of the class standing of a person or group. We will see that these indicators tend to be linked. A good income, for example, makes it possible to afford a house in a prestigious neighborhood and an exclusive education for one's children. In the sociological study of class, indicators such as income and education have had enormous value in revealing the outlines and influences of the class system.

The Class Structure of the United States: Growing Inequality

People think of the United States as a land of opportunity where one's class position matters less than individual effort. According to a recent survey, almost three-quarters of Americans think that hard work is the key to getting ahead in life. Compared to those in other countries, Americans are far more likely to believe in the importance of individual effort, a reflection of the cultural belief in individualism (Pew Research Global Attitudes Project 2014).

Despite these beliefs, class divisions in the United States are real, and inequality is growing. Perhaps this has become more apparent to people in recent years as the nation experienced a recession and a very fragile economic situation. Millions lost their homes and retirement savings and other investments. Many in the middle and working class feel that their way of life is slipping away. For the first time in our nation's history, only 17 percent of the public thinks that children

ZUMA Press, Inc./Alamy

College graduates who are facing an uncertain job market are experiencing some of the consequences of growing inequality in the United States.

davelogan/iStockphoto.com

© Vladimir Korostyshevskiy/Shutterstock.com

The class structure in the United States means very different living conditions for those of vast wealth and everyone else.

today will be better off than their parents; two-thirds no longer believe this (Pew Research Global Attitudes Project 2014; Rasmussen Reports 2009).

Even aside from the economic recession, the gap between the rich and the poor in the United States is greater than in other industrialized nations, and it is larger than at any time in the nation's history. Many analysts argue that this gap is the central problem of the age—contributing to crime and violence, political division, threats to democracy, and increased anxiety and frustration felt by large segments of the population (Piketty 2014; Noah 2012; Reich 2010).

Many factors have contributed to growing inequality in the United States, including the profound effects of national and global economic changes. Many think of the economic problems of the nation as stemming from individual greed on Wall Street, and this likely plays a role, but social inequality stems from systemic—that is, social structural—conditions, particularly what is called *economic restructuring.*

Economic restructuring refers to the decline of manufacturing jobs in the United States, the transformation of the economy by technological change, and the process of globalization. We examine economic restructuring more in Chapter 15 on the economy, but the point here is that these structural changes are having a profound effect on the life chances of people in different social classes. Many in the working class, for example, once largely employed in relatively stable manufacturing jobs with decent wages and good benefits, now likely work, if they work at all, in lower-wage jobs with fewer benefits, such as health care and pensions. Middle-class families have amassed large sums of debt, sometimes to support a middle-class lifestyle, but also perhaps to pay for their children's education.

The economic problems that produce inequality are not, however, purely economic: They are social, both in their origins and their consequences. Home ownership provides an example. For most Americans, owning one's own home is the primary means of attaining economic security—a central part of the American dream. Owning a home is also the key to other resources—good schools, cleaner neighborhoods, and an investment in the future. Similarly, losing your home is more than just a financial crisis—it reverberates through various aspects of your life. The odds of having a home—indeed, the odds of losing your home—are profoundly connected to social factors, such as your race and your gender.

Housing foreclosure is a trauma for anyone who experiences it, but foreclosure has hit some groups especially hard. The racial segregation of Hispanic and, especially, African American neighborhoods is a major contributing cause to the high rate of mortgage foreclosures (Rugh and Massey 2010). African Americans are almost twice as likely to experience foreclosure as White Americans (Bocian et al. 2010). Moreover, women are 32 percent more likely than men to have *subprime mortgages* (that is, mortgages with an interest rate *higher* than the prime lending rate). Black women earning double the region's median income were nearly five times more likely to receive subprime mortgages than White men with similar incomes (Fishbein and Woodall 2006).

Some might argue that foreclosures occur because individual people have made bad decisions—buying homes beyond their means. But, institutional lending practices also target particular groups, making them more vulnerable to the economic forces that can shatter individual lives. Lenders may see African Americans as a greater credit risk, but they also know that the value of real estate is less in racially segregated neighborhoods. Discriminatory practices in the housing market have also been well documented (Squires 2007; Oliver and Shapiro 2006).

The sociological point is that economic problems have a sociological dimension and cannot be explained by individual decisions alone. Economic policies also have different effects for different groups—sometimes intended, sometimes not. Wealthy people, as an example, typically pay a far lower tax rate than the middle class, because much of their money comes from investments, not income, and income is taxed at a much higher rate than investment income. Various tax loopholes (such as home mortgage deductions, tax shelters on real estate investments, or even offshore banking deposits) also significantly reduce the tax burden by those with the most resources.

Corporations benefit the most from the tax structure. The corporate tax rate in the United States is the highest in the world (35 percent), but many corporations pay much less than that, given the various loopholes, offshore investments, and tax subsidies that lessen one's tax obligation. A study of the Fortune 500 companies (those companies with the highest gross revenue in a given year) has found that most paid only about 20 percent in taxes. Many of these big companies paid no tax at all in some years (McIntyre et al. 2014).

The Distribution of Income and Wealth

Understanding inequality requires knowing some basic economic and sociological terms. Inequality is often presented as a matter of differences in income, one important measure of class standing. In addition to income inequality, there are vast inequalities in who owns what—that is, the wealth of different groups.

Income is the amount of money brought into a household from various sources (wages, investment income, dividends, and so on) during a given period. In recent years, income growth has been greatest for those at the top of the population; for everyone else, income (controlling for the value of the dollar) has either been relatively flat or grown at a far lesser rate. Inequality becomes even more apparent, however, when you consider both wealth and income.

Wealth is the monetary value of everything one actually owns. Wealth is calculated by adding all financial assets (stocks, bonds, property, insurance, savings, value of investments, and so on) and subtracting debts, resulting in one's **net worth**. Wealth allows you to accumulate assets over generations, giving advantages to subsequent generations that they might not have had on their own. Unlike income, *wealth is cumulative*—that is, its value tends to increase through investment. Wealth can also be passed on to the next generation, giving those who inherit wealth a considerable advantage in accumulating more resources.

To understand the significance of wealth compared to income in determining class location, imagine two college students graduating in the same year, from the same college, with the same major and same grade point average. Imagine further that both get jobs with the same salary in the same organization. In one case, parents paid all the student's college expenses and gave her a car upon graduation. The other student worked while in school and graduated with substantial debt from student loans. This student's family has no money with which to help support the new worker. Who is better off? Same salary, same credentials, but wealth (even if modest) matters, giving one person an advantage that will be played out many times over as the young worker buys a home, finances her own children's education, and possibly inherits additional assets.

Where is all the wealth? The wealthiest 1 percent own 35 percent of all net worth; the bottom half hold only 1.1 percent of all wealth (see ▲ Figure 8.1; Levine 2012).

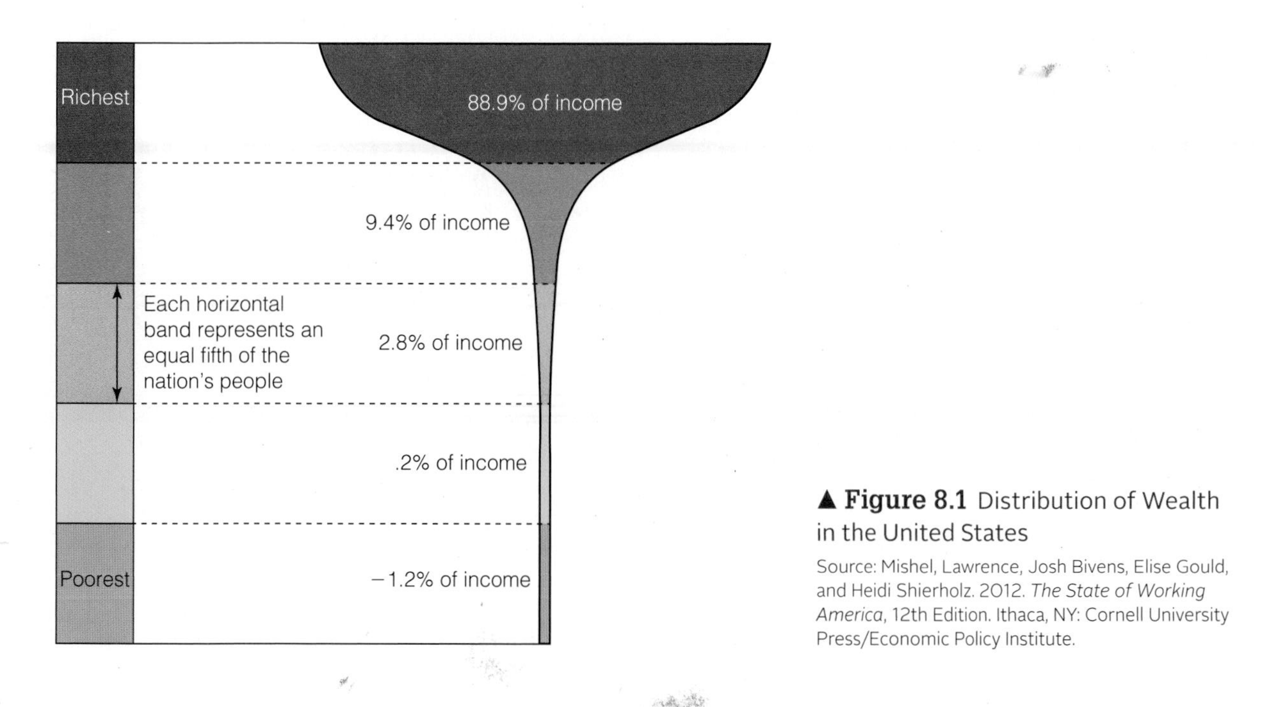

▲ **Figure 8.1** Distribution of Wealth in the United States

Source: Mishel, Lawrence, Josh Bivens, Elise Gould, and Heidi Shierholz. 2012. *The State of Working America*, 12th Edition. Ithaca, NY: Cornell University Press/Economic Policy Institute.

"It's a dog's life," or so the saying goes, but even dogs have their experiences shaped by the realities of social class.

Moreover, there has been an increase in the concentration of wealth since the 1980s, making the United States one of the most unequal nations in the world. The growth of wealth by a select few, though long a feature of the U.S. class system, has also reached historic levels. As just one example, John D. Rockefeller is typically heralded as one of the wealthiest men in U.S. history. Comparing Rockefeller with Bill Gates, controlling for the value of today's dollars, Gates has far surpassed Rockefeller's riches.

In contrast to the vast amount of wealth and income controlled by elites, a very large proportion of Americans have hardly any financial assets once debt is subtracted. Figure 8.4 shows the net worth of different parts of the population, and you can see that most of the population has very low net worth. One-fifth of the population has zero or negative net worth, usually because their debt exceeds their assets. The American dream of owning a home, a new car, taking annual vacations, and sending one's children to good schools—not to mention saving for a comfortable retirement—is increasingly unattainable for many. When you see the amount of income and wealth a small segment of the population controls, a sobering picture of class inequality emerges. Students themselves may be experiencing this burden, as levels of debt from student loans have escalated in recent years.

"I hope my parents can pay off their college loans before I go to college."

Despite the prominence of rags-to-riches stories in American legend, much of the wealth in this society

understanding **diversity**

The Student Debt Crisis

Numerous recent reports show that students are struggling over rising levels of debt from student loans. A record one in five households in the United States now has outstanding student debt. Not only is the number of those with student debt increasing, but so is the size of the indebtedness (see ▲ Figures 8.2 and 8.3 below; Lee 2013; Fry 2012). Leaving college or graduate school with large amounts of debt impedes one's ability to get financially established.

All students are at risk of accruing debt, given the rising cost of education and the higher interest rates now associated with student loans. Some groups though are more vulnerable than others, adding to the inequalities that accrue across different groups. Among those in the bottom fifth of income earners, student debt, on average, takes up 24 percent of all income; for the top fifth of earners, only 9 percent of income. For those in the middle, student debt consumes 12 percent of income (Fry 2012).

The highest amount of student debt is also among those under 35, those who are just beginning careers and, possibly, families. Race also matters. Black students are more likely to borrow money for college than other groups—and to borrow more; 80 percent of Black students have outstanding student loans, compared to 65 percent of Whites, 67 percent of Hispanics, and 54 percent of Asian students. Moreover, levels of debt are highest among Blacks—an average of $28,692, compared to $24,772 for Whites, $22,886 for Hispanics, and $21,090 for Asians (Demos 2014).

How does this reality of student debt influence the experience of those you see in your own environment? What are the sociological causes of this significant social problem?

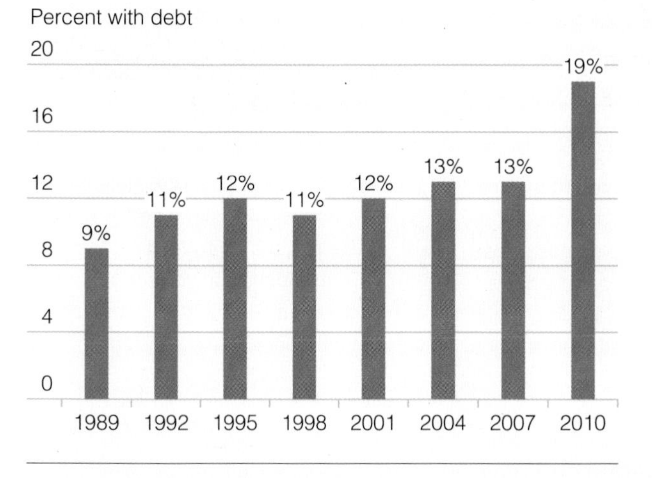

▲ **Figure 8.2** Households with Outstanding Student Debt

Source: Pew Research Center, Social and Demographic Trends Project. **www.pewsocialtrends.org/2012/09/26/**. Released September 26, 2012. *A Record One-in-Five Households Now Owe Student Loan Debt Burden Greatest on Young, Poor* by Richard Fry.

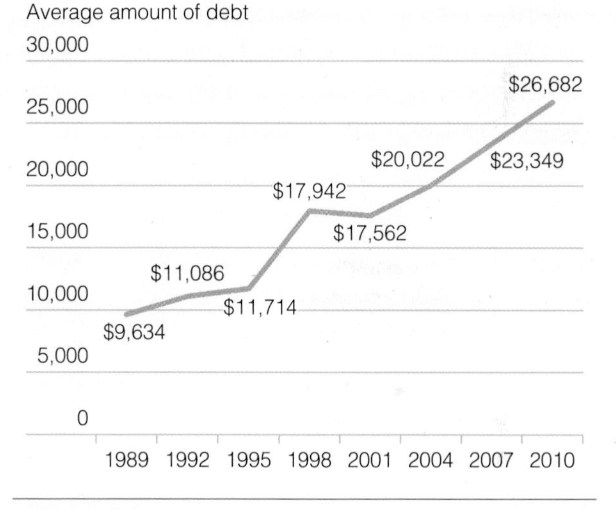

▲ **Figure 8.3** Average Amount of Household Student Debt

Source: Pew Research Center, Social and Demographic Trends Project. **www.pewsocialtrends.org/2012/09/26/**. Released September 26, 2012. *A Record One-in-Five Households Now Owe Student Loan Debt Burden Greatest on Young, Poor* by Richard Fry.

is inherited. In recent years, more of those who are very rich are "self-made"—that is, starting from modest origins: Bill Gates, a Harvard dropout; Mark Zuckerberg, founder of Facebook; and Oprah Winfrey come to mind. Such examples exist, although for many, if you scratch the surface of the rags-to-riches theme, you will find that they had a significant leg up. Among the now very rich, it has become more common for some to become amazingly rich, even though coming from modest origins. The technology boom has certainly helped, showing again how the historical context of one's life course can matter.

For most people, however, dramatically moving up in the class system remains highly likely. Children from low-income families have less than a 1 percent chance of reaching the top 5 percent of earners (T. Hertz 2006). Recently, even the modest wealth of those in the middle class has been significantly eroded by the impact of the recent economic recession; young and minority households have been especially hard hit by these changes, in large part because of being highly "leveraged"—that is, holding too much debt on their homes (Wolff 2014; Kochhar et al. 2011).

doing sociological research

The Fragile Middle Class

The hallmark of the middle class in the United States is its presumed stability. Home ownership, a college education for children, and other accoutrements of middle-class status (nice cars, annual vacations, an array of consumer goods) are the symbols of middle-class prosperity. The American middle class is not as secure as it has been presumed to be.

Personal bankruptcy has risen dramatically with more than a million nonbusiness filings for bankruptcy per year recently. How can this be happening in such a prosperous society? Sociologists Teresa Sullivan, Elizabeth Warren, and Jay Lawrence Westbrook have studied bankruptcy, and their research shows the fragility of the middle class in recent times.

Research Question: What is causing the rise of bankruptcy?

Research Method: This study is based on an analysis of official records of bankruptcy in five states, as well as on detailed questionnaires given to individuals who filed for bankruptcy.

Research Results: The research findings of Sullivan and her colleagues debunk the idea that bankruptcy is most common among poor people. Instead, they found bankruptcy is mostly a middle-class phenomenon representing a cross-section of those in this class (meaning that those who are bankrupt are matched on the demographic characteristics of race, age, and gender with others in the middle class). They also debunk the notion that bankruptcy is rising because it is so easy to file. Rather, they found many people in the middle class so overwhelmed with debt that they cannot possibly pay it off. Most people often file for bankruptcy as a result of job loss and lost wages. But divorce, medical problems, housing expenses, and credit card debt also drive many to bankruptcy court.

Conclusions and Implications: Sullivan and her colleagues explain the rise of bankruptcy as stemming from structural factors in society that fracture the stability of the middle class. The volatility of jobs under modern capitalism is one of the biggest factors, but add to this the "thin safety net"—no health insurance for many, but rising medical costs. Also, the American dream of owning one's own home means many are "mortgage poor"—extended beyond their ability to keep up.

The United States is also a credit-driven society. Credit cards are routinely mailed to people in the middle class, encouraging them to buy beyond their means. You can now buy virtually anything on credit: cars, clothes, doctor's bills, entertainment, groceries. You can even use one credit card to pay off other credit cards. Indeed, it is difficult to live in this society without credit cards. Increased debt is the result. Many are simply unable to keep up with compounding interest and penalty payments, and debt takes on a life of its own as consumers cannot keep up with even the interest payments on debt.

Sullivan, Warren, and Westbrook conclude that increases in debt and uncertainty of income combine to produce the fragility of the middle class. Their research shows that "even the most secure family may be only a job loss, a medical problem, or an out-of-control credit card away from financial catastrophe" (2000: 6).

Questions to Consider

1. Have you ever had a credit card? If so, how easy was it to get? Is it possible to get by without a credit card?
2. What evidence do you see in your community of the fragility or stability of different social class groups?

Source: T. A. Sullivan, E. Warren, and J. L. Westbrook, *The Fragile Middle Class: Americans in Debt*. Copyright © 2000 by Yale University Press.

Race also influences the pattern of wealth distribution in the United States. For every dollar of wealth White Americans hold, Black Americans have only 26 cents. At all levels of income, occupation, and education, Black families have lower levels of wealth than similarly situated White families (see ▲ Figure 8.4).

Being able to draw on assets during times of economic stress means that families with some resources can better withstand difficult times than those without assets. Even small assets, such as home ownership or a savings account, provide protection from crises such as increased rent, a health emergency, or unemployment. Because the effects of wealth are *intergenerational*—that is, they accumulate over time—just providing equality of opportunity in the present does not address the differences in class status that Black and White Americans experience (Oliver and Shapiro 2006).

What explains the disparities in wealth by race? Wealth accumulates over time. Thus government policies in the past have prevented Black Americans from being able to accumulate wealth. Discriminatory housing policies, bank lending policies, tax codes, and so forth have disadvantaged Black Americans, resulting in the differing assets Whites and Blacks in general hold now. Even though some of these discriminatory policies have ended, many continue. Either way, their effects persist, resulting in what sociologists Melvin Oliver and Thomas Shapiro call the *sedimentation of racial inequality*.

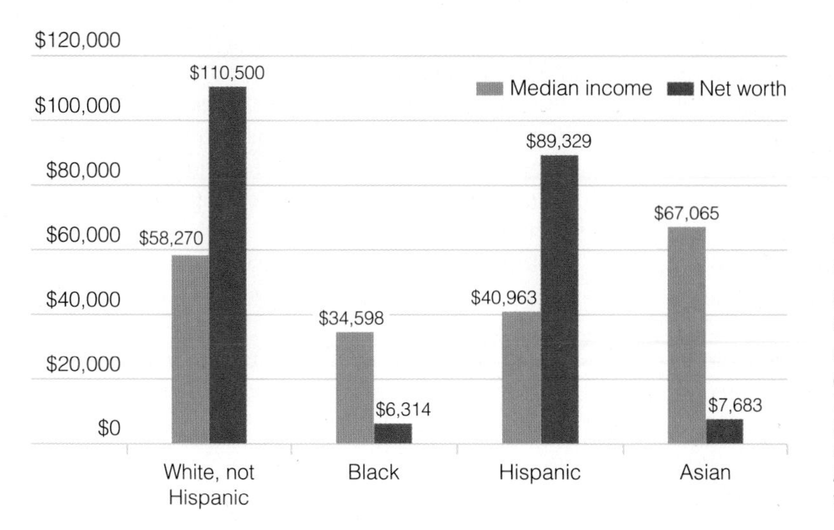

▲ **Figure 8.4** Median Income and Net Worth by Race

Note: Income data for 2013; net worth from 2011.

Source: U.S. Census Bureau. 2011. *Net Worth and Asset Ownership of Households: 2011, Table 1.* Washington, DC: U.S. Census Bureau; DeNavas-Walt, Carmen, and Bernadette Proctor. 2014. *Income and Poverty in the United States: 2013.* Washington, DC: U.S. Census Bureau. **www.census.gov**

Understanding the significance of wealth in shaping life chances for different groups also shows how important it is to understand diversity within the different labels used to define groups. Among Hispanics, for example, Cuban Americans and Spaniards are similar to Whites in their wealth holdings, whereas Mexicans, Puerto Ricans, Dominicans, and other Hispanic groups more closely resemble African Americans in measures of wealth and class standing. Without significant wealth holdings, families of any race are less able to transmit assets from previous generations to the next generation, one main support of *social mobility* (discussed later in the chapter).

Analyzing Social Class

The class structure of the United States is elaborate, arising from the interactions of race and gender inequality with class, the presence of old mixed with new wealth, the income and wealth gap between the haves and have-nots, a culture of entrepreneurship and individualism, and in recent times, accelerated globalization and high rates of immigration. Given this complexity, how do sociologists conceptualize social class?

Class as a Ladder

One way to think about the class system is as a ladder, with different class groups arrayed up and down the rungs, each rung corresponding to a different level in the class system. Conceptualized this way, social class is the common position groups hold in a status hierarchy (Lucal 1994; Wright 1979); class is indicated by factors such as levels of income, occupational standing, and educational attainment. People are relatively high or low on the ladder depending on the resources they have and whether those resources are education, income, occupation, or any of the other factors known to influence people's placement (or ranking) in the stratification system. Indeed, an abundance of sociological research has stemmed from the concept of **status attainment**, the process by which people end up in a given position in the stratification system. Status attainment research describes how factors such as class origins, educational level, and occupation produce class location.

The laddered model of class suggests that stratification in the United States is hierarchical but somewhat fluid. That is, the assumption is that people can move up and down different "rungs" of the ladder—or class system. In a relatively *open class system* such as the United States, people's achievements do matter, although the extent to which people rise rapidly and dramatically through the stratification system is less than the popular imagination envisions. Some people move down in the class system, but as we will see, most people remain relatively close to their class of origin. When people rise or fall in the class system, the distance they travel is usually relatively short, as we will see in a later section on social mobility.

The image of stratification as a laddered system, with different gradients of social standing, emphasizes that one's **socioeconomic status (SES)** is derived from certain factors. Income, occupational prestige, and education are the three measures of socioeconomic status that have been found to be most significant in determining people's placement in the stratification system.

The **median income** for a society is the midpoint of all household incomes. Half of all households earn more than the median income; half earn less. In 2013, median household income in the United States was $51,939 (DeNavas-Walt and Proctor 2014). To many, this may seem like a lot of money, but consider these facts: American consumers spend about one-third of their household budgets on housing; almost another 18 percent on transportation; 13 percent on food; and

11 percent on insurance and pensions (Bureau of Labor Statistics 2014a). If you do the calculations based on the median income level, you will see there is very little left for other living expenses (clothing, education, taxes, communication, entertainment, and so forth)—less than $1000 per month for all other expenses—hardly a lavish income, especially when you consider that half of Americans have less than this, given the definition of a median. (See also the "See for Yourself" exercise on household budgets later in this chapter.) Those bunched around the median income level are considered middle class, although sociologists debate which income brackets constitute middle-class standing because the range of what people think of as "middle class" is quite large. Nonetheless, income is a significant indicator of social class standing, although not the only one.

SEE for YOURSELF

Income Distribution: Should Grades Be the Same?

▲ Figure 8.5 shows the income distribution within the United States. Imagine that grades in your class were distributed based on the same curve. Let's suppose that after students arrived in class and sat down, different groups received their grades based on where they were sitting in the room and in the same proportion as the U.S. income distribution. Only students in the front receive A's; the back, D's and F's. The middle of the room gets the B's and C's. Write a short essay answering the following questions based on this hypothetical scenario:

1. How many students would receive A's, B's, C's, D's, and F's?
2. Would it be fair to distribute grades this way? Why or why not?
3. Which groups in the class might be more likely to support such a distribution? Who would think the system of grade distribution should be changed?
4. What might different groups do to preserve or change the system of grade distribution? What if you really needed an A, but got one of the F's? What might you do?
5. Are there circumstances in actual life that are beyond the control of people and that shape the distribution of income?
6. How is social stratification maintained by the beliefs that people have about merit and fairness?

Adapted from: Brislen, William, and Clayton D. Peoples. 2005. "Using a Hypothetical Distribution of Grades to Introduce Social Stratification." *Teaching Sociology* 33 (January): 74–80.

Occupational prestige is a second important indicator of socioeconomic status. **Prestige** is the value others assign to people and groups. **Occupational prestige** is the subjective evaluation people give to jobs. To determine occupational prestige, sociological researchers typically ask nationwide samples of adults to rank the general standing of a series of jobs. These subjective ratings provide information about how people perceive the worth of different occupations. People tend to rank professionals, such as physicians, professors, judges, and lawyers highly, with occupations such as electrician, insurance agent, and police officer falling in the middle. Occupations with low occupational prestige are maids, garbage collectors, and shoe shiners. These rankings do not reflect the worth of people within these positions but are indicative of the judgments people make about the worth of these jobs.

The final major indicator of socioeconomic status is **educational attainment**, typically measured as the total years of formal education. The more years of

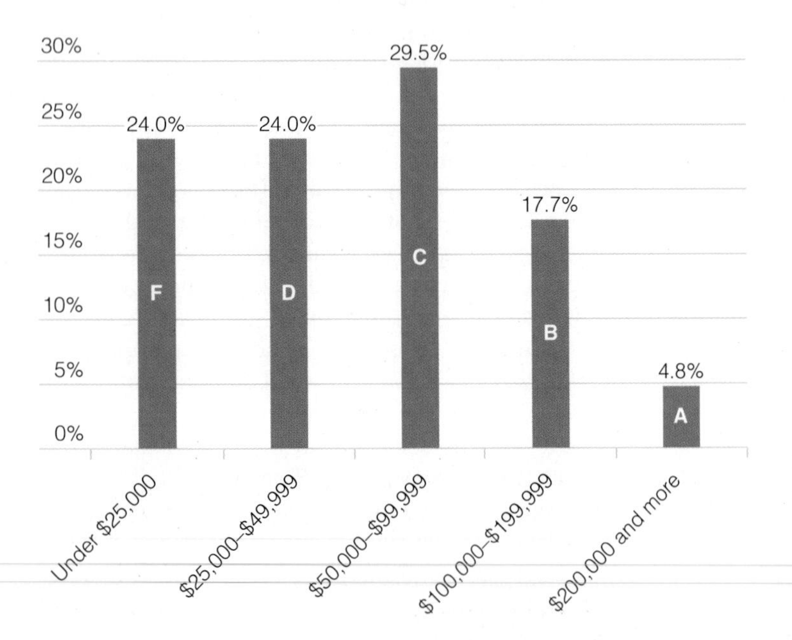

▲ **Figure 8.5** Income Distribution in the United States This graph shows the percentage of the total population that falls into each of five income groups. Would it be fair if course grades were distributed by the same percentages?

Source: DeNavas-Walt, Carmen and Bernadette Proctor. 2014. *Income and Poverty in the United States: 2013*. Washington, DC: U.S. Census Bureau. **www.census.gov**

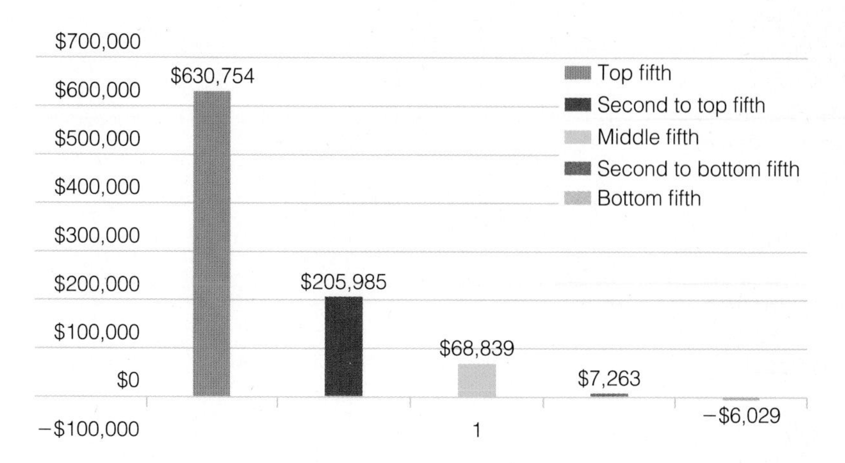

▲ **Figure 8.6** Median Net Worth by Income Quintile Recall that one's net worth is the value of everything owned minus one's debt. A quintile is one-fifth of a population, shown here for five different income brackets. You can see here the vast differences in wealth holdings by those in these different income brackets. How would one's wealth holdings affect your ability to withstand some sort of emergency—an illness, unemployment, a recession, and so forth?

Source: U.S. Census Bureau. 2012. *Wealth and Asset Ownership, 2011: Table A1.* Washington, DC: U.S. Census Bureau. **www.census.gov**

education attained, the more likely a person's class status. The prestige attached to occupations is strongly tied to the amount of education the job requires (Ollivier 2000; Blau and Duncan 1967).

Taken together, income, occupation, and education are good indicators of people's class standing. Using the laddered model of class, you can describe the class system in the United States as being divided into several classes: upper, upper middle, middle, lower middle, and lower class. The different classes are arrayed up and down, like a ladder, with those with the most money, education, and prestige on the top rungs and those with the least at the bottom.

In the United States, the *upper class* owns the major share of corporate and personal wealth (see ▲ Figure 8.6). The upper class includes those who have held wealth for generations as well as those who have recently become rich. Only a very small proportion of people actually constitute the upper class, but they control vast amounts of wealth and power in the United States. Those in this class are elites who exercise enormous control throughout society. Some wealthy individuals can wield as much power as entire nations (Friedman 1999).

Even the term *upper class*, however, can mask the degree of inequality in the United States. You might consider those in the top 10 percent as upper class, but within this class are the superrich, or those popularly known as the "one percent," so labeled by the Occupy America movement, in contrast to the remaining 99 percent. Since about 1980, the share of income (not to mention wealth) going to the top one percent has increased to levels not seen in the United States since 1920, a time labeled as the "Gilded Age" because of the concentration of wealth and income in the hands of a few. Income distribution now matches that of the Gilded Age and, given the trends, may well come to exceed it. Sociological research finds that this new concentration of income among the superrich is the result of several trends, including the lowest tax rates for high incomes, a more conservative shift in Congress, diminishing union membership, and asset bubbles in the stock and housing markets (Saez and Zucman 2014; Volscho and Kelly 2012).

How rich is rich? Each year, the business magazine *Forbes* publishes a list of the 400 wealthiest families and individuals in the country. By 2014, you had to have at least $1.5 billion to be on the list! Bill Gates and Warren Buffet are the two wealthiest people on the list—Gates with an estimated worth of $79.4 billion; Buffet, $67 billion. Even in the face of the massive economic downturn for so many in the United States, only two people in the top twenty of the group had less money than the year before. A substantial portion of those on the list describe themselves as "self-made," that is, living the American dream, but most of these were still able to borrow from parents, in-laws, or spouses. Although they may have built their fortunes, they did so with a head start on accumulation (Kroll and Dolan 2012). The best predictor of future wealth still remains the family into which you are born (McNamee and Miller 2009).

The upper class is overwhelmingly White, conservative, and Protestant. Members of this class exercise tremendous political power by funding lobbyists, exerting their social and personal influence on other elites, and contributing heavily to political campaigns (Domhoff 2013). They travel in exclusive social networks that tend to be open only to those in the upper class. They tend to intermarry, their children are likely to go to expensive schools, and they spend their leisure time in exclusive resorts.

Those in the upper class with newly acquired wealth are known as the *nouveau riche*. Luxury vehicles, high-priced real estate, and exclusive vacations may mark the lifestyle of the newly rich. Larry Ellison, who made his fortune as the founder of the software company Oracle, is the third wealthiest person in the United States. Ellis has a megayacht that is 482 feet long, five stories high, with 82 rooms inside. The megayacht also includes an indoor swimming pool, a cinema, a space for a private submarine, and a basketball court that doubles as a helicopter launch pad.

Social class influences many things, including the leisure time people experience. Few can even imagine having something like the yacht pictured on the left, owned by Larry Ellison, founder of Oracle.

The *upper-middle class* includes those with high incomes and high social prestige. They tend to be well-educated professionals or business executives. Their earnings can be quite high indeed, even millions of dollars a year. It is difficult to estimate exactly how many people fall into this group because of the difficulty of drawing lines between the upper, upper-middle, and middle classes. Indeed, the upper-middle class is often thought of as "middle class" because their lifestyle sets the standard to which many aspire, but this lifestyle is actually unattainable by most. A large home full of top-quality furniture and modern appliances, two or three relatively new cars, vacations every year (perhaps a vacation home), high-quality college education for one's children, and a fashionable wardrobe are simply beyond the means of a majority of people in the United States.

The *middle class* is hard to define in part because being "middle class" is more than just economic position. Half of all Americans identify themselves as middle class (Morin and Motel 2012), even though they vary widely in lifestyle and in resources at their disposal. The idea that the United States is an open class system leads many to think that the most have a middle-class lifestyle. The "middle class" is the ubiquitous norm, even though many who consider themselves middle class have a tenuous hold on this class position.

The *lower-middle class* includes workers in the skilled trades and low-income bureaucratic workers, some who may actually think of themselves as middle class. Also known as the *working class*, this class includes blue-collar workers (those in skilled trades who do manual labor) and many service workers, such as secretaries, hairstylists, food servers, police, and firefighters. A medium to low income, education, and occupational prestige define the lower-middle class relative to the class groups above it. The term *lower* in this class designation refers to the relative position of the group in the stratification system, but it has a pejorative sound to many people, especially to people who are members of this class, many who think of themselves as middle class.

The *lower class* is composed primarily of displaced and poor. People in this class tend to have little formal education and are often unemployed or working in minimum-wage jobs. People of color and women make up a disproportionate part of this class. The poor include the *working poor*—those who work at least twenty-seven hours a week but whose wages fall below the federal poverty level. Four percent of all people working full-time and 16 percent of those working part-time live below the poverty line, a proportion that has increased over time. Although this may seem a small number, it includes 9.1 million adults. Black and Hispanic workers are twice as likely to be among the working poor as White or Asian workers, and women are more likely than men to be so (U.S. Bureau of Labor Statistics 2014 d).

The concept of the **urban underclass** has been added to the lower class (W. Wilson 1987). The underclass includes those who are likely to be permanently unemployed and without much means of economic support. The underclass has little or no opportunity for movement out of the worst poverty. Rejected from the economic system, those in the underclass may become dependent on public assistance or illegal activities. Structural transformations in the economy have left large groups of people, especially urban minorities, in these highly vulnerable positions. The growth of the urban underclass has exacerbated the problems of urban poverty and related social problems (Wilson 2009, 1996, 1987).

Class Conflict

A second way of conceptualizing the class system is *conflict theory*. Conflict theory defines classes in terms of their structural relationship to other classes and their relationship to the economic system. The analysis of

those who have lived there the longest, even if newcomers arrive with more money. Although having power is typically related to also having high economic standing and high social status, this is not always the case, as you saw with the example of the drug dealer.

Finally, *party* (or what we would now call power) is the political dimension of stratification. Power is the capacity to influence groups and individuals even in the face of opposition. Power is also reflected in the ability of a person or group to negotiate their way through social institutions. An unemployed Latino man wrongly accused of a crime, for instance, does not have much power to negotiate his way through the criminal justice system. By comparison, business executives accused of corporate crime can afford expensive lawyers and thus frequently go unpunished or, if they are found guilty, serve relatively light sentences in comparatively pleasant facilities. Again, Weber saw power as linked to economic standing, but he did not think that economic standing was always the determining cause of people's power.

Marx and Weber explain different features of stratification. Both understood the importance of the economic basis of stratification, and they knew the significance of class for determining the course of one's life. Marx saw people as acting primarily out of economic interests. Weber refined the sociological analyses of stratification to account for the subtleties that can be observed when you look beyond the sheer economic dimension to stratification, stratification being the result of economic, social, and political forces. Together, Marx and Weber provide compelling theoretical grounds for understanding the contemporary class structure.

Functionalism and Conflict Theory: The Continuing Debate

Marx and Weber were trying to understand why differences existed in the resources that various groups in society hold. The question persists of why there is inequality. Two major frameworks in sociological theory—functionalist and conflict theory—take quite different approaches to understanding inequality (see ◆ Table 8.2).

The Functionalist Perspective on Inequality. Functionalist theory views society as a system of institutions organized to meet society's needs (see Chapter 1). The functionalist perspective emphasizes that the parts of society are in basic harmony with each other; society is held together by cohesion, consensus, cooperation, stability, and persistence (Eitzen and Baca Zinn 2012; Merton 1957; Parsons 1951a). Different parts of the social system complement one another. To explain stratification, functionalists see the roles filled by the upper classes—such as governance, economic innovation, investment, and management—are essential for a cohesive and smoothly running society. The upper classes are then rewarded in proportion to their contribution to the social order (Davis and Moore 1945).

◆ Table 8.2 Functionalist and Conflict Theories of Stratification

Interprets	Functionalism	Conflict Theory
Inequality	The purpose of inequality is to motivate people to fill needed positions in society.	Inequality results from a system where those with the most resources exploit and control others.
Reward system	Greater rewards are attached to higher positions to ensure that people will be motivated to train for functionally important roles in society.	Inequality prevents the talents of those at the bottom from being discovered and used.
Classes	Some groups are rewarded because their work requires the greatest degree of talent and training.	Classes conflict with each other as they vie for power and economic, social, and political resources.
Elites	The most talented are rewarded in proportion to their contribution to the social order.	The most powerful reproduce their advantage by distributing resources and controlling the dominant value system.
Class consciousness/ ideology	Beliefs about success and failure confirm the status of those who succeed.	Elites shape societal beliefs to make their unequal privilege appear to be legitimate and fair.
Poverty	Poverty serves economic and social functions in society.	Poverty is inevitable because of the exploitation built into the system.
Social policy	Because the system is basically fair, social policies should only reward merit.	Because the system is basically unfair, social policies should support disadvantaged groups.

According to the functionalist perspective, social inequality serves an important purpose in society: It motivates people to fill the different positions in society that are needed for the survival of the whole. Functionalists think that some positions in society are more important than others and require the most talent and training. The rewards attached to those positions (such as higher income and prestige) ensure that people will make the sacrifices needed to acquire the training for functionally important positions (Davis and Moore 1945). Higher class status thus comes to those who acquire what is needed for success (such as education and job training). In other words, functionalist theorists see inequality as based on a reward system that motivates people to succeed.

The Conflict Perspective on Inequality. Conflict theory also sees society as a social system, but unlike functionalism, conflict theory interprets society as being held together through conflict and coercion. From a conflict-based perspective, society comprises competing interest groups, some with more power than others. Groups struggle over societal resources and compete for social advantage. Conflict theorists argue that those who control society's resources also hold power over others. The powerful are also likely to act to reproduce their advantage and try to shape societal beliefs to make their privileges appear to be legitimate and fair. In sum, conflict theory emphasizes the friction in society rather than the coherence, and sees society as dominated by elites.

From the perspective of conflict theory, social stratification is based on class conflict and blocked opportunity. Conflict theorists see stratification as a system of domination and subordination in which those with the most resources exploit and control others. They also see the different classes as in conflict with each other, with the unequal distribution of rewards reflecting the class interests of the powerful, not the survival needs of the whole society (Eitzen and Baca Zinn 2012). According to the conflict perspective, inequality provides elites with the power to distribute resources, make and enforce laws, and control value systems. Elites then use these powers to reproduce their own advantage. Others in the class structure, especially the working class and the poor, experience blocked mobility.

From a conflict point of view, the more stratified a society, the less likely that society will benefit from the talents of its citizens. Inequality limits the life chances of those at the bottom, preventing their talents from being discovered and used.

The Debate between Functionalist and Conflict Theories. Implicit in the argument of each perspective is criticism of the other perspective. Functionalism assumes that the most highly rewarded jobs are the most important for society, whereas conflict theorists argue that some of the most vital jobs in society—those that sustain life and the quality of life, such as farmers, mothers, trash collectors, and a wide range of other laborers—are usually the least rewarded. Conflict theorists also criticize functionalist theory for assuming that the most talented get the greatest rewards. They point out that systems of stratification tend to devalue the contributions of those left at the bottom and to underutilize the diverse talents of all people (Tumin 1953). In contrast, functionalist theorists contend that the conflict view of how economic interests shape social organization is too simplistic. Conflict theorists respond by arguing that functionalists hold too conservative a view of society and overstate the degree of consensus and stability that exists.

The debate between functionalist and conflict theorists raises fundamental questions about how people view inequality. Is inequality inevitable? How is inequality maintained? Do people basically accept it? This debate is not just academic. The assumptions made from each perspective frame public policy debates. Whether the topic is taxation, poverty, or homelessness, if people believe that anyone can get ahead by ability alone, they will tend to see the system of inequality as fair and accept the idea that there should be a differential reward system. Those who tend toward the conflict view of the stratification system are more likely to advocate programs that emphasize public responsibility for the well-being of all groups and to support programs and policies that result in more of the income and wealth of society going toward the needy.

Poverty

Despite the relatively high average standard of living in the United States, poverty afflicts millions of people. There are now more than 45 million poor people in the United States—a whopping 14.5 percent of the population. Even more startling is the large number of people living in very deep poverty, or what experts define as *extreme poverty* (the U.S. measure being living on two dollars or less per day; the world measure of extreme poverty is $1.25 per day or less; see also Chapter 9). Extreme poverty in the United States includes 3.5 million children who are living with virtually no income—a shocking fact for such a rich nation (Shaefer and Edin 2014).

Poverty deprives people of basic human needs—food, shelter, and safety from harm. It is also the basis for many of our nation's most intractable social problems. Failures in the education system; crime and violence; inadequate housing and homelessness; poor health care—all are related to poverty. Who is poor, and why is there so much poverty in an otherwise affluent society?

Defining Poverty

The federal government has established an official definition of poverty used to determine eligibility for government assistance and to measure the extent of poverty in the United States. The **poverty line** is the amount of money needed to support the basic needs of a household, as determined by government; below this line, one is considered officially poor. To determine the poverty line, the Social Security administration takes a low-cost food budget (based on dietary information provided by the U.S. Department of Agriculture) and multiplies it by a factor of three, assuming that a family spends approximately one-third of its budget on food. The resulting figure is the official poverty line, adjusted slightly each year for increases in the cost of living. In 2013, the official poverty line for a family of four (including two children) was $23,624. Although a cutoff point is necessary to administer antipoverty programs, this definition of poverty can be misleading. A person or family earning $1 above the cutoff point would not be officially categorized as poor.

There are numerous problems with the official definition of poverty. To name a few, it does not account for regional differences in the cost of living; it does not reflect changes in the cost of housing nor changes in the cost of modern standards of living that were not imagined in the 1930s, when the definition was established (Meyer and Sullivan 2012). Experts have argued that the government should develop alternative poverty measures, such as *shelter poverty*—a measure that would account for the cost of housing in different regions (Stone 1993). To date, Congress has resisted changing the official definition of poverty—a change that would likely increase the reported rate of poverty and potentially increase the cost of federal antipoverty programs.

SEE for YOURSELF

Using the current federal poverty line ($23,264 for a family of four, including two children), develop a monthly budget that does not exceed this income level and that accounts for all of your family's needs. Base your budget on the actual costs of such things in your locale (rent, food, transportation, utilities, clothing, and so forth). Don't forget to account for taxes (state, federal, and local), health care expenses, your children's education, and so on. What does this exercise teach you about those who live below the poverty line?

Who Are the Poor?

After the 1950s, poverty declined in the United States. The poverty rate has generally been increasing since 2000, from about 11 to nearly 15 percent of the population. The majority of the poor are White, although there are disproportionately high rates of poverty among Asian Americans, Native Americans, Black Americans, and Hispanics. Of those considered poor, 27 percent are Native Americans, 27.2 percent are African Americans, 23.5 percent are Hispanics, 10.5 percent are Asian Americans, and 9.6 percent are non-Hispanic Whites (DeNavas-Walt and Proctor 2014; McCartney et al. 2013). Among Hispanics, there are further differences among groups. Puerto Ricans—the Hispanic group with the lowest median income—have been most likely to suffer increased poverty, probably because of their concentration in the poorest segments of the labor market and their high unemployment rates (Hauan et al. 2000; Tienda and Stier 1996). Asian American poverty has also increased substantially in recent years, particularly among the most recent immigrant groups, including Laotians, Cambodians, Vietnamese, Chinese, and Korean immigrants; Filipino, Japanese, and Asian Indian families have lower rates of poverty (White House 2012).

The vast majority of the poor have always been women and children, but the percentage of women and children considered to be poor has increased in recent years. The term **feminization of poverty** refers to the large proportion of the poor who are women and children. This trend results from several factors, including the dramatic growth of female-headed households, a decline in the proportion of the poor who are elderly (not matched by a decline in the poverty of women and children), and continuing wage inequality between women and men. The large number of poor women is associated with a commensurate large number of poor children. By 2013, 20 percent of all children in the United States (those under age 18) were poor,

Steve Debenport/Getty Images

Child poverty in the United States is higher than one would expect for such an otherwise materially well-off nation.

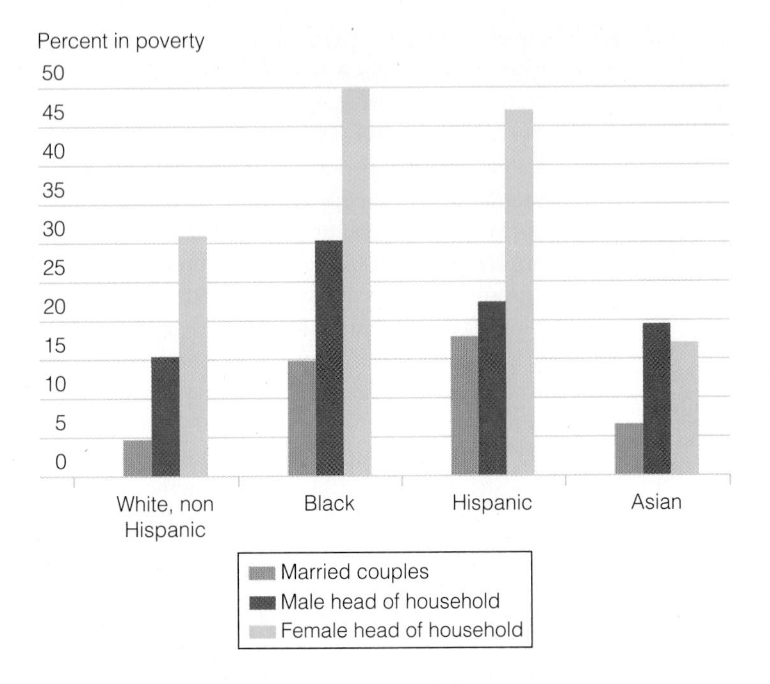

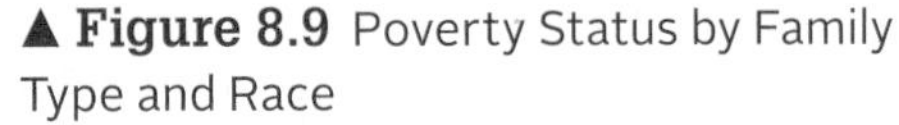

▲ **Figure 8.9** Poverty Status by Family Type and Race

Note: Families with children under 18 present.

Source: U.S. Census Bureau. 2014. *Historical Income Tables, Table POV-02, People in Families, by Family Structure, Age, and Sex, Iterated by Income-to-Poverty Ratio and Race: 2013.* Washington, DC: U.S. Census Bureau. **www.census.gov**

including 9.9 percent of non-Hispanic White children, 36.7 percent of Black children, 30 percent of Hispanic children, and 9.4 percent of Asian American children (DeNavas-Walt and Proctor 2014).

One-third of all families headed by women are poor (see ▲ Figure 8.9). In recent years, wages for young workers have declined; because most unmarried mothers are quite young, there is a strong likelihood that their children will be poor. Because of the divorce rate and generally little child support provided by men, women are also increasingly likely to be without the contributing income of a spouse and for longer periods of their lives. Women are more likely than men to live with children and to be financially responsible for them. However, women without children also suffer a high poverty rate, compounded in recent times by the fact that women now live longer than before and are less likely to be married than in previous periods.

DEBUNKING Society's Myths

Myth: Marriage is a good way to reduce women's dependence on welfare.

Reality: Although it is true that married-couple households are less likely to be poor than single-headed households, forcing women to marry encourages women's dependence on men and punishes women for being independent. Research indicates that poor women place a high value on marriage and want to be married, but also understand that men's unemployment and instability makes their ideal of marriage unattainable. In addition, large numbers of women receiving welfare have been victims of domestic violence (Edin and Kefalas 2005; Scott et al. 2002).

The poor are not a one-dimensional group. They are racially diverse, including Whites, Blacks, Hispanics, Asian Americans, and Native Americans. They are diverse in age, including not just children and young mothers, but also men and women of all ages, and especially a substantial number of the elderly, many of whom live alone. The poor are also geographically diverse, to be found in areas east and west, south and north, urban and rural.

As ■ Map 8.1 shows, poverty rates are generally higher in the South and Southwest. What the map cannot show, however, is concentrated poverty. **Concentrated poverty** means that there are areas of counties, cities, or states where larger percentages of people are poor. Such areas then have higher

Mario Tama/Getty Images

Although most people associate poverty with urban areas, poverty rates outside of metropolitan areas are actually higher than you might expect.

map 8.1

Mapping America's Diversity: Poverty in the United States

This map shows regional differences in poverty rates (that is, the percentage of poor in different counties). As you can see, poverty is highest in the South, Southwest, and some parts of the upper Midwest. This reflects the higher rates of poverty among Native Americans, Latinos, and African Americans, especially in rural areas. What the map does not show is the concentration of poverty in particular urban areas. According to this map, how much poverty is there in your region? Is there poverty that the map does not show?

Source: U.S. Census Bureau. 2010. "Poverty in the United States." **www.census.gov**

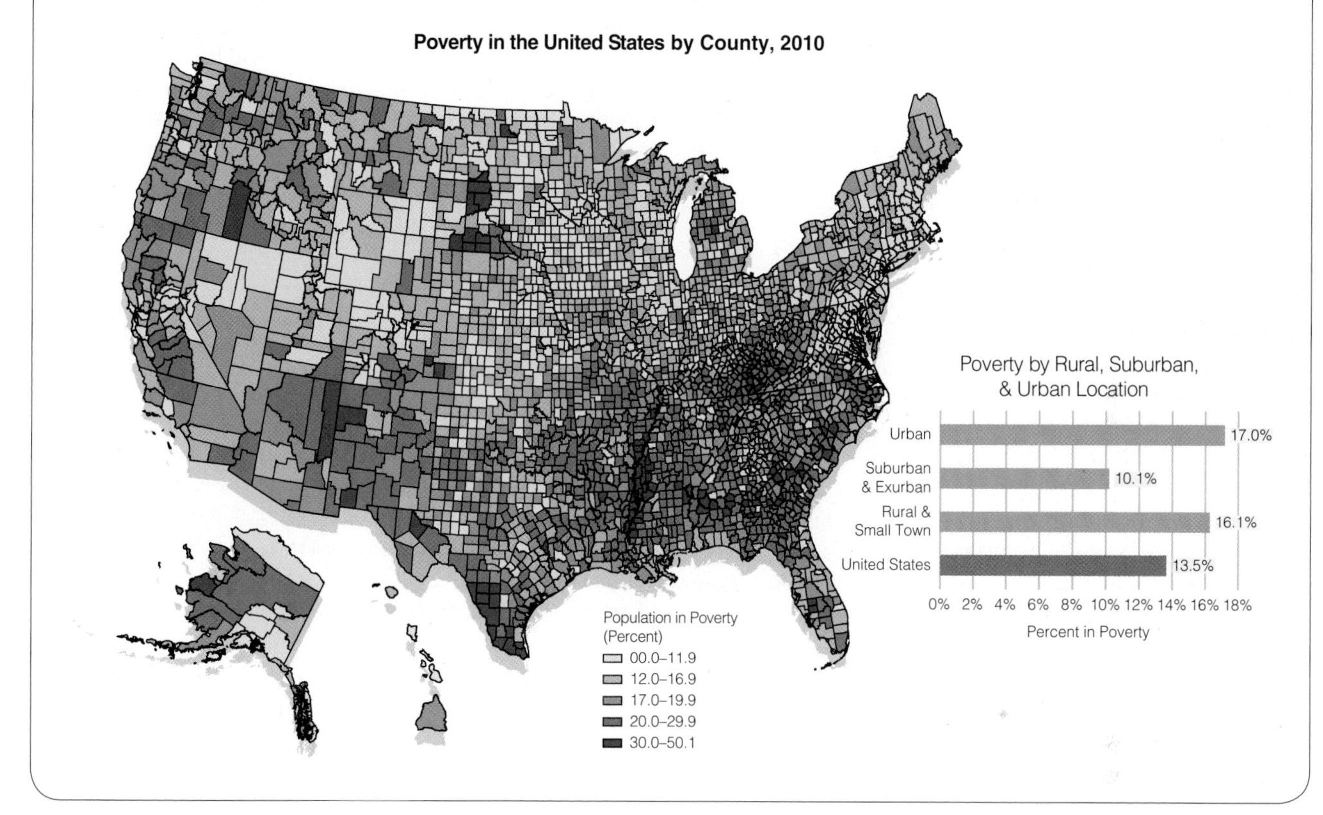

rates of crime, poor schools, few job opportunities, poor health and housing, and less access to services. Concentrated poverty is highest among African Americans and American Indians (including Alaska natives). Among these groups, 10 percent of people live in areas where 40 percent or more of the population is poor, compared to 3 percent of poor White Americans and 7 percent of Hispanics (Bishaw 2011).

One marked change in poverty is the growth of poverty in suburban areas. One-third of the nation's poor are now found in suburbs where poverty is growing twice as fast as in center cities (Kneebone and Holmes 2014). Rural poverty also persists in the United States, even though people tend to think of poverty as an urban phenomenon. The truth is that the poverty rate is actually higher outside of metropolitan areas than inside (DeNavas-Walt and Proctor 2014).

Despite the idea that the poor "milk" the system, government supports for the poor are limited. So-called welfare is now largely in the form of food stamps, not cash assistance. Half of households receiving food stamps are those with children present; one-fifth of recipients are disabled, and 17 percent are elderly people. Considering that the average monthly coupon value is $134, it is hard to understand why federal support is reviled as overly generous (U.S. Census Bureau 2012a).

Among the poor are thousands of homeless. Depending on how one defines and measures homelessness, estimates of the number of homeless people vary widely. If you count the number of homeless on any given night, there may be anywhere between 444,000 to 842,000; over an entire year, estimates are that between 2.3 and 3.5 million people experience homelessness,

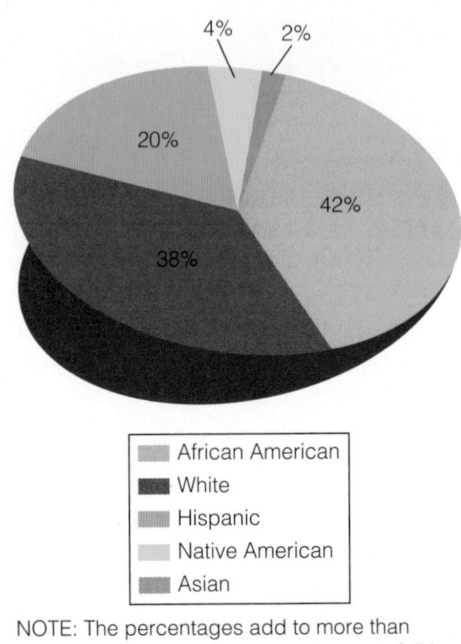

NOTE: The percentages add to more than 100 percent because some groups may fall into more than one race.

▲ **Figure 8.10** Who Are the Homeless?

Source of data: National Coalition for the Homeless. 2014. **www.nationalhomeless.org**

though not necessarily for an entire year (National Coalition for the Homeless 2014). The transient nature of this population makes accurate estimates of the extent of homelessness impossible.

Whatever the actual numbers of homeless people, there has been an increase in homelessness over the past two decades. Families are the fastest-growing segment of the homeless—40 percent—and, children are also 40 percent of the homeless. Moreover, half of the women with children who are homeless have fled from domestic violence (National Coalition against Domestic Violence 2001; Zorza 1991). A shocking number of the homeless are veterans (about 11 percent), including those returning from Iraq and Afghanistan (National Coalition for the Homeless 2014; see also ▲ Figure 8.10).

There are many reasons for homelessness. The great majority of the homeless are on the streets because of a lack of affordable housing and an increase in poverty, leaving many people with no choice but to live on the street. Add to that problems of inadequate health care, domestic violence, and addiction, and you begin to understand the factors that create homelessness. Some of the homeless have mental illness (about 16 percent of single, homeless adults); the movement to relocate patients requiring mental health care out of institutional settings has left many without mental health resources that might help them (National Coalition for the Homeless 2014).

Causes of Poverty

Most agree that poverty is a serious social problem. There is far less agreement on what to do about it. Public debate about poverty hinges on disagreements about its underlying causes. Two points of view prevail: Some blame the poor for their own condition, and some look to social structural causes to explain poverty. The first view, popular with the public and many policymakers, is that poverty is caused by the cultural habits of the poor. According to this point of view, behaviors such as crime, family breakdown, lack of ambition, and educational failure generate and sustain poverty, a syndrome to be treated by forcing the poor to fend for themselves. The second view is more sociological, one that understands poverty as rooted in the structure of society, not in the morals and behaviors of individuals.

DEBUNKING Society's Myths

Myth: The influx of unskilled immigrants raises the poverty level by taking jobs away from U.S. citizens who would otherwise be able to find work and lift themselves from poverty.

Sociological Perspective: A state-by-state analysis of poverty and immigration in recent years finds that immigrants actually improve local economies by increasing the supply of workers, generating labor market expansion, and promoting entrepreneurship—in other words, stimulating the economy. In some regions, immigration has actually lessened poverty (Peri 2014).

Blaming the Victim: The Culture of Poverty. Blaming the poor for being poor stems from the myth that success requires only individual motivation and ability. Many in the United States adhere to this view and hence have a harsh opinion of the poor. This attitude is also reflected in U.S. public policy concerning poverty, which is rather ungenerous compared with other industrialized nations. Those who blame the poor for their own plight typically argue that poverty is the result of early childbearing, drug and alcohol abuse, refusal to enter the labor market, and crime. Such thinking puts the blame for poverty on individual choices, not on societal problems. In other words, it blames the victim, not the society, for social problems (Ryan 1971).

The **culture of poverty** argument attributes the major causes of poverty to the absence of work values and the irresponsibility of the poor (Lewis 1969, 1966). In this light, poverty is seen as a dependent way of life that is transferred, like other cultural values, from generation to generation. Policymakers have adapted the culture of poverty argument to argue that the actual causes of poverty are found in the breakdown of major institutions, including the family, schools, and churches.

Is the culture of poverty argument true? To answer this question, we might ask: Is poverty transmitted across generations? Researchers have found only mixed support for this assumption. Many of those who are poor remain poor for only one or two years; only a small percentage of the poor are chronically poor. More often, poverty results from a household crisis, such as divorce, illness, unemployment, or parental death. People tend to cycle in and out of poverty. The public stereotype that poverty is passed through generations is thus not well supported by the facts.

A second question is: Do the poor want to work? The persistent public stereotype that they do not is central to the culture of poverty thesis. This attitude presumes that poverty is the fault of the poor, that poverty would go away if they would only change their values and adopt the American work ethic. What is the evidence for these claims?

Detailed studies of the poor simply find no basis for the assumption that the poor hold different values about work compared to every one else (Lakso 2013; Lee and Anat 2008). They simply find that work is difficult to find. Several other facts also refute this popular claim.

Most of the able-bodied poor *do* work, even if only part-time. As we saw previously, the number of workers who constitute the *working poor* has actually increased. You can see why this is true when you calculate the income of someone working full-time for minimum wage. Someone working forty hours per week, fifty-two weeks per year, at minimum wage will have an income far below the poverty line. This is the major reason that many have organized a *living wage campaign*, intended to raise the federal minimum wage to provide workers with a decent standard of living.

Current policies that force those on welfare to work also tend to overlook how difficult it is for poor people to retain the jobs they get. Prior to welfare reform in the mid-1990s, poor women who went off welfare to take jobs often found they soon had to return to welfare because the wages they earned were not enough to support their families. Leaving welfare often means losing health benefits, yet incurring increased living expenses. The jobs that poor people find often do not lift them out of poverty. In sum, attributing poverty to the values of the poor is both unproven and a poor basis for public policy.

Structural Causes of Poverty. From a sociological point of view, the underlying causes of poverty lie in the economic and social transformations taking place in the United States. Careful scholars do not attribute poverty to a single cause. There are many causes. Two of the most important are: (1) the restructuring of the economy, which has resulted in diminished earning power and increased unemployment; and, (2) the status of women in the family and the labor market, which has contributed to women being overrepresented among the poor. Add to these underlying conditions the federal policies in recent years that have diminished social support for the poor in the form of welfare, public housing, and job training. Given these reductions in federal support, it is little wonder that poverty is so widespread.

The restructuring of the economy has caused the disappearance of manufacturing jobs, traditionally an avenue of job security and social mobility for many workers, especially African American and Latino workers (Wilson 1996). The working class has been especially vulnerable to these changes. Economic decline in those sectors of the economy where men have historically received good pay and good benefits means that fewer men are the sole support for their families. Most families now need two incomes to achieve a middle-class way of life. The new jobs that are being created fall primarily in occupations that offer low wages and few benefits; they also tend to be filled by women, especially women of color, leaving women poor and men out of work. Such jobs offer little chance to get out of poverty. New jobs are also typically located in neighborhoods far away from the poor, creating a mismatch between the employment opportunities and the residential base of the poor.

Declining wage rates caused by transformations taking place within the economy fall particularly hard on young people, women, and African Americans and Latinos, the groups most likely to be among the working poor. The high rate of poverty among women is also strongly related to women's status in the family and the labor market. Divorce is one cause of poverty, because without a male wage in the household, women are more likely to be poor. Women's child-care responsibilities make working outside the home on marginal incomes difficult. Many women with children cannot manage to work outside the home, because it leaves them with no one to watch their children. More women now depend on their own earnings to support themselves, their children, and other dependents. Whereas unemployment has always been considered a major cause of poverty among men, low wages play a major role for women.

The persistence of poverty also increases tensions between different classes and racial groups. William Julius Wilson, one of the most noted analysts of poverty and racial inequality, has written, "The ultimate basis for current racial tension is the deleterious effect of basic structural changes in the modern American economy on Black and White lower-income groups, changes that include uneven economic growth, increasing technology and automation, industry relocation, and labor market segmentation" (1978: 154). Wilson's comments demonstrate the power of sociological thinking by convincingly placing the causes of both poverty and racism in their societal context, instead of the individualistic thinking that tends to blame the poor for their plight.

Welfare and Social Policy

The 1996 Personal Responsibility and Work Opportunity Reconciliation (PRWOR) Act governs current welfare policy. This federal policy eliminated the long-standing welfare program titled Aid to Families with Dependent Children (AFDC), which was created in 1935 as part of the Social Security Act. Implemented during the Great Depression, AFDC was meant to assist poor mothers and their children. This program acknowledged that some people are victimized by economic circumstances beyond their control and deserve assistance. For much of its lifetime, this law supported mostly White mothers and their children. Not until the 1960s did welfare come to be identified with Black families.

The new welfare policy gives block grants to states to administer their own welfare programs through the program called **Temporary Assistance for Needy Families (TANF)**. TANF stipulates a lifetime limit of five years for people to receive aid and requires all welfare recipients to find work within two years—a policy known as *workfare*. Those who have not found work within two years of receiving welfare can be required to perform community service jobs for free.

In addition, welfare policy denies payments to unmarried teen parents under age 18 unless they stay in school and live with an adult. It also requires unmarried mothers to identify the fathers of their children or risk losing their benefits (Edin and Kefalas 2005; Hays 2003). These broad guidelines are established at the federal level, but individual states can be more restrictive, as many have been. At the heart of public beliefs about support for the poor is the idea that public assistance creates dependence, discouraging people from seeking jobs. The very title of current welfare policy emphasizes "personal responsibility and work," suggesting that poverty is the fault of the poor. Low-income women, for example, are stereotyped as just wanting to have babies to increase the size of their welfare checks. Low-income men are also stereotyped as shiftless and irresponsible, even though research finds no support for either idea (Edin and Nelson 2013; Edin and Kefalas 2005).

Is welfare reform working? Many claim that welfare reform is working because, since passage of the new law, the welfare rolls have shrunk. Since 1996, the year that welfare reform was passed, the number receiving welfare support has declined from twelve million to four million (U.S. Census Bureau 2012a). Having fewer people on welfare does not, however, mean that poverty is reduced. In fact, as we have seen, extreme poverty has actually increased since passage of welfare reform. Having fewer people on the rolls can simply mean that people are without a safety net. Holes in the safety net makes those already vulnerable to economic distress even more so. Although not limited to them, single women with children and people of color have been those most negatively affected by the 1996 welfare reforms (Shaefer and Edin 2014).

Research done to assess the impact of a changed welfare policy is relatively recent. Politicians brag that welfare rolls have shrunk, but reduction in the welfare rolls is a poor measure of the true impact of welfare reform because this would be true simply if people are denied benefits. Because welfare has been decentralized to the state level, studies of the impact of current law must be done on a state-by-state basis. Such studies are showing that those who have gone into workfare programs most often earn wages that keep them below the poverty line. Although some states report that family income has increased following welfare reform, the increases are slight. More people have been evicted because of falling behind on rent. Families also report an increase in other material hardships, such as phones and utilities being cut off. Marriage rates among former recipients have not changed, although more now live with nonmarital partners, most likely as a way of sharing expenses. The number of children living in families without either parent has also increased, probably because parents had to relocate to find work. In some states, the numbers of people neither working nor receiving aid also increased (Acker et al. 2002; Bernstein 2002). Many studies also find that low-wage work does not lift former welfare recipients out of poverty (Shaefer and Edin 2014; Hays 2003). Forcing welfare recipients

Is It True?*

	True	False
1. Income growth has been greatest for those in the middle class in recent years.		
2. The average American household has most of its wealth in the stock market.		
3. Social mobility is greater in the United States than in any other Western nation.		
4. Poor teen mothers do not have the same values about marriage as middle-class people.		
5. Old people are the most likely to be poor.		
6. Poverty in U.S. suburbs is increasing.		

*The answers can be found at the end of this chapter.

to work provides a cheap labor force for employers and potentially takes jobs from those already employed.

The public debate about welfare rages on, often in the absence of informed knowledge from sociological research and almost always without input from the subjects of the debate, welfare recipients themselves. Although stigmatized as lazy and not wanting to work, those who have received welfare actually believe that it has negative consequences for them, but they say they have no other viable means of support. They typically have needed welfare when they could not find work or had small children and needed child care. Most were forced to leave their last job because of layoffs or firings or because the work was only temporary. Few left their jobs voluntarily.

Welfare recipients also say that the welfare system makes it hard to become self-supporting because the wages one earns while on welfare are deducted from an already minimal subsistence. Furthermore, there is not enough affordable day care for mothers to leave home and get jobs. The biggest problem they face in their minds is lack of money. Contrary to the popular image of the conniving "welfare queen," welfare recipients want to be self-sufficient and provide for their families, but they face circumstances that make this very difficult to do. Indeed, studies of young, poor mothers find that they place a high value on marriage, but they do not think they or their boyfriends have the means to achieve the marriage ideals they cherish (Edin and Kefalas 2005; Hays 2003).

Another popular myth about welfare is that people use their welfare checks to buy things they do not need. Research finds that when former welfare recipients find work, their expenses actually go up. Although they may have increased income, their expenses (in the form of child care, clothing, transportation, lunch money, and so forth) increase, leaving them even less disposable income. Moreover, studies find that low-income mothers who buy "treats" for their children (brand-name shoes, a movie, candy, and so forth) do so because they want to be good mothers (Edin and Lein 1997).

Other beneficiaries of government programs have not experienced the same kind of stigma. Social Security supports virtually all retired people, yet they are not stereotyped as dependent on federal aid, unable to maintain stable family relationships, or insufficiently self-motivated. Spending on welfare programs is also a pittance compared with the spending on other federal programs. Sociologists conclude that the so-called welfare trap is not a matter of learned dependency, but a pattern of behavior forced on the poor by the requirements of sheer economic survival (Edin and Kefalas 2005; Hays 2003).

Chapter Summary

What different kinds of stratification systems exist?

Social stratification is a relatively fixed hierarchical arrangement in society by which groups have different access to resources, power, and perceived social worth. All societies have systems of stratification, although they vary in composition and complexity. *Estate systems* are those in which a single elite class holds the power and property; in *caste systems*, placement in the stratification is by birth; in *class systems*, placement is determined by achievement.

How do sociologists define class?

Class is the social structural position that groups hold relative to the economic, social, political, and cultural resources of society. Class is highly significant in determining one's *life chances*.

How is the class system structured in the United States?

Social class can be seen as a hierarchy, like a ladder, where income, occupation, and education are indicators of class. *Status attainment* is the process by which people end up in a given position in this hierarchy. *Prestige* is the value others assign to people and groups within this hierarchy. Classes are also organized around common interests and exist in conflict with one another.

Is there social mobility in the United States?

Social mobility is the movement between class positions. Education gives some boost to social mobility, but social mobility is more limited than people believe; most people end up in a class position very close to their class of origin. *Class consciousness* is both the perception that a class structure exists and the feeling of shared identification with others in one's class. The United States has not been a particularly class-conscious society because of the belief in upward mobility.

What analyses of social stratification do sociological theorists provide?

Karl Marx saw class as primarily stemming from economic forces; Max Weber had a multidimensional view of stratification, involving economic, social, and political dimensions. Functionalists argue that social inequality motivates people to fill the different positions in society that are needed for the survival of the whole, claiming that the positions most important for society require the greatest degree of talent or training and are thus most rewarded. Conflict theorists see social stratification as based on class conflict and blocked opportunity, pointing out that those at the bottom of the stratification system are least rewarded because they are subordinated by dominant groups.

How do sociologists explain why there is poverty in the United States?

Culture of poverty is the idea that poverty is the result of the cultural habits of the poor that are transmitted from generation to generation, but sociologists see poverty as caused by social structural conditions, including unemployment, gender inequality in the workplace, and the absence of support for child care for working parents.

What current policies address the problem of poverty?

Current welfare policy, adopted in 1996, provides support through individual states, but recipients are required to work after two years of support and have a lifetime limit of five years of support.

Is It True? (Answers)

1. FALSE. Income growth has been highest in the top 5 percent of income groups (DeNavas-Walt and Proctor 2014).
2. FALSE. Eighty percent of all stock is owned by a small percentage of people. For most people, home ownership is the most common financial asset (Oliver and Shapiro 2006).
3. FALSE. The United States has lower rates of social mobility than Canada, Sweden, and Norway, and ranks near the middle in comparison to other Western nations (Jäntti 2006).
4. FALSE. Research finds that poor teen mothers value marriage and want to be married, but associate marriage with economic security, which they do not think they can achieve (Edin and Kefalas 2005).
5. FALSE. Although those over age 65 used to be the most likely to be poor, poverty among the elderly has declined; the most likely to be poor are children (DeNavas-Walt and Proctor 2014).
6. TRUE. Although most of the poor live inside metropolitan areas, poverty in the suburbs has been increasing (DeNavas-Walt and Proctor 2014).

Key Terms

caste system **172**
class consciousness **187**
class system **172**
concentrated poverty **192**
conspicuous consumption **173**
culture of poverty **194**
economic restructuring **174**
educational attainment **180**
estate system **172**
false consciousness **187**
feminization of poverty **191**
ideology **188**
income **175**
life chances **172**
median income **179**
meritocracy **186**
net worth **175**
occupational prestige **180**
poverty line **191**
prestige **180**
social class **172**
social mobility **186**
social stratification **171**
socioeconomic status (SES) **179**
status attainment **179**
Temporary Assistance for Needy Families (TANF) **196**
urban underclass **182**
wealth **175**

9

Gideon Mendel/Historical/Corbis

Global Stratification

"It takes a village to raise a child," the saying goes, but it also seems to take a world to make a shirt—or so it seems from looking at the global dimensions of the production and distribution of goods. Try this simple experiment: Look at the labels on your clothing. (If you do this in class, try to do so without embarrassing yourself and others!) What do you see? "Made in Indonesia," "Made in Vietnam," "Made in Malawi"—all indicating the linkage of the United States to clothing manufacturers around the world. The popular brand Nike, as just one example, contracts with factories all over the world. Most of Nike's products are made in hundreds of factories throughout Asia.

Taking your experiment further, ask yourself: Who made your clothing? A young person trying to lift his or her family out of poverty? Might it have been a child? The International Labour Organization (ILO) distinguishes *child labor* (those under age 17) from *employed children* (such as a teenager holding a part-time job or babysitting). *Child labor* specifically refers to "work that deprives children of their childhood, their potential, and their dignity and that is also harmful to mental and physical development" (International Labour Organization 2014). The ILO estimates that about 168 million children around the world are trapped in child labor, almost half of whom are involved in dangerous work and many of whom are separated from their families and possibly held in slavery (International Labour Organization 2014). This does not mean that a child necessarily made your clothing. In fact, most child labor occurs in agricultural work, although a significant component (25 percent) is in service and manufacturing work.

Data on child labor indicate that our global systems of work are deeply connected to inequality. Especially in the poorest countries, trying to survive forces people into forms of work—or lack of work—that produce some of the world's greatest injustices—both for children and for adult women and men. As we will see in this chapter, nations are interlocked in a system of global inequality, in which the status of the people in one country is intricately linked to the status of the people in others.

Recall from Chapter 1 that C. Wright Mills identified the task of sociology as seeing the social forces that exist beyond individuals. This is particularly important when studying global inequality. A person in the United States (or western Europe

in this chapter, you will learn to:

- Define global stratification and describe its components
- Compare and contrast different explanations of global stratification
- Describe the various consequences of global stratification
- Explain the causes and consequences of global poverty
- Summarize the impact of globalization for social change

or Japan) who thinks he or she is expressing individualism by wearing the latest style is actually part of a global system of inequality. The adornments available to that person result from a whole network of forces that produce affluence in some nations and poverty in others.

The United States and other wealthy nations are dominant in the system of global stratification. Those at the top of the global stratification system have enormous power over the fate of other nations. Although world conflict stems from many sources, including religious differences, cultural conflicts, and struggles over political philosophy, the inequality between rich and poor nations causes much hatred and resentment. One cannot help but wonder what would happen if the differences between the wealth of some nations and the poverty of others were smaller. In this chapter, we examine the dynamics and effects of global stratification.

Global Stratification

In the world today, there are not only rich and poor people but also rich and poor countries. Some countries are well off, some countries are doing so-so, and a growing number of countries are poor and getting poorer. There is, in other words, a system of **global stratification** in which the units are countries, much like a system of stratification within countries in which the units are individuals or families.

Just as we can talk about the upper-class or lower-class individuals within a country, we can also talk of the equivalent upper-class or lower-class countries in this world system. One manifestation of global stratification is the great inequality in life chances that differentiates nations around the world. Simple measures of well-being (such as life expectancy, infant mortality, access to education and health, and measures of environmental quality) reveal the consequences of global inequality. The gap between rich and poor people is also sometimes greatest in nations where poverty rates are highest. No longer can nations be understood without considering the global system of stratification of which they are a part.

The effects of the global economy on inequality have become increasingly evident, as witnessed by public concerns about jobs being sent overseas. Unions, environmentalists, and other groups have coalesced to protest global trade policies that they think threaten U.S. jobs, erode workers' rights, and contribute to environmental degradation. The global economy has also further spread McDonaldization, bringing this form of production and consumption throughout the world (see Chapter 6). Popular stores such as Gap and Niketown often have been targets of political protests because they symbolize the expansion of global capitalism. Protestors see the growth of such stores as eroding local cultural values and spreading the values of unfettered consumerism around the globe. A student-based movement has also emerged to protest the sweatshop labor that is often used by companies that manufacture college logo apparel.

The relative affluence of the United States means that U.S. consumers have access to goods produced around the world. A simple thing, such as a child's toy, can represent this global system. For many young girls in the United States, Barbie is the ideal of fashion and romance. Young girls may have not just one Barbie, but several, each with a specific role and costume. Cheaply bought in the United States, but produced overseas, Barbie is manufactured by those probably not much older than the young girls who play with her and who would need all of their monthly pay to buy just one of the dolls that many U.S. girls collect by the dozens (Press 1996: 12).

The manufacturing of toys and clothing is an example of the global stratification that links the United States and other parts of the world. *Global outsourcing* locates jobs overseas even while supporting U.S.-based businesses. Many of the jobs that have been outsourced in this way are semiskilled jobs, such as data entry, medical transcription, and so forth. Increasingly, outsourced jobs are also found in high-tech industries, software design, market research, and product research activities. Although it is difficult to measure the extent of global outsourcing, it has become a common phenomenon—something you experience when, for example, you engage in a telephone or Internet transaction, such as getting help for your computer or arranging a trip. India, China, and Russia have been major players in the economy of global outsourcing, but other nations, such as Ireland, South Africa, Poland, and Hungary, among others, are increasingly playing an important role. The consequences can be very positive for the economies of the host nations. The practice of outsourcing also lowers personnel costs for U.S.-based companies, given the lower wages in nations where jobs flow. Outsourcing can be at the expense of jobs for workers in the United States, however (Rajan and Srivastava 2007). The practice of global outsourcing increasingly links the economies and social systems of nations around the world.

Alice Hartley

Some nations have so much wealth that they actually serve gold as food, such as these real gold leaves on a dessert in Dubai!

Rich and Poor

One dimension of stratification between countries is wealth. Enormous differences exist between the wealth of the countries at the top of the global stratification system and the wealth of the countries at the bottom. As you can see in ▲ Figures 9.1 and 9.2, a very small proportion of the world's population receives a vast share of all income—a visual reminder of the inequality that characterizes our world.

You will recall that we looked at the "champagne glass" of inequality within the United States in Chapter 8. A similar image can show you the inequality of income worldwide (see Figure 9.2). As you can see, a small percentage of the world's population has a very disproportionate share of world income.

There are different ways to measure the wealth of nations, but the most common is to use the per capita **gross national income (GNI)**. The GNI measures the total output of goods and services produced by residents of a country each year plus the income from nonresident sources, divided by the size of the population. The GNI does not truly reflect what individuals or families receive in wages or pay; it is simply each person's annual share of their country's income if income were shared equally. You can use this measure to get a picture of global stratification (see ■ Map 9.1).

Per capita GNI is reliable only in countries that are based on a cash economy. It does not measure informal exchanges or bartering in which resources are exchanged without money changing hands. These noncash transactions are not included in the GNI calculation, but they are common in developing countries. As a result, measures of wealth based on the GNI, or other statistics that count cash transactions, are less reliable among the poorer countries and may underestimate the wealth of the countries at the lower end of the economic scale.

The per capita GNI of the United States, one of the wealthier nations in the world (though not the wealthiest on a per capita basis), was $53,860

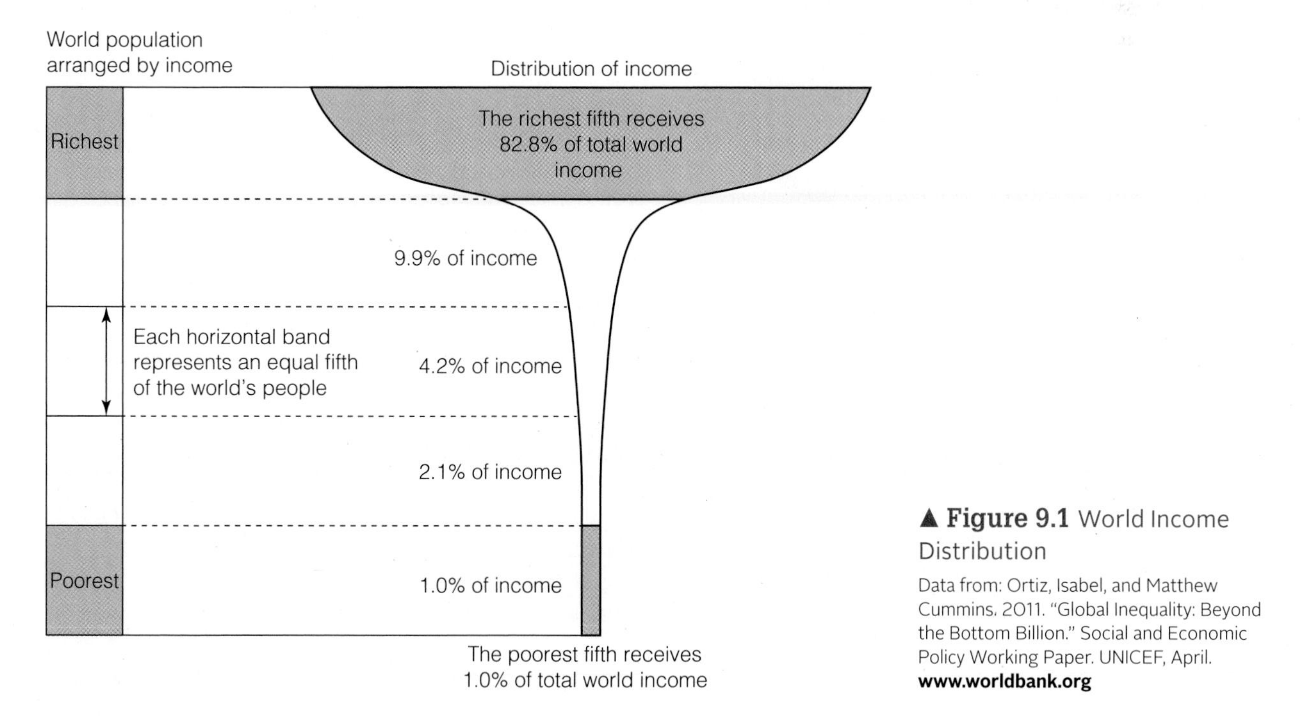

▲ **Figure 9.1** World Income Distribution

Data from: Ortiz, Isabel, and Matthew Cummins. 2011. "Global Inequality: Beyond the Bottom Billion." Social and Economic Policy Working Paper. UNICEF, April. **www.worldbank.org**

map 9.1

Viewing Society in Global Perspective: Rich and Poor

Most nations are linked in a world system that produces wealth for some and poverty for others. The GNI (gross national income), depicted here on a per capita basis for most nations in the world, is an indicator of the wealth and poverty of nations.

Source: The World Bank. 2011. Reprinted by permission. **www.worldbank.org**

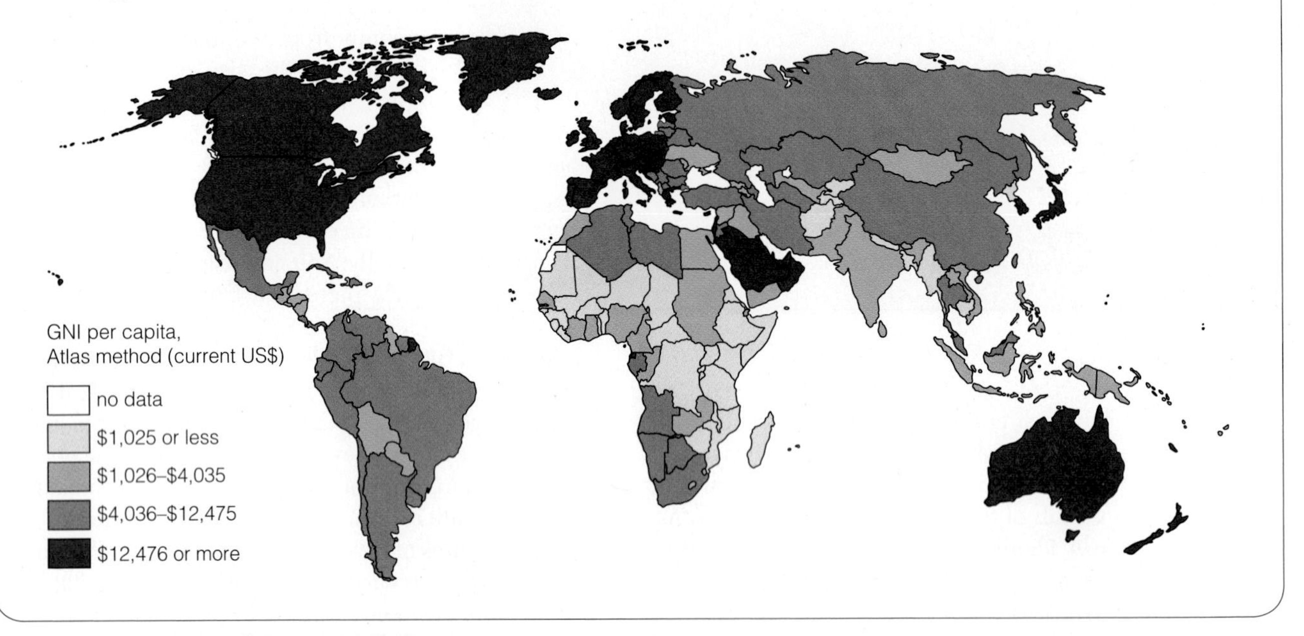

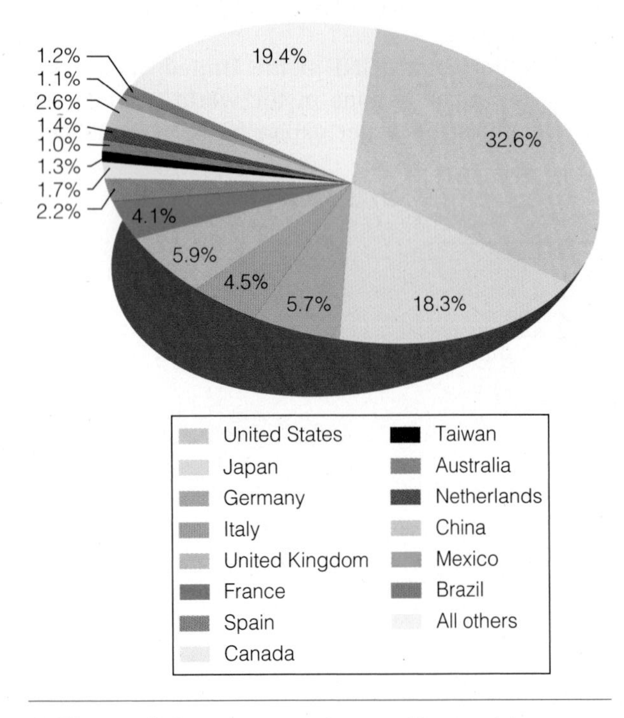

▲ **Figure 9.2** Who Owns the World's Wealth?

Data from: Davies, James B., Susanna Sandstrom, Anthony Shorrocks, and Edward N. Wolff. 2008. "The World Distribution of Household Wealth." UNU-WIDER, World Institute for Development Economics Research. Helsinki, Finland.

in 2013. The per capita GNI in Burundi, one of the poorest countries in the world, was $280. Even compared to other well-to-do, industrialized nations, the United States' per capita GNI shows us to be one of the most affluent nations in the world; GNI per capita is $46,140 in Japan, $46,100 in Germany, $39,140 in the United Kingdom, and only $6520 in China (World Bank 2014b).

Which are the wealthiest nations? ▲ Figure 9.3 shows the ten richest and the ten poorest countries in the world (measured by the annual per capita GNI). Monaco is the richest nation in the world on a per capita basis. Of course, Monaco has a tiny population compared with the United States. The poorest country in the world is Malawi, closely tied with Burundi, but note how many of the poorest nations are in sub-Saharan Africa, one of the poorest regions of the world. We will return to this fact in the discussion of world poverty later in this chapter.

The poorest nations are largely rural, have high fertility rates, large populations, and still depend heavily on subsistence agriculture. In very poor countries, the life of an average citizen is meager. Often poor nations are rich with natural resources but are exploited for such resources by more powerful nations. Still, they rank at the bottom of the global stratification system.

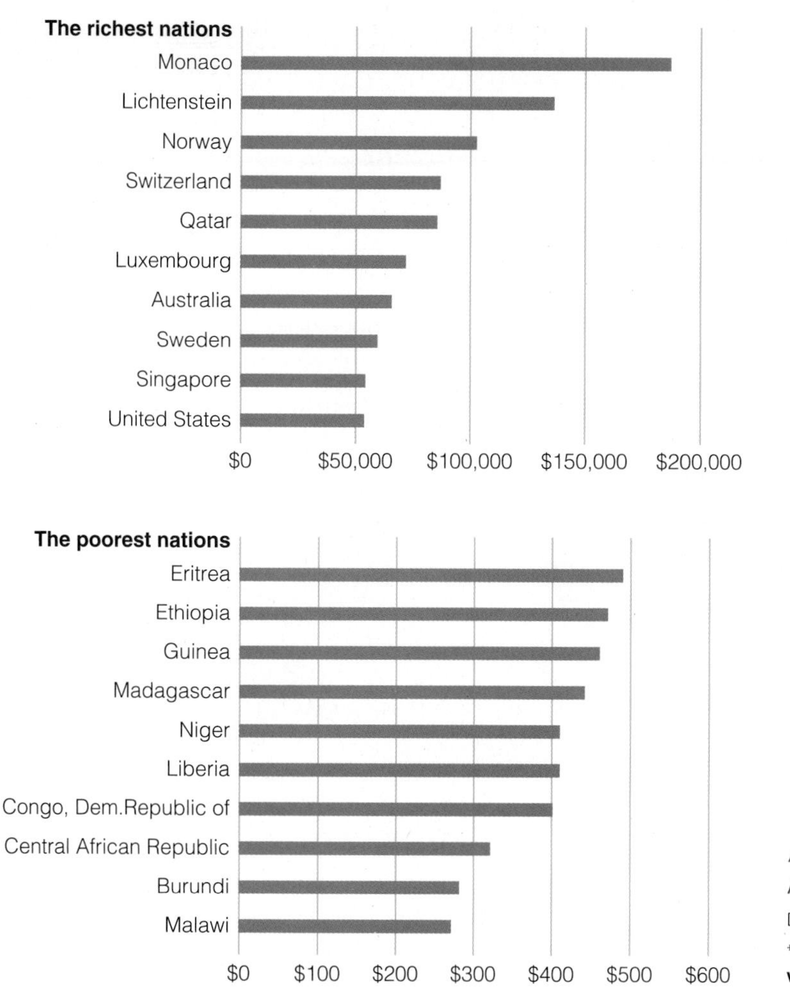

▲ **Figure 9.3** The Rich and the Poor: A World View*

Data from: The World Bank. 2014b.

*Measured by GNI per capita, in U.S. dollars, for 2011.
www.worldbank.org

Because the poorest nations suffer from extreme poverty, there is terrible human suffering in these places. This also produces instability and the potential for violence, as well as risks to human health. We witnessed this in the Ebola virus outbreak that devastated some of the countries of western Africa. The nations most affected are among some of the poorest in the world—the Central African Republic with a GNI of only $320; Burundi, $280; and Malawi, $270. Each of these nations suffers from very high poverty rates, short life expectancy, and poor water facilities. We will look more closely at the nature and causes of such world poverty later in this chapter.

The wealthiest countries, you will see, are largely industrialized nations or those that are oil-rich. These countries represent the equivalent of the upper class. Simply being one of the wealthiest nations in the world does not mean that all of the nation's population is well off. Some, especially the Scandinavian countries, have low degrees of inequality within. Others, including the United States, have great inequality within, as we have seen in the previous chapter on class inequality. Moreover, inequality between nations has to be seen in relative terms. For example, a very wealthy person in India may have the income of someone in the bottom 5 percent of income earners in the United States, but within India, this can afford the person an expensive mansion and a highly lavish lifestyle relative to other Indian people (Milanovic 2010).

Inequality within nations is measured by something called the Gini coefficient. The **Gini coefficient** is a measure of income distribution within a given population or nation. The figure ranges from zero to one, with zero representing a population where there is perfect equality and one indicating a population where just one person has all the money—in other words, the greatest inequality. South Africa has the highest Gini coefficient in the world; the Scandinavian countries, among the lowest. But the United States ranks very high in the degree of internal inequality among other industrialized nations, as you can see from ▲ Figure 9.4. ■ Map 9.2 also gives you a visual image, based on the Gini coefficient, of inequality within nations throughout the world.

map 9.2

Viewing Society in Global Perspective: The Gini Coefficient

Source: Central Intelligence Agency. 2009. *World Factbook.* **www.cia.org**

Global Networks of Power and Influence

Global stratification involves nations in a large and integrated network of economic and political relationships. *Power*—that is, the ability of a country to exercise control over other countries or groups of countries—is a significant dimension of global stratification. Countries can exercise several kinds of power over other countries, including military, economic, and political power. The **core countries** have the most power in the world economic system. These countries control and profit

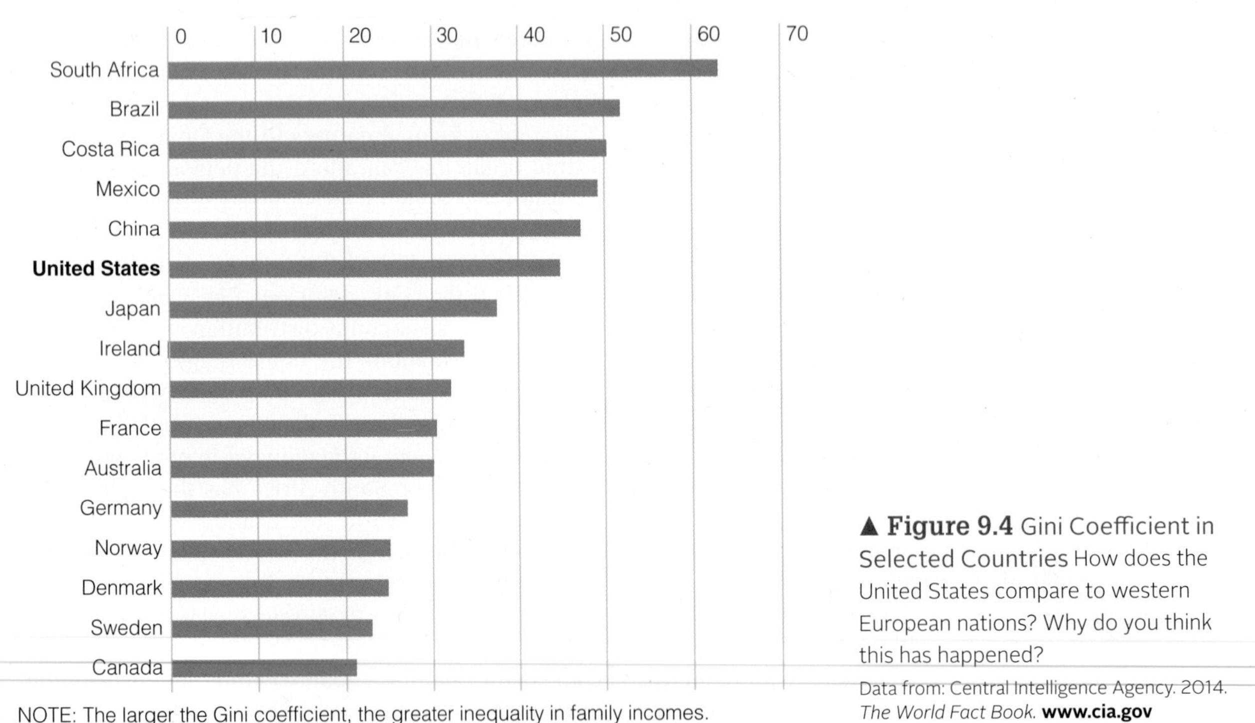

▲ **Figure 9.4** Gini Coefficient in Selected Countries How does the United States compare to western European nations? Why do you think this has happened?

Data from: Central Intelligence Agency. 2014. *The World Fact Book.* **www.cia.gov**

doing sociological research

Servants of Globalization: Who Does the Domestic Work?

Research Question: International migration is becoming an increasingly common phenomenon. Women are one of the largest groups to experience migration, often leaving poor nations to become domestic workers in wealthier nations. What are these women's experiences in the context of global stratification? This is what Rhacel Salazar Parreñas wanted to know.

Research Method: Parreñas studied two communities of Filipina women, one in Los Angeles and one in Rome, Italy, conducting her research through extensive interviewing with Filipina domestic workers in these two locations. She supplemented the interviews with participant observation in church settings, after-work social gatherings, and in employers' homes. The interviews were conducted in a mixture of both cities.

Research Results: Parreñas found that Filipina domestics experienced many status inconsistencies. They were upwardly mobile in terms of their home country but were excluded from the middle-class Filipino communities in the host nation. Thus they experienced feelings of social exclusion in addition to being separated from their own families.

Conclusions and Implications: The women Parreñas studied are part of a new social pattern for *transnational families*—that is, families whose members live across the borders of nations. These Filipinas provide the labor for more affluent households while their own lives are disrupted by these new global forces. As global economic restructuring evolves, it may be that more families will experience this form of family living.

Questions to Consider

1. Are there domestic workers in your community who provide child care and other household work for middle- and upper-class households? What are the race, ethnicity, nationality, and gender of these workers? What does this tell you about the division of labor in domestic work and its relationship to global stratification?
2. Why do you think domestic labor is so underpaid and undervalued? Are there social changes that might result in a reevaluation of the value of this work?

Sources: Parreñas, Rhacel Salazar. 2001. *Servants of Globalization: Women, Migration and Domestic Work.* Stanford, CA: Stanford University Press.

the most from the world system, and thus they are the "core" of the world system. These include the powerful nations of Europe, the United States, Australia, and, increasingly, East Asia.

Surrounding the core countries, both structurally and geographically, are the **semiperipheral countries** that are semi-industrialized and, to some degree, represent a kind of middle class (such as Spain, Turkey, and Mexico). They play a middleman role, extracting profits from the poor countries and passing those profits on to the core countries.

At the bottom of the world stratification system are the **peripheral countries**. These are the poor, largely agricultural countries of the world. Even though they

The gap between the rich and poor worldwide can be staggering. At the same time that many struggle for mere survival, others enjoy the pleasantries of a gentrified lifestyle.

are poor, they often have important natural resources that are exploited by the core countries. Exploitation, in turn, keeps them from developing and perpetuates their poverty. Often these nations are politically unstable. Political instability within poor nations can create a crisis for core nations that depend on their resources. Military intervention by the United States or European nations is often the result.

SEE for YOURSELF

The Global Economy of Clothing

Look at the labels in your clothes and note where your clothing was made. Where are the products bearing your college logos manufactured and sold? Who profits from the distribution of these goods? What does this tell you about the relationship of *core*, *semiperipheral*, and *peripheral* countries within world systems theory? What further information would reveal the connections between the country where you live and the countries where your clothing is made and distributed?

To explore this further, you can read the account by Kelsey Timmerman who wanted to know where his clothes came from. Timmerman traveled to Honduras, Bangladesh, Cambodia, and China, talking to factory workers and their families about the experiences in making the clothes that others wear. His journey can teach you a lot about global production and consumption (Timmerman 2012).

Further resources: www.whereamiwearing.com

Race and Global Inequality

Along with class inequality, there is a racial component to world inequality. In the richest nations, the population is largely White; in the poorest countries of the world, mostly in Africa, the populations are people of color. Exploitation of the human and natural resources of regions populated by people of color has characterized the history of Western capitalism, with people of color being dominated by Western imperialism and colonialism. The inequities that have resulted are enormous, including malnutrition and hunger.

How did this racial inequality come about? On the surface, global capitalism is not explicitly racist, as were earlier forms of industrial capitalism. Yet, the rapid expansion of the global capital system has led to an increase in racial inequality between nations. A new **international division of labor** has emerged that is not tied to particular places but seeks cheap labor, usually in non-Western countries. The exploitation of cheap labor has created a poor and dependent workforce comprised mostly of people of color. Profits accrue to wealthy owners, who are mostly White, resulting in a racially divided world. Some argue that the exploitation of the poor peripheral nations by multinational capitalists has forced an exodus of unskilled workers from the impoverished nations to the rich nations. The flood of third-world refugees into the industrialized nations is thereby increasing racial tensions, fostering violence, and destroying worker solidarity (Sirvananadan 1995).

The meaning of race, however, varies in different national settings. South Africa, the United States, and Brazil each developed different sets of racial categories. Although all three countries have many people of mixed descent, race is defined differently in each place. In South Africa, the particular history of Dutch and English colonialism led to strongly drawn racial categories that defined people in four separate categories: "White," "Coloured" (including indigenous Khoi and San people, as well as people of mixed descent), "Black," and "Indian." Black South Africans had no political representation under apartheid. There were three separate parliaments—one for each of the other groups.

Brazil is yet a different case. The Brazilian elite declared Brazil a racial democracy at the early stages of national development. Racial differences were thought not to matter. Yet, instead of creating an egalitarian society free of racism, Afro-Brazilians were still of lower social status and Euro-Brazilians remain at the highest social status, suggesting that color itself stratifies people—a sociological phenomenon sometimes referred to as "colorism" (Telles 2004; Fredrickson 2003; Marx 1997).

Theories of Global Stratification

How did world inequality occur? Sociological explanations of world stratification generally fall into three camps: modernization theory, dependency theory, and world systems theory (see ◆ Table 9.1).

Modernization Theory

Modernization theory views the economic development of countries as stemming from technological change. According to this theory, a country becomes more "modernized" by increased technological development.

Modernization theory sees economic development as a process by which traditional societies become more complex. For economic development to occur, modernization theory predicts, countries must change their traditional attitudes, values, and institutions. Economic achievement is thought to derive from attitudes and values that emphasize hard work, saving, efficiency, and enterprise. Modernization theory suggests that nations remain underdeveloped when traditional customs and culture discourage individual achievement and kin relations dominate.

As an outgrowth of functionalist theory, modernization theory derives some of its thinking from the work

◆ **Table 9.1** Theories of Global Stratification

	Modernization Theory	Dependency Theory	World Systems Theory
Economic Development	Arises from relinquishing traditional cultural values and embracing new technologies and market-driven attitudes and values	Exploits the least powerful nations to the benefit of wealthier nations that then control the political and economic systems of the exploited countries	Has resulted in a single economic system stemming from the development of a world market that links core, semiperipheral, and peripheral nations
Poverty	Results from adherence to traditional values and customs that prevent societies from competing in a modern global economy	Results from the dependence of low-income countries on wealthy nations	Is the result of core nations extracting labor and natural resources from peripheral nations
Social Change	Involves increasing complexity, differentiation, and efficiency	Is the result of neocolonialism and the expansion of international capitalism	Leads to an international division of labor that increasingly puts profit in the hands of a few while exploiting those in the poorest and least powerful nations

of Max Weber. In *The Protestant Ethic and the Spirit of Capitalism* (1958/1904), Weber saw the economic development that occurred in Europe during the Industrial Revolution as a result of the values and attitudes of Protestantism. The Industrial Revolution took place in England and northern Europe, Weber argued, because the people of this area were hardworking Protestants who valued achievement and believed that God helped those who helped themselves.

Modernization theory can partially explain why some countries have become successful. Japan and China are examples of countries that have made huge strides in economic development, in part because of a national work ethic. Work ethic alone, however, does not explain Japan's success. Modernization theory may partially explain the cultural context in which some countries become successful and others do not, but it is not a substitute for explanations that also look at the economic and political context of national development. Cultural attitudes may impede economic development in some cases, but you have to be careful not to assume that developed nations have superior values compared to others. Blaming the cultural values of a poor nation overlooks the fact that a nation's status in the world may be outside their control. Whether a country develops or remains poor is often the result of other countries exploiting the less powerful. Modernization theory does not sufficiently take into account the interplay and relationships between countries that can affect a country's economic or social condition.

Developing countries, modernization theory says, are better off if they let the natural forces of competition guide world development. Free markets, according to this perspective, will result in the best economic order. As critics argue, markets do not develop independently of government's influence. Governments can spur or hinder economic development, especially as they work with private companies to enact export strategies, restrict imports, or place embargoes on the products of nonfavored nations.

Dependency Theory

Although market-oriented theories may explain why some countries are successful, they do not explain why some countries remain in poverty or why some countries have not developed. It is necessary to look at issues outside the individual countries and to examine the connections between them. Keep in mind that many of the poorest nations are former colonies of European powers. This focuses your attention on colonization and imperialism as causes of global stratification.

Dependency theory holds that the poverty of the low-income countries is a direct result of their political and economic dependence on the wealthy countries. Specifically, dependency theory argues that the poverty of many countries is a result of exploitation by powerful countries. This theory is derived from the work of Karl Marx, who foresaw that a capitalist world economy would create an exploited class of dependent countries, just as capitalism within countries had created an exploited class of workers.

Dependency theory begins with understanding the historical development of this system of inequality. As the European countries began to industrialize in the 1600s, they needed raw materials for their factories and places to sell their products. To accomplish this, the European nations colonized much of the world, including most of Africa, Asia, and the Americas. **Colonialism** is a system by which Western nations became wealthy by taking raw materials from colonized societies and reaping profits from products finished in the homeland. Colonialism worked best for the industrial countries when the colonies were kept undeveloped to avoid competition with the home country. For example, India was a British colony from 1757 to 1947. During that time, Britain bought cheap cotton from India, made it into cloth in British mills, and then sold the cloth back to India, making large profits. Although India was able to make cotton into cloth at a much cheaper cost than Britain, and very fine cloth at that, Britain nonetheless did not allow India to develop its cotton industry. As long as India was dependent on Britain, Britain became wealthy and India remained poor.

Under colonialism, dependency was created by the direct political and military control of the poor countries by powerful developed countries. Most colonial powers were European countries, but other countries, particularly Japan and China, had colonies as well. Colonization came to an end soon after the Second World War, largely because of protests by colonized people and the resulting movement for independence. As a result, according to dependency theory, the powerful countries turned to other ways to control the poor countries and keep them dependent. The powerful countries still intervene directly in the affairs of the dependent nations by sending troops or, more often, by imposing economic or political restrictions and sanctions. But other methods, largely economic, have been developed to control the dependent poor countries, such as price controls, tariffs, and, especially, the control of credit. Indeed, the level of debt that some nations accrue is a major source of global inequality.

The rich industrialized nations, according to dependency theory, are able to set prices for raw materials produced by the poor countries at very low levels so that the poor countries are unable to accumulate enough profit to industrialize. As a result, the poor, dependent countries must borrow from the rich countries. However, debt creates only more dependence. Many poor countries are so deeply indebted to the major industrial countries that they must follow the economic edicts of the rich countries that loaned them the money, thus increasing their dependency. This form of international control has sometimes been called **neocolonialism**, a form of control of the poor countries by the rich countries but without direct political or military involvement.

Multinational corporations are companies that draw a large share of their profits from overseas investments and that conduct business across national borders. They play a role in keeping the dependent nations poor, dependency theory suggests. Although their executives and stockholders are from the industrialized countries, multinational corporations recognize no national boundaries and pursue business where they can best make a profit. Multinationals buy resources where they can get them cheapest, manufacture their products where production and labor costs are lowest, and sell their products where they can make the largest profits.

Many critics fault companies for perpetuating global inequality by taking advantage of cheap overseas labor to make large profits for U.S. stockholders. Companies are, in fact, doing what they should be doing in a market system: trying to make a profit. Nonetheless, dependency theory views the practices of multinationals as responsible for maintaining poverty in the poor parts of the world.

One criticism of dependency theory is that many poor countries (for example, Ethiopia) were never colonies. Some former colonies have also done well. Two of the greatest postwar success stories of economic development are Singapore and Hong Kong. Both of these places were British colonies—Hong Kong until 1997—and were clearly dependent on Britain, yet they have had successful economic development precisely because of their dependence on Britain. Other former colonies are also improving economically, such as India.

World Systems Theory

Modernization theory examines the factors internal to an individual country, and dependency theory looks to the relationship between countries or groups of countries. Another approach to global stratification is called **world systems theory**. Like dependency theory, this theory begins with the premise that no nation in the world can be considered in isolation. Each country, no matter how remote, is tied in many ways to the other countries in the world. However, unlike dependency theory, world systems theory argues that there is a world economic system that must be understood as a single unit, not in terms of individual countries or groups of countries. This theoretical approach derives to some degree from the work of dependency theorists and is most closely associated with the work of Immanuel Wallerstein in *The Modern World System* (1974) and *The Modern World System II* (1980). According to this theory, the level of economic development is explained by understanding each country's place and role in the world economic system.

This world system has been developing since the sixteenth century. The countries of the world are tied together in many ways, but of primary importance are the economic connections in the world markets of goods, capital, and labor. All countries sell their products and services on the world market and buy products and services from other countries. However, this is not a market of equal partners. Because of historical and strategic imbalances in this economic system, some countries are able to use their advantage to create and maintain wealth, whereas other countries that are at a disadvantage remain poor. This process has led to a global system of stratification in which the units are not people but countries.

World systems theory sees the world divided into three groups of interrelated nations: core or first-world countries, semiperipheral or second-world countries, and peripheral or third-world countries. This world economic system has resulted in a modern world in which some countries have obtained great wealth and other countries have remained poor. The core countries control and limit the economic development in the peripheral countries so as to keep the peripheral countries from developing and competing with them on the world market; thus the core countries can continue to purchase raw materials at a low price.

Although world systems theory was originally developed to explain the historical evolution of the world system, modern scholars now focus on the international division of labor and its consequences. This approach is an attempt to overcome some of the shortcomings in world systems theory by focusing on the specific mechanism by which differential profits are attached to the production of goods and services in the world market. A tennis shoe made by Nike is designed in the United States; uses synthetic rubber made from petroleum from Saudi Arabia; is sewn in Indonesia; is transported on a ship registered in Singapore, which is run by a Korean management firm using Filipino sailors; and is finally marketed in Japan and the United States. At each of these stages, profits are taken, but at very different rates.

THINKING Sociologically

What are the major industries in your community? In what parts of the world do they do business, including where their product is produced? How does the *international division of labor* affect jobs in your region?

World systems theorists call this global production process a **commodity chain**, the network of production and labor processes by which a product becomes a finished commodity. By following a commodity through its production cycle and seeing where the profits go at each link of the chain, one can identify which country is getting rich and which country is being exploited. As an example, the Gap hoodie that you buy in the United States for about $30 was likely produced from cotton grown in Uzbekistan where workers are paid 2 cents a pound, cut and sewn by workers in Russia who are paid between 39 and 69 dollars a month, and then distributed and sold in the United States (Gordon and Designs 2001).

World systems theory also helps explain the growing phenomenon of international migration. An *international division of labor* means that the need for cheap labor in some of the industrial and developing nations draws workers from poorer parts of the globe. International migration is also the result of refugees seeking asylum from war-torn parts of the world or from countries where political oppression, often against particular ethnic groups, forces some to leave.

The development of a world economy is also resulting in large changes in the composition of populations around the globe. **World cities**, that is, cities that are closely linked through the system of international commerce, have emerged. Within these cities, families and their surrounding communities often form *transnational communities*, communities that may be geographically distant but socially and politically close. Linked through various communication and transportation networks, transnational communities share information, resources, and strategies for coping with the problems of international migration.

International migration, sometimes legal, sometimes not, has radically changed the racial and ethnic composition of populations not only in the United States but also in many European and Asian nations. Over 200 million people now live outside the country of their birth, some of whom moved because of war and persecution, but many of whom move as work moves around the globe (Eitzen 2009). Many such migrants work in the lowest segments of the labor force. The work of these low-wage laborers is critical to the world economy, but they are often treated with hostility and suspicion, discriminated against, and stereotyped as undeserving and threatening. The United States receives the most international migrants of any nation, but they are common in western Europe, Saudi Arabia, Iran, and other parts of the world (Koser 2007). In many nations, the presence of migrants can lead to political tensions over immigration, even though international faces in world cities are now a major feature of the urban landscape.

It is useful to see the world as an interconnected set of economic ties between countries, and to understand that these ties often result in the exploitation of poor countries. This process of globalization means that countries that were once at the center of this world system—England, for example—no longer occupy such a lofty position. Peripheral countries can also improve their standard of living with investment by core countries, although the benefits do not accrue equally to groups within such nations and investment by outsiders can also put receiving nations in debt, thus harming them in the long run. Low-wage factories may benefit managers, but not the working class. Even core countries can be hurt by the world system, such as when jobs move overseas. Who benefits from this world system is differentiated—in all countries—by one's placement, not just in the global class system but

Feng Li/ Getty Images Entertainment/ Getty Images
Peter Cade/Stone/Getty Images

Global stratification often means that consumption in the more affluent nations is dependent on cheap labor in other less affluent nations.

also in the class system internal to each country within this global system. World systems theory has provided a powerful tool for understanding global inequality.

Consequences of Global Stratification

It is clear that some nations are wealthy and powerful and some are poor and powerless. What are the consequences of this world stratification system? Basic indicators of national well-being include such things as infant mortality, literacy levels, access to safe water, and the status of women. There are, as we will see, considerable differences in the quality of life based on these indicators in different places in the world.

Population

One of the biggest differences in rich and poor nations is population. The poorest countries have the highest birthrates and the highest death rates. The total *fertility rate*—how many live births a woman will have over her lifetime at current fertility rates—shows that women in the poorest countries have on average almost five children. Because of this high fertility rate, the populations of poor countries are growing faster than the populations of wealthy countries. Poor countries therefore also have a high proportion of young children.

In contrast, the richest countries have a total population of approximately one billion people—only 15 percent of the world's population. The populations of the richest countries are not growing nearly as fast as the populations of the poorest countries. In the richest countries, women have about two children over their lifetime, and the populations of these countries are growing by only 1.2 percent. Many of the richest countries, including most of the countries of Europe, are actually experiencing population declines. With a low fertility rate, the rich countries have proportionately fewer children, but they also have proportionately more elderly, which can also be a burden on societal resources. Different from the poorest nations, the richest ones are largely urban.

Rapid population growth as a result of high fertility rates can make a large difference in the quality of life of the country. Countries with high birthrates are faced with the challenge of having too many children and not enough adults to provide for the younger generation. Public services, such as schools and hospitals, are strained in high-birthrate countries, especially because these countries are poor to begin with. Very low birthrates, as many rich countries are now experiencing, can also lead to other problems. In countries with low birthrates, there often are not enough young people to meet labor force needs, and workers must be imported from other countries.

Scholars disagree about the relationship between the rate of population growth and economic development. Some theorize that rapid population growth and high birthrates lead to economic stagnation and that too many people keep a country from developing, thus miring the country in poverty (Ehrlich 1968). Yet, some countries with very large populations have become developed: China and India come to mind; both are showing significant economic development. The United States has the third largest population in the world at 318 million people, yet it is one of the richest and most developed nations in the world. Scholars now believe that even though large population and high birthrates can impede economic development in some situations, in general, fertility levels are affected by levels of industrialization, not the other way around. That is, as countries develop, their fertility levels decrease and their population growth levels off (Hirschman 1994; Watkins 1987).

Health and the Environment

Significant differences are also evident in the basic health standards of countries, depending on where they are in the global stratification system. The high-income countries have lower childhood death rates, higher life expectancies, and fewer children born underweight. People born today in wealthy countries can expect to live about seventy-seven years, and women outlive men by several years. Except for some isolated or poor areas of the rich countries, almost all people have access to clean water and acceptable sewer systems.

In the poorest countries, the situation is completely different. Many children die within the first five years of life, people live considerably shorter lives, and fewer

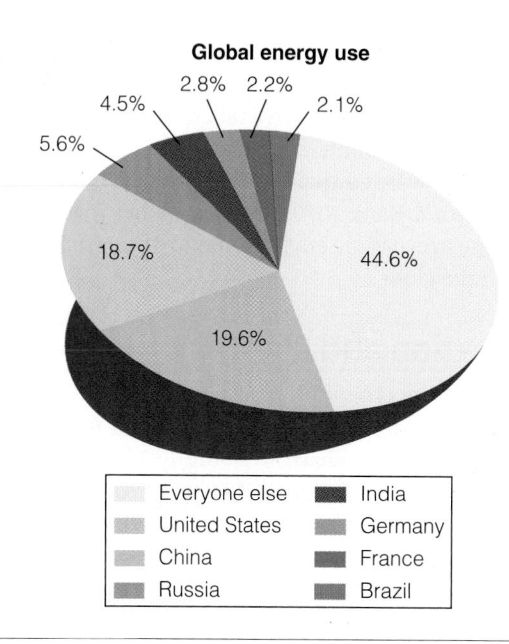

▲ **Figure 9.5** Who Uses the World's Energy?
Data source: U.S. Energy Information Administration. 2012. **www.eia.gov**

Eric Nathan / Alamy

There can be innovative solutions to reduce world poverty, such as this solar panel delivering energy in southern Mozambique.

people have access to clean water and adequate sanitation. In the low-income countries, the problems of sanitation, clean water, childhood death rates, and life expectancies are all closely related. In many of the poor countries, drinking water is contaminated from poor or nonexistent sewage treatment. This contaminated water is then used to drink, to clean eating utensils, and to make baby formula. For adults, waterborne illnesses such as cholera and dysentery sometimes cause severe sickness but seldom result in death. Children under age 5 and, especially, those under the age of 1 are highly susceptible to the illnesses carried in contaminated water. A common cause of childhood death in countries with low incomes is dehydration brought on by the diarrhea contracted from drinking contaminated water.

Degradation of the environment is a problem that affects all nations, which are linked in one vast environmental system, but global stratification also means that some nations suffer at the hands of others. Overdevelopment is resulting in deforestation, and high population and the dependency on agriculture in the poorest nations contribute to the depletion of natural resources. In the most industrial nations where the most energy is used, the overproduction of "greenhouse gas"—emission of carbon dioxide from burning fossil fuels—is resulting in numerous threats to our environment, including climate change (see also Chapter 16).

Although high-income countries have only 15 percent of the world population, together they use more than half of the world's energy. The United States alone uses 20 percent of the world's energy, although it holds only 4 percent of the world's population (see ▲ Figure 9.5). Safe water is also crucial; more than 700 million people in 43 different countries are experiencing what the World Bank calls "water stress"—that is, inadequate access to water. The dwindling of water supplies will only be exacerbated by population growth and economic development. The World Bank has, in fact, warned that we are facing a "global water crisis" (World Bank 2010). Clearly, global stratification has some irreversible environmental effects that are felt around the globe.

Education and Illiteracy

In the high-income nations of the world, education is almost universal, and the vast majority of people have attended school, at least at some level. Literacy and school enrollment are now taken for granted in the high-income nations, although people in these wealthy nations who do not have a good education stand little chance of success. In the middle- and lower-income nations, the picture is quite different. Elementary school enrollment, virtually universal in wealthy nations, is less common in the middle-income nations and even less common in the poorest nations.

How do people survive who are not literate or educated? In much of the world, education takes place outside formal schooling. Just because many people in the poorer countries never go to school does not mean that they are ignorant or that they are uneducated. Most of the education in the world takes place in family settings, in religious congregations, or in other settings where elders teach the next generation the skills and knowledge they need to survive. This type of informal education often includes basic literacy and math skills that people in these poorer countries need for their daily lives.

The disadvantage of this informal and traditional education is that, although it prepares people for their traditional lives, it often does not give them the skills and knowledge needed to operate in the modern world. In an increasingly technological world, this can perpetuate the underdeveloped status of some nations.

Gender Inequality

The position of a country in the world stratification system also affects gender relations within different countries. Poverty is usually felt more by women than by men. Although gender equality has not been achieved in the industrialized countries, compared with women in other parts of the world, women in the wealthier countries are much better off.

The United Nations (UN) is one of the organizations that carefully monitors the status of women globally. The UN has developed an index to assess the progress of women in nations around the world. Called the **gender inequality index**, the measure is a composite of three key components of women's lives: reproductive health, empowerment, and labor market status. Each of these three major components is then measured by particular facts about women's status, such as maternal mortality, educational attainment, and labor force participation (see ▲ Figure 9.6). Given how this index is computed, nations with the lowest gender inequality index have the greatest equality between women and men (see ◆ Table 9.2). Based on this index, the United Nations has concluded that, around the world, reproductive health—or lack thereof—is the greatest contributor to gender inequality (United Nations 2010).

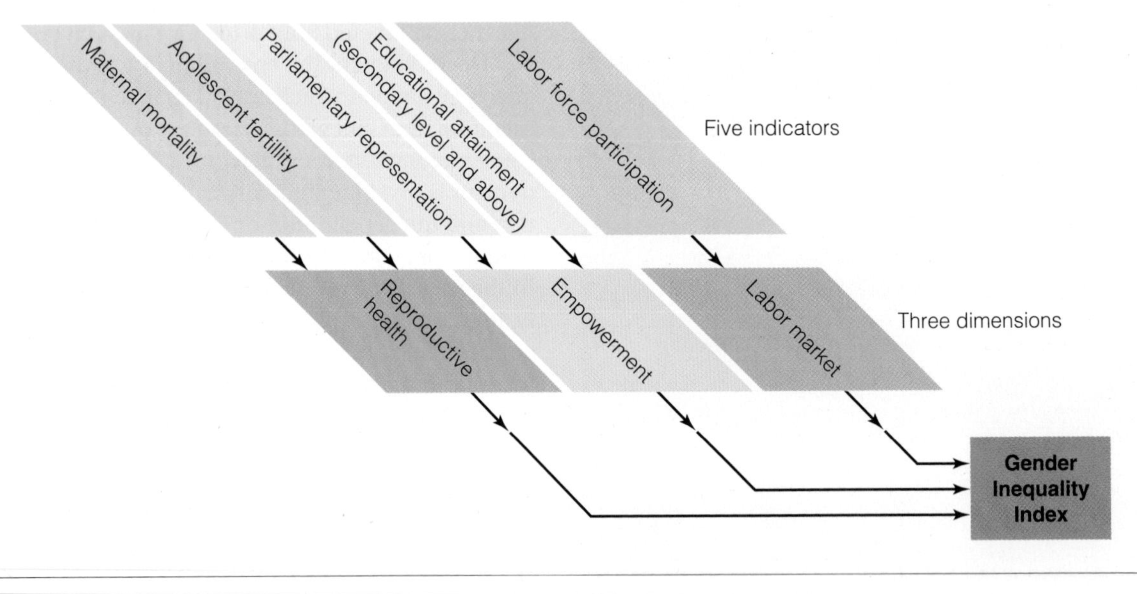

▲ **Figure 9.6** The Gender Inequality Index

Source: United Nations Development Programme. 2010. "Components of the Gender Inequality Index." http://hdr.undp.org. Reprinted with permission.

what would a sociologist say?

Human Trafficking

The U.S. State Department estimates that 12.3 million people worldwide are enslaved in human trafficking. This includes sexual servitude, forced labor, forced child labor, and other forms of coercive treatment. Human trafficking is a modern form of slavery in which people are used for commercial gain through the use of force, coercion, or fraud (U.S. State Department 2012).

There are many ways to think about human trafficking—including as a moral wrong, as a criminal act by corrupt individuals, and as a human rights issue. As a sociological issue, human trafficking is a complex social structure that is integrally connected to international trade, the social structure of tourism, and the racial, class, and gender inequality that crosses national borders.

Sexual trafficking is a particular form of human trafficking in which women and, often, young girls are bought and sold in an international system of prostitution. Sociologists argue that the male-dominated character of state institutions plays a part in the tolerance of sexual trafficking. Sexual trafficking and sexual tourism are part of a culture in which women's bodies are treated as a commodity. Racial and ethnic inequality also play a part as women of color are sexually exploited based on the racial/gender stereotypes that define them as exotic but also available for the pleasure of men.

An antitrafficking movement has developed that involves a coalition of feminists, various voluntary organizations, the United Nations, some politicians, and others who have organized to stop this practice (Limoncelli 2010).

Aurora Photos / Alamy

Sex trafficking, particularly of young girls, is a common part of the system of global stratification.

production, the need for labor in certain industries has declined. Even though new technologies provide new job opportunities, they also create new forms of illiteracy because many people have neither the access nor the skills to use information technology. In sub-Saharan Africa, the poor live in marginal areas where poor soil, erosion, and continuous warfare have created extremely harsh conditions. Political instability and low levels of economic productivity also contribute to the high rates of poverty in some nations. Solutions to world poverty in these different regions require sustainable economic development, as well as an understanding of the diverse regional factors that contribute to high levels of poverty.

Women and Children in Poverty

There is no country in the world in which women are treated as well as men. As with poverty in the United States, women bear a larger share of the burden of world poverty. Some have called this *double deprivation*—in many of the poor countries, women suffer because of their gender and because they disproportionately carry the burden of poverty. For instance, in situations of extreme poverty, women have the burden of taking on much of the manual labor because the men in many cases have left to find work or food. The United Nations concludes that strengthening women's economic security through better work is essential for reducing world poverty.

Because of their poverty, women tend to suffer greater health risks than men. Although women outlive men in most countries, the life expectancy gap between women and men is *less* in the poorest countries. This is explained by several factors that put women at special risks. For one, fertility rates are higher in poor countries. Giving birth is a time of high risk for women, and women in poor countries with poor nutrition, poor maternal care, and the lack of trained birth attendants are at higher risk of dying during and after the birth.

Dan Vincent/Alamy

When children are poor, they may turn to child labor to help support their families. Such is the case with this child collecting trash for potential sale or use in a municipal dump in Phnom Penh.

High fertility rates are also related to the degree of women's empowerment in society—an often neglected aspect of the discussion between fertility and poverty. Societies where women's voices do not count for much tend to have high fertility rates as well as other social and economic hardships for women, including lack of education, job opportunities, and information about birth control. Empowering women through providing them with employment, education, property, and voting rights can have a strong impact on reducing the fertility rate (Sen 2000).

Women also suffer in some poor countries because of traditions and cultural norms. Most (though not all) of the poor countries are patriarchal, meaning that men control the household. As a result, in some situations of poverty, the women eat after the men, and boys are fed before girls. In conditions of extreme poverty, baby boys may also be fed before baby girls because boys have higher status than girls. As a result, female infants have a lower rate of survival than male infants.

A distressing number of children in the world are also poor, including in the most industrialized and affluent nations. As you can see in ▲ Figure 9.9,

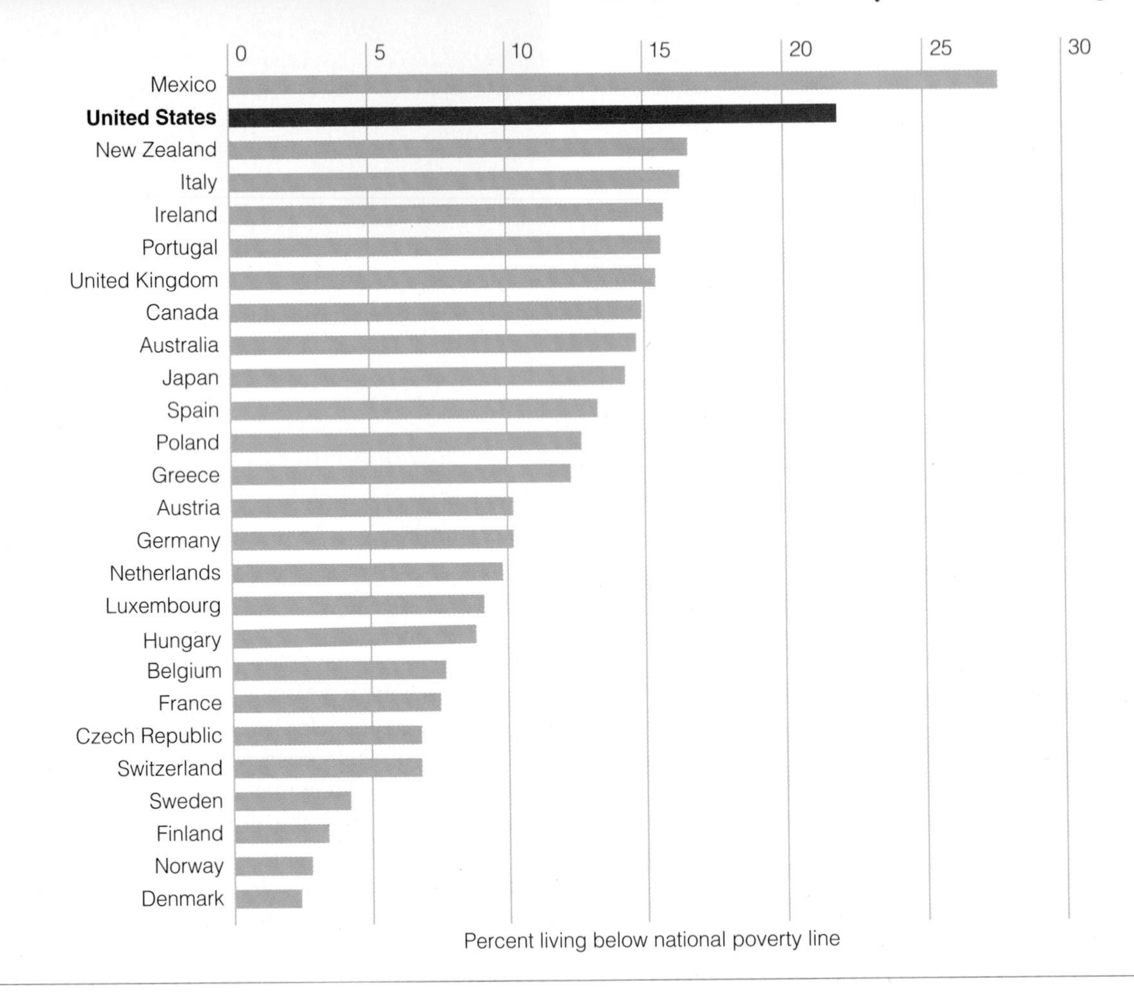

▲ **Figure 9.9** Child Poverty in the Wealthier Nations

Source: UNICEF. 2000. Child Poverty in Rich Countries 2005. Florence, Italy: United Nations Children's Fund. **www.unicef.org.**

poverty among children in the United States exceeds that of other industrialized nations and is second only to Mexico. Children in poverty seldom have the luxury of an education. Schools are usually few or nonexistent in poor areas of the world, and families are so poor that they cannot afford to send their children to school. Children from a very early age are required to help the family survive by working or performing domestic tasks such as fetching water. In extreme situations, even very young children may work as beggars (Boo 2012). Young boys and girls may end up working in sweatshops. Families may have to sell young girls into prostitution. This may seem unusually cruel and harsh by Western standards, but it is difficult to imagine the horror of starvation and the desperation that many families in the world must feel that would force them to take such measures to survive. In poor countries, families feel they must have more children for their survival, yet having more children perpetuates the poverty.

Estimates are that 168 million children under age 17 are in the labor force throughout the world. Most of the children are in Asia, though some are also in sub-Saharan Africa (International Labour Organization 2014). Many of these children work long hours in difficult conditions and enjoy few freedoms, making products (soccer balls, clothing, and toys, for example) for those who are much better off.

Another problem in the very poor areas of the world is homeless children (Mickelson 2000). In many situations, families are so poor that they can no longer care for their children, and the children must go without education or be out on their own, even at young ages. Many of these homeless children end up in the streets of the major cities of Asia and Latin America. In Latin America, it is estimated that there are thirteen million street children, some as young as six years old. Alone, they survive through a combination of begging, prostitution, drugs, and stealing. They sleep in alleys or in makeshift shelters. Their lives are harsh, brutal, and short.

DEBUNKING Society's Myths

Myth: There are too many people in the world, and there is simply not enough food to go around.

Sociological Perspective: Growing more food will not end hunger. If systems of distributing the world's food were more just, hunger could be reduced.

Poverty and Hunger

Poverty is also directly linked to malnutrition and hunger because people in poverty cannot find or afford food. It is estimated that about 805 million people (12 percent of the world population) are malnourished. Experts attribute the increase to a number of factors, including poverty, lack of agricultural development, displacement and war, instability of economic markets, climate and weather, and wasting food (United Nations World Food Programme 2014).

Hunger stifles the mental and physical development of children and leads to disease and death. Although the food supply is plentiful in the world and is actually increasing faster than the population, the rate of malnutrition remains dangerously high.

Is there not enough food to feed all the people in the world? In fact, plenty of food is grown in the world. The world's production of wheat, rice, corn, and other grains is sufficient to adequately feed all the people in the world. Much grain grown in the United States is stored and not used. The problem is that the surplus food does not get to the truly needy. The people who are starving lack what they need for obtaining adequate food, such as arable land or a job that would pay a living wage. In the past, people in most cases grew food crops and were able to feed themselves, but much of the best land today has been taken over by agribusinesses that grow cash crops, such as tobacco or cotton, and subsistence farmers have been forced onto marginal lands on the flanks of the desert where conditions are difficult and crops often do not grow.

Clearly, poverty is a cause of malnutrition, but there are other causes as well. Violence and war within a nation can displace people, leading to large numbers of refugees crowding places where food may not be available to all. Disasters can leave people without food and water—a situation complicated when a nation is already poor. Even climate change can threaten to create hunger, especially if farming practices cannot adjust to drought, floods, and extreme changes in weather patterns.

Causes of World Poverty

What causes world poverty, and why are so many people so desperately poor and starving? More to the point, why is poverty decreasing in some areas but increasing in others? We do know what does *not* cause poverty. Poverty is not necessarily caused by rapid population growth, although high fertility rates and poverty are related. In fact, many of the world's most populous countries—India and China, for instance—have large segments of their population that are poor, but even with very large populations, these countries have begun to reduce poverty levels. Poverty is also not caused by people being lazy or disinterested in working. People in extreme poverty work tremendously hard just to survive, and they would work hard at a job if they had one. It is not that they are lazy; it is that there are no jobs for them.

understanding **diversity**

War, Childhood, and Poverty

"In 2003, surgeons were forced to amputate both of Ali Ismaeel Abbas's arms after an errant U.S. bomb slammed into his Baghdad home during the opening phase of the Iraq war. Pictures of the twelve-year-old, who lost his parents in the attack, soon appeared on TV screens and in newspapers around the world. Since then, Abbas, who was treated in Kuwait, has come to represent a grim reality: All too often the victims of war are innocent children" (McClelland 2003: 20).

UNICEF estimates that over two million children have died in war, with even more injured, disabled, orphaned, or forced into refugee camps (Machel 1996). One estimate is that of all the victims of war, 90 percent are civilian—half of those, children (McClelland 2003).

Because children are seen as innocent, the trauma of war for children seems especially tragic to people. When children witness the violence of war or are forced to leave their homes, they experience a range of difficulties: loss, depression, injury, and other troubling results. In the aftermath of war, children are also highly vulnerable to outbreaks of disease (Cook and Wall 2011).

The United Nations has passed resolutions prohibiting the use of children under age 18 in combat. It has linked the threats to children from violence with high rates of poverty around the world. Although reducing poverty would not eliminate the threat of war, it would go a long way toward improving children's lives in war-torn regions.

MOHAMMED ABED/Getty Images

War, though it may seem remote to some, affects millions in both the United States and in war-torn countries. Many of those most affected are children.

Poverty is a result of a mix of causes. For one, the areas where poverty is increasing have a history of unstable governments or, in some cases, virtually no effective government to coordinate national development or plans that might alleviate extreme poverty and starvation. World relief agencies are reluctant to work in or send food to countries where the national governments cannot guarantee the safety of relief workers or the delivery of food and aid to where it should go. Food convoys may be hijacked or roads blocked by bandits or warlords.

In many countries with high proportions of poverty, the economies have collapsed and the governments have borrowed heavily to remain afloat. As a condition of these international loans, lenders, including the World Bank and the International Monetary Fund, have demanded harsh economic restructuring to increase capital markets and industrial efficiency. These economic reforms may make good sense for some and can improve the standard of living in these countries, but imposed reforms have also harmed people, especially the poor, when reforms also call for drastically reduced government spending on human services.

Poverty is also caused by changes in the world economic system. Although poverty has been a long-term problem and has many causes, increases in poverty and starvation in Africa can be attributed in part to the changes in world markets that have favored Asia economically but put sub-Saharan Africa at a disadvantage. As the price of products declined with more industrialization in places such as India, China, Indonesia, South Korea, Malaysia, and Thailand, commodity-producing nations in Africa and Latin America suffered. In Latin America, the poor have flooded to the cities, hoping to find work, whereas they did the opposite in Africa, fleeing to the countryside hoping to be able to grow subsistence crops. Governments often had to borrow to provide help to their citizens. Some governments collapsed or find themselves in such great debt that they are unable to help their own people. This has created massive amounts of poverty and starvation.

An often unrecognized cause of poverty is war. War disrupts the infrastructure of a society—its roads, utility systems, water, sanitation, even schools. For countries already struggling economically, this can be devastating. Food production may be disrupted and commerce can be threatened as it may be difficult, even impossible, to move goods in and out of a country. The loss of life and major injury can mean that there are fewer productive citizens who can work, thus threatening family and community

Race and Ethnicity

You might expect a society based on the values of freedom and equality, such as the United States, not to be deeply afflicted by racial–ethnic conflict, but think of the following situations:

- In 2009, James von Brunn, an eighty-eight-year-old self-proclaimed White supremacist, gunned down and killed a security guard at the U.S. Holocaust Memorial Museum in Washington, DC. Federal authorities knew James von Brunn was affiliated with various hate groups. The shooter left an anti-Semitic letter in his car parked outside the museum, charging that "Obama was created by Jews." The guard who was shot and killed, Stephen T. Johns, was a thirty-nine-year-old African American guard who worked at the museum.
- A sorority at a major East Coast university posted a photo of their group dressed in sombreros, ponchos, and fake mustaches, also carrying signs that said, "Will mow lawn for weed and beer." Such denigrating and offensive "racial theme parties" are common on college campuses (Cabrera 2014), including the March 2015 event in which a chapter of Sigma Alpha Epsilon at the University of Oklahoma was closed after a busload of fraternity brothers chanted a highly racist tune: "There will never be a ni**** at SAE. You can hang him from a tree, but he can never sign with me. There will never be a ni**** at SAE" (cnn.com).
- A thirty-eight-year-old American man of East Indian descent and vice president of a major bank was attacked on a Lake Tahoe beach as his attackers called him a "terrorist," "relative of Osama bin Laden," and "Indian garbage." The attack broke his eye socket and he will have dizzy spells for the rest of his life.

These are all ugly incidents. They all have one thing in common—racial–ethnic prejudice and overt racism. Race and ethnicity have fundamental importance in human social interaction and are integral parts of the social institutions in the United States. Unfortunately, ethnic prejudice and racism are also integral to American society.

Of course, racial and ethnic groups do not always interact as enemies, and interracial tension is not always obvious. It can be as subtle as a White person who simply does not initiate

in this chapter, you will learn to:

- Define race as a social construction
- Define and give examples of stereotype interchangeability
- Understand the difference between prejudice and racism
- Define and distinguish the different forms of racism
- Show how sociological theory broadens our understanding of prejudice and discrimination
- Discuss at least one common thread present in the histories of Blacks, Mexican Americans, and Japanese Americans in the United States
- Compare and contrast the different social change strategies toward the attainment of racial–ethnic equality and freedom in the United States

interactions with African Americans and Latinos, or an elderly White man who almost imperceptibly leans backward at a cocktail party as a Japanese American man approaches him.

In everyday human interaction, as African American philosopher Cornel West has cogently argued, *race matters* and still matters a lot (West 2004; 1994). What is race, and what is ethnicity? Why does society treat racial and ethnic groups differently, and why is there social inequality—stratification—between these groups? Racial and ethnic inequality is so strong and persistent in American society that sociologists reject the notion that we are a "postracial" society. As this chapter will show, race and ethnicity remain two of the most important axes of social stratification in the United States.

Race and Ethnicity

Within sociology, the terms *ethnicity, race, minority,* and *dominant group* have very specific meanings, different from their meanings in common usage. These concepts are important in developing a sociological perspective on race and ethnicity.

Ethnicity

An **ethnic group** is a social category of people who share a common culture, for example, a common language or dialect, a common nationality, a common religion, and common norms, practices, customs, and history. Ethnic groups have a consciousness of their common cultural bond—a "consciousness of kind." Italian Americans, Japanese Americans, Arab Americans, Polish Americans, Greek Americans, Mexican Americans, and Irish Americans are all examples of ethnic groups in the United States. Ethnic groups are also found in other societies, such as the Pashtuns in Afghanistan or the Shiites and Sunnis in Iraq, whose ethnicity is based on religious differences.

An ethnic group does not exist only because of the common national or cultural origins of a group, however. Ethnic groups develop also because of their unique historical and social experiences. These experiences become the basis for the group's *ethnic identity,* meaning the definition the group has of itself as sharing a common cultural bond. Prior to immigration to the United States, Italians, for example, did not necessarily think of themselves as a distinct group with common interests and experiences. Originating from different villages, cities, and regions of Italy, Italian immigrants identified themselves by their family background and community of origin. However, the process of immigration and the experiences Italian Americans faced as a group in the United States, including discrimination, created a new identity for the group, who subsequently began to define themselves as "Italians" (Waters and Levitt 2002; Alba 1990; Waters 1990).

The social and cultural basis of ethnicity allows ethnic groups to develop more or less intense ethnic identification at different points in time. Ethnic identification may grow stronger when groups face prejudice or hostility from other groups. Perceived or real threats and perceived competition from other groups may unite an ethnic group around common political and economic interests, which as you may recall was an idea advanced by early sociological theorist Emile Durkheim (see Chapter 1). Ethnic unity can develop voluntarily, or it may be involuntarily imposed when more powerful groups exclude ethnic groups from certain residential areas, occupations, or social clubs. Exclusionary practices strengthen ethnic identity.

Defining Race

Like ethnicity, race is primarily, though not exclusively, a socially constructed category. A **race** is a group treated as distinct in society based on certain characteristics, some of may be biological, that have been assigned or attributed social importance. Because of presumed biologically or culturally inferior characteristics (as defined by powerful groups in society), a race is often singled out for differential and unfair treatment. It is not the biological characteristics per se that define racial groups but *how groups have been treated and labeled historically and socially* (Higginbotham and Andersen 2012).

Society assigns people to racial categories, such as Black, White, and so on, not because of science, logic, or fact, but because of opinion and social experience. In other words, how groups are defined racially is a *social* process. This is what is meant when one says that race is "socially constructed." Although the meaning of race begins with alleged biological/genetic differences between groups (such as differences in physical characteristics like skin color, lip form, and hair texture), on closer examination, the assumption that racial differences are purely biological breaks down. In fact, biologists have pointed out that there is *little correspondence* between races as defined biologically/genetically and the actual naming of the races (Taylor 2012; Morning 2011; Ledger 2009; Lewontin 1996).

DEBUNKING Society's Myths

Myth: Racial differences are fixed, biological categories.
Sociological Perspective: Race is a social construct, one in which certain physical or cultural characteristics take on social meanings that become the basis for racism and discrimination. The definition of race varies across cultures within a society, across different societies, and at different times in the history of a given society (Graves 2004).

The social categories used to divide groups into races are not fixed. They vary from society to society and at different times in the history of a given society (Morning 2011, 2008; Washington 2011). Within the United States, laws defining who is Black have historically varied from state to state. North Carolina and Tennessee law historically defined people as Black if they have even one great-grandparent who was Black (thus being one-eighth Black—called "octoroon" in the 1890 Census; see Table 10.1). In other southern states, having any Black ancestry at all defined one as a Black person—the so-called *one-drop rule,* that is, one drop of Black blood (Washington 2011; Broyard 2007; Malcomson 2000). This one-drop rule still applies to a great extent today in the United States, even though its use for defining one's race has eroded somewhat.

This is even more complex when we consider the meaning of race in other countries. In Brazil, a light-skinned Black person could well be considered White, especially if the person is of high socioeconomic status. This demonstrates that one's race in Brazil is in part actually defined by one's social class. Thus, in parts of Brazil, it is often said that "money lightens" (*o dinheiro embranquence*). In this sense, a category such as social class can become *racialized.* In fact, people in Brazil are considered Black only if they are clearly of African descent and have little or no discernible White ancestry at all. A large percentage of U.S. Blacks would not be considered Black in Brazil (Telles et al. 2011; Telles 2004). Although Brazil is often touted as being a utopia of race "mixing" and racial social equality, nonetheless, as sociologist Edward Telles notes, light-skinned Brazilians continue to be privileged and continue to hold a disproportionate share of the wealth and power. Brazilians of darker skin color have significantly lower earnings, occupational status, and lower access to education (Telles et al. 2011; Villareal 2010; Telles 2004, 1994).

Racialization is a process whereby some social category, such as a social class or nationality, takes on what *society perceives* to be racial characteristics (Omi and Winant 2014; Harrison 2000; Malcomson 2000). The experiences of Jewish people provide a good example of what it means to say that race is a socially constructed category. Jews are more accurately called an *ethnic group* because of common religious and cultural heritage, but in Nazi Germany, Hitler defined Jews as a "race." An ethnic group thus became racialized. Jews were presumed to be biologically inferior to the group Hitler labeled the Aryans—white-skinned, blonde, tall, blue-eyed people. On the basis of this definition—which was supported through Nazi law, taught in Nazi schools, and enforced by the Nazi military—Jewish people were brutally mistreated. They were segregated, persecuted, and systematically murdered in what has come to be called the Holocaust during the Second World War.

Mixed-race people defy the biological categories that are typically used to define race. Is someone who is the child of an Asian mother and an African American father Asian or Black? Reflecting this issue, the U.S. Census's current practice is for people to have the option of checking several racial categories rather than just one, thus defining one's self as "biracial" or "multiracial" (Spencer 2011; Waters 1990). As ◆ Table 10.1 shows, the decennial U.S. census (taken every ten years) has dramatically changed its racial and ethnic classifications since 1890, reflecting the fact that society's thinking about racial and ethnic categorization has not remained constant through time (Spencer 2012; Saulny 2011; Washington 2011; Rodriguez 2000; Lee 1993).

Opposition to the multiple categorization of races has arisen upon both scholarly and political grounds. Some (Spencer 2012) have argued that advocating simultaneous multiple categorization of races tends to downplay the rich cultural traditions in the case of Blacks in the United States, including but not limited to language ("Ebonics"), music (jazz, blues, rock, hip-hop, and so on), dance, a vast literature, and many, many others. Some people have argued that multiracial classification will ultimately lead to a "postracial" society and thus a solution of sorts to race problems in the United States. Wiping out single-race categorization will not, and has not, however, led to less discrimination against minorities of color.

The "postracial dream" conflicts with the hard realities of housing discrimination, higher foreclosure rates during the recession, racial discrimination in education and in standardized testing, differential access to medical care, a lower life expectancy, and many other forms of discrimination (Bobo 2012; Rugh and Massey 2010). Some of these forms of racial discrimination actually *increased* even within the last decade! At least one sociological analyst concluded that "Those who proclaim that multiracial identity will destroy racial distinctions are living a lie" (Spencer 2012: 70).

◆ Table 10.1 Comparison of U.S. Census Classifications, 1890–2010

Census Date	White	African American	Native American	Asian American	Other Categories
1890	White	Black, Mulatto, Quadroon, Octoroon	Indian	Chinese, Japanese	
1900	White	Black	Indian	Chinese, Japanese	
1910	White	Black, Mulatto	Indian	Chinese, Japanese	Other
1990	White	Black or Negro	Indian (American) Eskimo Aleut	Chinese, Japanese, Filipino, Korean, Asian, Indian, Vietnamese	Hawaiian, Guamanian, Samoan, Asian or Pacific Islander, Other
2000 and 2010[a, b]	White	Black or African American	American Indian, Alaskan Native	Chinese, Japanese, Filipino, Korean, Asian, Indian, Vietnamese	Native Hawaiian, Pacific Islander, Other

[a] In 2000, for the first time ever, and again in 2010, individuals could select more than one racial category. In 2010, only 5 percent actually did so.
[b] Hispanics were included under "Other" in 1910 and 1920. In 1930 and subsequent years, the category "Mexican" was listed in addition to the category "Other."

Sources: Lee, Sharon. 1993. "Racial Classification in the U.S. Census: 1890–1990." *Ethnic and Racial Studies* 16(1): 75–94; U.S. Census Bureau. 2003. "Racial and Ethnic Classification Used in Census 2000 and Beyond." Rodriguez, Clara E. 2000. *Changing Race: Latinos, the Census, and the History of Ethnicity*. New York: New York University Press; Silver, Alexandra. 2010. "Brief History of the U.S. Census." *Time* (February 8): 16; Washington, Scott. 2011. "Who Isn't Black? The History of the One-Drop Rule." PhD dissertation, Department of Sociology, Princeton University, Princeton, NJ.

Scott Olson/Getty Images News/Getty Images

This is Barack Obama, the first African American ever to be elected U.S. president, and for two terms. His father is Black African (Kenyan) and his mother is White American. Why is his race African American?

The Significance of Defining Race. The biological characteristics that have been used to define different racial groups vary considerably both within and between groups. Many Asians, for example, are actually lighter skinned than many Europeans and White Americans but, regardless of their skin color, have been defined in racial terms as "yellow." Some light-skinned African Americans are also lighter in skin color than some White Americans. Developing racial categories overlooks the fact that human groups defined as races are—biologically speaking—much more alike than they are different (Graves 2004).

The biological differences that are presumed to define different racial groups are somewhat arbitrary. Why, for example, do we differentiate people based on skin color and not some other characteristic such as height or hair color? You might ask yourself how a society based on the presumed racial inferiority of red-haired people would compare to other racial inequalities in the United States. The likelihood is that if a powerful group defined another group as inferior because of some biological characteristics and they used their power to create social institutions that treated this group unfairly, a system of racial inequality would result. In fact, very few biological differences exist between racial groups. As we already noted, most of the variability in almost all biological characteristics, even blood type and various bodily chemicals, is *within* and not between racial groups.

Different groups use different criteria to define racial groups. To American Indians, being classified as an American Indian depends upon proving one's

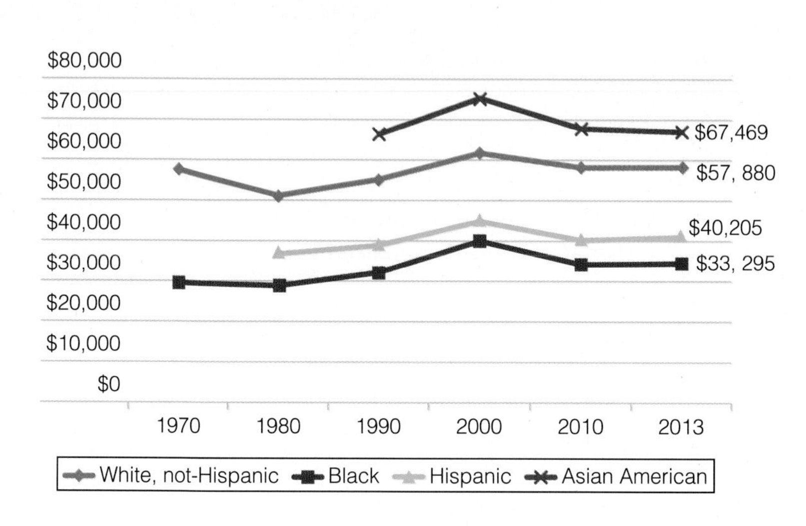

▲ **Figure 10.1** The Income Gap by Race, 1970–2013 Be careful about interpreting this chart. The U.S. Census Bureau changed how it counted different racial–ethnic groups over this period of time. Asian Americans, for example, included Pacific Islanders prior to 2002, which inflates the overall median income. Nonetheless, you can observe persistent gaps in median income by looking at the income status of these different groups over time. Note, however, that with the changing way that the census counted race and ethnicity, data are not available for some groups at earlier points in time.

Source: DeNavas-Walt, Carmen, and Bernadette Proctor. 2014. *Income and Poverty in the United States: 2013.* Washington, DC: U.S. Census Bureau. **www.census.gov**

group or stratum solely because of their membership in that group or stratum. Prejudice is an attitude; discrimination is overt behavior. *Racial-ethnic discrimination* is unequal treatment of a person on the basis of race or ethnicity.

Discrimination occurs in many sites and it can be seen in studies. Audit studies take two people identical in nearly all respects (age, education, gender, social class, and other characteristics) who then present themselves as potential tenants or employees. If one is White and the other is a minority, the minority person will often be refused housing or employment by landlords and employers. Audit studies have found that discrimination occurs far more frequently than most people imagine (Feagin 2007).

The discrimination affecting the nation's minorities takes a number of forms—for example, income discrimination, as you can see in ▲ Figure 10.1. Although the median income of Black and Hispanic families has increased since 1970, the size of the income gap between these two groups and Whites has remained much the same over time.

Median income figures tell only part of the story though. In addition to annual income, the net worth of White families has consistently grown faster than that of Black families (Oliver and Shapiro 2008). Net worth may well be a better indicator of economic inequality than annual income. Poverty among Blacks has also decreased since the 1950s, but is now close to the same level as in 1970. As ▲ Figure 10.2 shows, the current poverty rate is highest for African Americans and Hispanics compared with Whites or Asians. Note that it is greater for Asians than for Whites. In all these racial groups, children have the highest rate of poverty.

Discrimination is illegal under U.S. law. Nonetheless, for many years, various discriminatory processes have continued. Take housing as an example. Even very recently, banks and mortgage companies have withheld mortgage loans from minorities based on "redlining," an illegal practice in which entire minority neighborhoods are designated as "ineligible for loans." Racial segregation may also be fostered by *gerrymandering,* the calculated redrawing of election districts, school districts, and similar political boundaries to maintain racial segregation.

Segregation is the spatial and social separation of racial and ethnic groups. Although desegregation has been mandated by law (thus eliminating *de jure segregation,* or legal segregation), *de facto segregation*—segregation in fact—still exists, particularly in housing and education. Segregation has contributed to the creation of an **urban underclass**, a grouping of people, largely minorities and the poor, who live at the absolute bottom of the socioeconomic ladder in urban areas

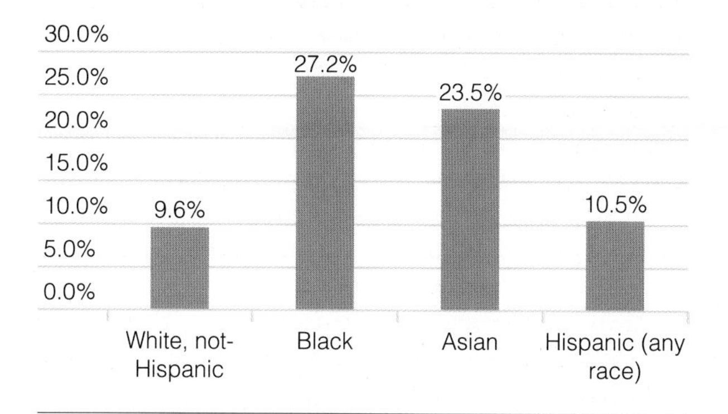

▲ **Figure 10.2** People in Poverty by Race, 2013 This shows the overall percentage of people living below the official poverty line in 2013. Given what you learned in Chapter 8 about how poverty is measured, what might you conclude from this chart?

Source: DeNavas-Walt, Carmen, and Bernadette Proctor. 2014. *Income and Poverty in the United States: 2013.* Washington, DC: U.S. Census Bureau. **www.census.gov**

doing sociological research

American Apartheid

The term *apartheid* was used to describe the society of South Africa prior to the election of Nelson Mandela in 1994. It refers to the rigid separation of the Black and White races. Sociological researchers Massey and Denton argue that the United States is now under a system of apartheid and that it is based on a very rigid residential segregation in the country.

Research Question: What is the current state of residential segregation? Massey and Denton note that the terms *segregation* and *residential segregation* practically disappeared from the American vocabulary in the late 1970s and early 1980s. These terms were spoken little by public officials, journalists, and even civil rights officials. This was because the ills of race relations in the United States were at the time attributed, though erroneously, to other causes such as a "culture of poverty" among minorities, or inadequate family structure among Blacks, or too much welfare for minority groups. The Fair Housing Act was passed in 1968, and the problem of segregation and discrimination in housing was declared solved. Yet nothing could be farther from the truth.

Research Methods and Results: Researchers Massey and Denton amassed a large amount of survey data demonstrating that residential segregation not only has persisted in American society but also that it has actually increased since the 1960s. Most Americans vaguely realize that urban America is still residentially segregated, but few appreciate the depth of Black and Hispanic segregation or the degree to which it is maintained by ongoing institutional arrangements and contemporary individual actions. Urban society is thus hypersegregated, or characterized by an extreme form of residential and educational segregation.

Conclusions and Implications: Massey and Denton find that most people think of racial segregation as a faded notion from the past, one that is decreasing over time. Today, theoretical concepts such as the culture of poverty, institutional racism, and welfare are widely debated, yet rarely is residential segregation considered to be a major contributing cause of urban poverty and the underclass. Massey and Denton argue that their purpose is to redirect the focus of public debate back to race and racial segregation.

Questions to Consider

1. Think about the degree of residential segregation in the neighborhood in which you grew up. How racially and/or ethnically segregated was it? If it was segregated at all, what racial–ethnic groups were living there?
2. Do you think the problem of racial–ethnic residential segregation in the United States is largely "solved"? How so or why not?
3. What are the consequences of residential segregation—for example, for education? For employment?

Sources: Massey, Douglas S., and Nancy A. Denton. 1993. *American Apartheid: Segregation and the M.aking of the Underclass*. Cambridge, MA: Harvard University Press; Massey, Douglas S. 2005. *Strangers in a Strange Land: Humans in an Urbanizing World*. New York: Norton; Rugh, Jacob S., and Douglas S. Massey. 2010. "Racial Segregation and the American Foreclosure Crisis." *American Sociological Review* 75 (5): 629–651.

(Wilson 2009, 1987; Massey and Denton 1993). Indeed, the level of housing segregation is so high for some groups, especially poor African Americans and Latinos, that it has been termed **hypersegregation**, referring to a pattern of extreme segregation (Rugh and Massey 2010; Massey 2005; Massey and Denton 1993). Currently, the rate of segregation of Blacks and Hispanics in U.S. cities is actually increasing, thus allowing for less and less interaction between White and Black children and White and Hispanic children (Schmitt 2001; Massey and Denton 1993; see the box "Doing Sociological Research: American Apartheid"). In education, the extraordinary realization is that schools are also becoming more segregated, a phenomenon called *resegregation,* because American schools are now more segregated than they were even in the 1980s (Frankenberg and Lee 2002).

Residential segregation and other forms of special segregation are so pervasive that people have psychologically internalized notions of "the White space" in neighborhoods, workplaces, restaurants, and other formally nonsegregated places. As a consequence, in restaurants as an example, both White and Black patrons recognize—only semiconsciously—that certain seating sections are "for Blacks." As a result, such areas tend to be avoided by White patrons and even actively sought out by Black patrons (Anderson 2011).

Racism

Racism includes both attitudes and behaviors. A negative attitude taken toward someone simply because he or she belongs to a racial or ethnic group is a prejudice, as has already been discussed. **Racism** is the perception and treatment of a racial or ethnic group, or member of that group, as intellectually, socially, and culturally inferior to one's own group. It is more than an attitude; it is institutionalized in society.

There are different forms of racism. Researchers have often called obvious, overt racism, such as physical assaults, lynchings, and overt expressions of prejudice,

Jim Crow racism. This form of racism has declined somewhat in our society since the 1950s, though it certainly has not disappeared (Bobo 2004).

Racism can also be subtle, covert, and nonobvious; this is known as **aversive racism** (Jones et al. 2013). Consistently avoiding interaction with someone of another race or ethnicity is an example of aversive racism. This form of racism is quite common and has remained at roughly the same level for more than sixty years, with perhaps a slight increase (Jones et al. 2013). Even when overt forms of racism dissipate, aversive racism tends to persist, because it is less visible than overt racism; people can believe racism has diminished when it has not (Gaertner and Dovidio 2005).

Another subtle nonobvious form of racism, akin to aversive racism, is what researcher Jennifer Eberhardt calls **implicit bias**. It is a largely nonconscious form of racism, where individuals make unconscious associations, say between race and crime. Culture forces such associations on individuals. Starting with childhood socialization, Blackness is mentally associated with criminality; Whiteness is not. Eberhardt's research concretely demonstrates that the association between race and crime directly impacts how individuals behave and make decisions (Eberhardt 2010).

Eberhardt's research finds, for example, that in a court trial, holding other things constant, a defendant's skin color and hair texture correlate with the sentencing decisions of jurors: Black defendants are more likely to receive the death penalty than are otherwise similar White defendants. Eberhardt attributes such findings to implicit bias, a subtle form of racism that individuals internalize and carry with them always. This bias also carries over to police officers, who mistakenly identify Black faces as "criminal faces" relative to White faces.

Laissez-faire racism is another form of racism (Bobo 2006). Laissez-faire racism includes several elements:

1. The subtle but persistent negative stereotyping of minorities, particularly Black Americans, especially in the media
2. A tendency to blame Blacks themselves for the gap between Blacks and Whites in socioeconomic standing, occupational achievement, and educational achievement
3. Clear resistance to meaningful policy efforts (such as affirmative action, discussed later) designed to ameliorate racially oppressive social conditions and practices in the United States.

A close relative of laissez-faire racism is Eduardo Bonilla-Silva's (2013) concept of **color-blind racism**—so named because individuals affected by this type of racism prefer to ignore legitimate racial-ethnic, cultural, and other differences and insist that the race problems in the United States will go away if only race is ignored altogether. Accompanying this belief is the opinion that race differences in the United States are merely an illusion and that race is not real. Simply refusing to perceive any differences at all between racial groups (thus being color-blind) is in itself a form of racism (Bonilla-Silva 2013; Gallagher 2013; Bonilla-Silva and Baiocchi 2001). This will come as a surprise to many. Here is an example of color-blind racism: Have you heard someone say, as they attempt to show that they are "not racist": "I don't care if you are white, black, yellow, or green! I am not prejudiced!"

Such people insist that they are only objective and fair people, people who do not notice skin color. By definition, this is color-blind racism—a form of racism, for sure. This ignores the reality of race and its significance in society.

Color-blindness hides what is called **White privilege** behind a mask: It allows Whites to define themselves as politically and racially tolerant as they proclaim adherence to a belief system that does not see or judge individuals by "the color of their skin." They think of skin color as irrelevant, but it is not "irrelevant." Racial domination—that is, white privilege, is structured into society. Color-blind racism gives the false impression that racial barriers have fallen when they have not (Gallagher 2013; Kristof 2008).

Institutional racism as a form of racism is the negative treatment and oppression of one racial or ethnic group by society's existing institutions based on the presumed inferiority of the oppressed group. Institutional racism exists at the level of social structure. In Durkheim's sense, it is external to individuals—thus institutional. It is then possible to have "racism without racists," as sociologist Bonilla-Silva has shown (Bonilla-Silva 2013). It is a purely sociological notion: Institutional racism can exist apart from the individual personality or personalities. Key to understanding institutional racism is seeing that dominant groups have the economic and political power to subjugate the minority group, even if they do not have the explicit intent of being prejudiced or discriminating against others.

Racial profiling is an example of institutional racism in the criminal justice system. African American and Hispanic people are arrested—and serve longer sentences—considerably more often than are Whites and Asians. In fact, an African American or Hispanic wrongdoer is more likely to be arrested than a White person who commits the exact same crime, even when the White person shares the same age, socioeconomic environment, and prior arrest record as the Black or Hispanic. These kinds of disparities are very prominent in traffic arrests, where police officers often report that the arrested person "fit the profile" (Doermer and Demuth 2010; Eberhardt 2010).

As most people now know, in a heavily Black suburb of St. Louis, in Ferguson, Missouri, a Black teenager,

Michael Brown, was shot and killed in the summer of 2014 by a White policeman (Darren Wilson) for allegedly physically attacking Wilson. Massive public demonstrations followed. Many Black observers of the shooting reported that Brown did not attack Wilson, was unarmed, and had his hands up. "Hands up/don't shoot" became a rallying cry of the public demonstrations.

The institutional racism in this incident is shown by the data. Ferguson, Missouri, is a predominantly Black suburb (55 percent Black), yet 90 percent of all traffic arrests are of Black people. Second, Blacks in Ferguson were more likely than Whites to say that race was definitely a factor in Wilson's fatal shooting of Michael Brown: Eighty percent of a total of Blacks interviewed in a later survey thought so, whereas only 23 percent of Whites interviewed thought so (Pew Research Center 2014). The cry of Whites that "race had nothing to do with it" is all too familiar in such situations.

Consider this: Even if every White person in the country lost all of his or her personal prejudices, and even if he or she stopped engaging in individual acts of discrimination, institutional racism would still persist for some time. Over the years, it has become so much a part of U.S. institutions (hence, the term *institutional racism*) that discrimination can occur even when no single person is deliberately causing it. To sum up, racism is a characteristic of the institutions and not necessarily of the individuals within the institution. *This is why institutional racism can exist even without prejudice being the cause.*

Daryl Lang/Shutterstock.com

Minorities are more likely to be arrested than Whites for the same offense. Does this reflect institutional racism in addition to possible individual prejudice of the arresting police officer?

DEBUNKING Society's Myths

Myth: The primary cause of racial inequality in the United States is the persistence of prejudice.

Sociological Perspective: Prejudice is one dimension of racial problems in the United States, but institutional racism can flourish even while prejudice is on the decline. Prejudice is an attribute of the individual, whereas institutional racism is an attribute of social structure.

Theories of Prejudice and Racism

Why does racial inequality persist? Sociological theory provides insight into this question.

Assimilation theory examines the process by which a minority becomes socially, economically, and culturally absorbed within the dominant society. This theory assumes that to become fully fledged members of society, minority groups must adopt as much of the dominant society's culture as possible, particularly its language, mannerisms, and goals for success, and thus give up much of its own culture. Assimilation stands in contrast to racial cultural **pluralism**—the separate maintenance and persistence of one's culture, language, mannerisms, practices, art, and so on.

Many Americans believe that with enough hard work and loyalty to the dominant White culture of the country, any minority can make it and thus "assimilate" into American society. It is the often heard argument that African Americans, Hispanics, and Native Americans need only to pull themselves up "by their own bootstraps" to become a success.

THINKING Sociologically

Write down your own racial–ethnic background and list one thing that people from this background have positively contributed to U.S. society or culture. Also list one experience (current or historical) in which people from your group have been victimized by society. Discuss how these two things illustrate the fact that racial–ethnic groups have both been victimized and have made positive contributions to this society. Share your comments with others: What does this reveal to you about the connections between different groups of people and their experiences as racial–ethnic groups in the United States?

Assimilationists believe that to overcome adversity and oppression, minority people need only imitate the dominant White culture as much as possible. In this sense, minorities must assimilate "into" White culture and White society. The general assumption is that with inferior race that did not deserve social, educational, or political equality. This is an example of the *racial formation process,* as noted earlier in this chapter (Omi and Winant 2014). Anglos believed that Mexicans were lazy, corrupt, and cowardly, yet violent, which launched stereotypes that would further oppress Mexicans. These stereotypes were used to justify the lower status of Mexicans and Anglo control of the land that Mexicans were presumed to be incapable of managing (Telles and Ortiz 2008; Moore 1976).

During the twentieth century, advances in agricultural technology changed the organization of labor in the Southwest and West. Irrigation allowed year-round production of crops and a new need for cheap labor to work in the fields. Migrant workers from Mexico were exploited as a cheap source of labor. Migrant work was characterized by low earnings, poor housing conditions, poor health, and extensive use of child labor. The wide use of Mexican migrant workers as field workers, domestic servants, and other kinds of poorly paid work continues, now throughout the United States.

Puerto Ricans. The island of Puerto Rico was ceded to the United States by Spain in 1898. In 1917, the Jones Act extended U.S. citizenship to Puerto Ricans, although it was not until 1948 that Puerto Ricans were allowed to elect their own governor. In 1952, the United States established the Commonwealth of Puerto Rico, with its own constitution. Following the Second World War, the first elected governor launched a program known as Operation Bootstrap, which was designed to attract large U.S. corporations to the island of Puerto Rico by using tax breaks and other concessions. This program contributed to rapid overall growth in the Puerto Rican economy, although unemployment remained high and wages remained low. Seeking opportunity, unemployed farmworkers began migrating to the United States. These migrants were interested in seasonal work, and thus a pattern of temporary migration characterized the Puerto Ricans' entrance into the United States (Amott and Matthaei 1996; Rodriguez 1989).

Unemployment in Puerto Rico became so severe that the U.S. government even went so far as to attempt a reduction in the population by some form of population control. Pharmaceutical companies experimented with Puerto Rican women in developing contraceptive pills, and the U.S. government actually encouraged the sterilization of Puerto Rican women. More than 37 percent of the women of reproductive age in Puerto Rico had been sterilized by 1974 (Roberts 1997). More than one-third of these women have since indicated that they regret sterilization because they were not made aware at the time that the procedure was irreversible.

AP Images/Richard Drew

Activities such as this Puerto Rican Day Parade in New York City reflect pride in one's group culture and result in greater cohesiveness of the group.

Cubans. Cuban migration to the United States is recent in comparison with the many other Hispanic groups. The largest migration has occurred since the revolution led by Fidel Castro in 1959. Between then and 1980, more than 800,000 Cubans—one-tenth of the entire island population—migrated to the United States. The U.S. government defined this as a political exodus, facilitating the early entrance and acceptance of these migrants. Many of the first migrants had been middle- and upper-class professionals and landowners under the prior dictatorship of Fulgencio Batista, but they had lost their land during the Castro revolution. In exile in the United States, some worked to overthrow Castro, often with the support of the U.S. federal government. Many other Cuban immigrants were of more modest means who, like other immigrant groups, came seeking freedom from political and social persecution and an escape from poverty.

A second wave of Cuban immigration came in 1980, when the Cuban government, still under Castro, opened the Port of Mariel to anyone who wanted to leave Cuba. In the five months following this action, 125,000 Cubans came to the United States—more than the combined total for the preceding eight years. The arrival of people from Mariel has produced debate and tension, particularly in Florida, a major center of Cuban migration. The Cuban government had previously labeled the people fleeing from Mariel as "undesirable"; some had been incarcerated in Cuba before leaving. They were actually not much different from previous refugees such as the "golden exiles," who were professional and high-status refugees (Portes and Rumbaut 1996). But because the refugees escaping from Mariel had been labeled (stereotyped) as undesirables, and because they were forced to live in primitive camps for long periods after their arrival, they have been unable to achieve much social and economic mobility in the United States—thus ironically reinforcing the initial perception that they were "lazy" and "undesirable." In contrast, the earlier Cuban migrants, who were on average more educated and much more settled, have enjoyed a fair degree of success (Portes and Rumbaut 2001, 1996; Amott and Matthei 1996; Pedraza 1996).

Asian Americans

Like Hispanic Americans, Asian Americans are from many different countries and diverse cultural backgrounds; they cannot be classified as the single cultural rubric of Asians. Asian Americans include migrants from China, Japan, the Philippines, Korea, and Vietnam, as well as more recent immigrants from Southeast Asia.

Chinese. Attracted by the U.S. demand for labor, Chinese Americans began migrating to the United States during the mid-nineteenth century. In the early stages of this migration, the Chinese were tolerated because they provided cheap labor. They were initially seen as good, quiet citizens, but racial stereotypes turned hostile when the Chinese came to be seen as competing with White California gold miners for jobs. Thousands of Chinese laborers worked for the Central Pacific Railroad from 1865 to 1868. They were relegated to the most difficult and dangerous work, worked longer hours than the White laborers, and for a long time were paid considerably less than the White workers.

The Chinese were virtually expelled from railroad work near the turn of the twentieth century (in 1890–1900) and settled in rural areas throughout the western states. As a consequence, anti-Chinese sentiment and prejudice ran high in the West. This ethnic antagonism was largely the result of competition between the White and Chinese laborers for scarce jobs. In 1882, the federal government passed the Chinese Exclusion Act, which banned further immigration of unskilled Chinese laborers. Like African Americans, the Chinese and Chinese Americans were legally excluded from intermarriage with Whites (Takaki 1989). During this period, several Chinatowns were established by those who had been forcibly uprooted and who found strength and comfort within enclaves of Chinese people and culture (Nee 1973).

Japanese. Japanese immigration to the United States took place mainly between 1890 and 1924, after which passage of the Japanese Immigration Act forbade further immigration. Most of these first-generation immigrants, called issei, were employed in agriculture or in small Japanese businesses. Many issei were from farming families and wished to acquire their own land, but in 1913, the Alien Land Law of California stipulated that Japanese aliens could lease land for only three years and that lands already owned or leased by them could not be bequeathed to heirs. The second generation of Japanese Americans, or nisei, were born in the United States of Japanese-born parents. They became better educated than their parents, lost their Japanese accents, and in general became more "Americanized," that is, culturally assimilated. The third generation, called sansei, became even better educated and assimilated, yet still met with prejudice and discrimination, particularly where Japanese Americans were present in the highest concentrations, as on the West Coast from Washington to southern California (Takaki 1989; Glenn 1986).

The Japanese suffered the complete indignity of having their loyalty questioned when the federal government, thinking they would side with Japan after the Japanese attack on Pearl Harbor in December 1941, herded them into concentration camps. By executive order of president Franklin D. Roosevelt, much of the West Coast Japanese American population (a great many—perhaps most—of them loyal second- and third-generation Americans) had their assets frozen and their real estate confiscated by the government. A media campaign immediately followed, labeling Japanese Americans "traitors" and "enemy aliens." Virtually

During World War II, Japanese Americans, who were full American citizens, were forced into concentration camps. A noon food ("mess") line at one of these camps, Manzanar, is shown here.

11

Hoang Dinh Nam/AFP/Getty Images

of some journalists, the United States is by no means headed toward a "postracial" society.

What are some of the approaches to attaining racial and ethnic equality?
Approaches include Reverend Martin Luther King's nonviolent civil rights strategy, radical social change, and movements such as the Black Power movement, La Raza Unida, and the American Indian Movement (AIM), all of which directly addressed *institutional racism. Affirmative action* policies, which are race-specific rather than race-blind programs, continue to be changed and modified through Supreme Court cases.

Key Terms

affirmative action **250**
anti-Semitism **246**
assimilation theory **238**
aversive racism **237**
color-blind racism **237**
contact theory **239**
discrimination **234**
dominant group **232**
ethnic group **228**
ethnocentrism **234**
hypersegregation **236**
implicit bias **237**
institutional racism **237**
intersection perspective **240**
laissez-faire racism **237**
minority group **232**
pluralism **238**
prejudice **234**
race **228**
racial formation **231**
racial profiling **237**
racialization **229**
racism **236**
residential segregation **236**
salience principle **232**
segregation **235**
stereotype **232**
stereotype interchangeability **233**
stereotype threat **234**
urban underclass **235**
White privilege **237**

and Middle Easterners, nor have the recent genuinely terrorist attacks of ISIS against Middle Easterners and some Anglos in Mid-East locations.

White Ethnic Groups

The story of White ethnic groups in the United States begins during the colonial period. White Anglo Saxon Protestants (WASPs), who were originally immigrants from England and to some extent Scotland and Wales, settled in the New World (what is now North America). They were the first ethnic group to come into contact—most often hostile contact—on a large scale with those people already here, namely, Native American Indians. WASPs came to dominate the newly emerging society earlier than any other White ethnic group.

In the late 1700s, the WASPs regarded the later immigrants from Germany and France as "foreigners" with odd languages, accents, and customs, and applied derogatory labels ("krauts" for Germans; "frogs" for the French) to them. Again, this demonstrates that virtually every immigrant group, except WASPs themselves, was discriminated against and subject to racist name-calling. Tension between the "old stock" and the "foreigners" continued through the Civil War era until around 1860, when the national origins of U.S. immigrants began to change (Kibria et al. 2013; Handlin 1951).

Of all racial and ethnic groups in the United States during that time and since, only WASPs do not think of themselves as an ethnic nationality. The WASPs came to think of themselves as the "original" Americans despite the prior presence of Native American Indians, whom the WASPs in turn described and stereotyped as savages. As immigrants from northern, western, eastern, and southern Europe began to arrive, particularly during the mid- to late-nineteenth century, WASPs began to direct prejudice and discrimination against many of these newer groups.

There were two waves of migration of White ethnic groups in the mid- and late-nineteenth century. The first stretched from about 1850 through 1880, and included northern and western Europeans: English, Irish, Germans, French, and Scandinavians. The second wave of immigration occurred from 1890 to 1914, and included eastern and southern European populations: Italians, Greeks, Poles, Russians, and other eastern Europeans, in addition to more Irish. The immigration of Jews to the United States extended for well over a century, but the majority of Jewish immigrants came to the United States during the period from 1880 to 1920.

The Irish arrived in large numbers in the mid-nineteenth century and after, as a consequence of food shortages and massive starvation in Ireland. During the latter half of the nineteenth century and in the early twentieth century, the Irish in the United States were abused, attacked, and viciously stereotyped. It is instructive to remember that the Irish, particularly on the East Coast and especially in Boston, underwent a period of ethnic oppression of extraordinary magnitude. A frequently seen sign posted in Boston saloons during that time proclaimed "no dogs or Irish allowed." The sign was not intended as a joke. German immigrants were similarly stereotyped, as were the French and the Scandinavians. It is easy to forget that virtually all immigrant groups have gone through times of oppression and prejudice, although these periods were considerably longer for some groups than for others. As a rule, where the population density of an ethnic group in a town, city, or region was greatest, so too was the amount of prejudice, negative stereotyping, and discrimination to which that group was subjected.

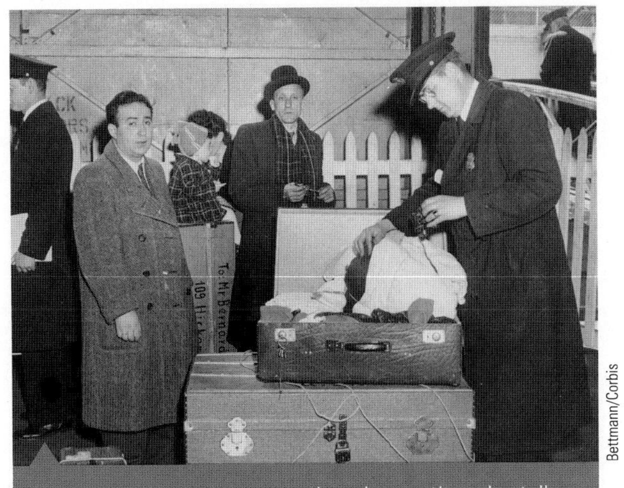

Bettmann/Corbis

Jewish immigrants were questioned, sometimes brutally, at Ellis Island, the point of entry to the United States for many early European immigrants.

More than 40 percent of the world's Jewish population lives in the United States, making it the largest community of Jews in the world. Most of the Jews in the United States arrived between 1880 and the First World War, originating from the eastern European countries of Russia, Poland, Lithuania, Hungary, and Romania. Jews from Germany arrived in two phases; the first wave came just prior to the arrival of those from eastern Europe, and the second came as a result of Hitler's ascension to power in Germany during the late 1930s. Because many German Jews were professionals who also spoke English, they assimilated more rapidly than those from the eastern European countries. Jews from both parts of Europe underwent lengthy periods of anti-Jewish prejudice, **anti-Semitism** (defined as the hatred of Jewish people), and discrimination, particularly on Manhattan's Lower East Side. Significant anti-Semitism still exists in the United States (Ferber 1999; Essed 1991). In 1924, the National Origins all Japanese Americans in the United States had been removed from their homes by August 1942, and some were forced to stay in relocation camps until as late as 1946. Relocation destroyed numerous Japanese families and ruined them financially (Takaki 1989; Glenn 1986; Kitano 1976).

In 1986, the U.S. Supreme Court allowed Japanese Americans the right to file suit for monetary reparations. In 1987, legislation was passed, awarding $20,000 to each person who had been relocated and offering an official apology from the U.S. government. One is motivated to contemplate how far this paltry sum and late apology could go in righting what many have argued was the "greatest mistake" the United States has ever made as a government.

Filipinos. The Philippine Islands in the Pacific Ocean fell under U.S. rule in 1899 as a result of the Spanish-American War, and for a while Filipinos could enter the United States freely. By 1934, the islands became a commonwealth of the United States, and immigration quotas were imposed on Filipinos. More than 200,000 Filipinos immigrated to the United States between 1966 and 1980, settling in major urban centers on the West and East Coasts. More than two-thirds of those arriving were professional workers; their high average levels of education and skill have eased their assimilation. By 1985, more than one million Filipinos were in the United States. They are now the second largest Asian American population.

Koreans. Many Koreans entered the United States in the late 1960s, after amendments to the immigration laws in 1965 raised the limit on immigration from the Eastern Hemisphere. The largest concentration of Koreans is in Los Angeles. As much as half of the adult Korean American population is college educated, an exceptionally high proportion. Many of the immigrants were successful professionals in Korea; upon arrival in the United States, though, they have been forced to take on menial jobs, thus experiencing downward social mobility and status inconsistency. This is especially true of those Koreans who migrated to the East Coast. However, nearly one in eight Koreans in the United States today owns a business; many own small greengrocer businesses. Many of these stores are located in predominantly African American communities and have become one among several sources of ongoing conflict between some African Americans and Koreans. This has fanned negative feeling and prejudice on both sides—among Koreans against African Americans and among African Americans against Koreans (Chen 1991).

Vietnamese. Among the more recent groups of Asians to enter the United States have been the South Vietnamese, who began arriving following the fall of South Vietnam to the communist North Vietnamese at the end of the Vietnam War in 1975. These immigrants, many of them refugees who fled for their lives, numbered about 650,000 in the United States in 1975. About one-third of the refugees settled in California. Many faced prejudice and hostility, resulting in part from the same perception that has dogged many immigrant groups before them—that they were in competition for scarce jobs. A second wave of Vietnamese immigrants arrived after China attacked Vietnam in 1978. As many as 725,000 arrived in the United States, only to face discrimination in a variety of locations. Tensions became especially heated when the Vietnamese became a substantial competitive presence in the fishing and shrimping industries in the Gulf of Mexico on the Texas shore. Since that time, however, many communities have welcomed them, and many Vietnamese heads of households have become employed full-time (Kim 1993; Winnick 1990).

Middle Easterners

Since the mid-1970s, immigrants from the Middle East have been arriving in the United States. They have come from countries such as Syria, Lebanon, Egypt, and Iran, and more recently, especially Iraq. Contrary to popular belief, the immigrants speak no single language and follow no singular religion and thus are ethnically diverse. Some are Catholic, some are Coptic Christian, and many are Muslim. Many are from working-class backgrounds, but many were professionals—teachers, engineers, scientists, and other such positions—in their homelands. About 65 percent of those residing in this country were born outside the United States; about half are college-educated (Kohut 2007). Like immigrant populations before them, Middle Easterners have formed their own ethnic enclaves in the cities and suburbs of this country as they pursue the often elusive American dream (Abrahamson 2006).

Since the terrorist attacks on the World Trade Center and the Pentagon on September 11, 2001, many male Middle Easterners of several nationalities have become unjustly suspect in this country and are subjected to severe harassment; racially motivated physical attacks; and as already noted, out-and-out racial profiling, if only because they had dark skin and—as with some—wore a turban of some sort on their heads. Most of these individuals, of course, probably had no discernible connection at all with the terrorists. A survey shows that most Muslims in this country believe that the September 11, 2001, terrorist attacks were indeed the cause of subsequent increased racial harassment and violence against them (Kohut 2007). Finally, the U.S. wars with Iraq and Afghanistan have not helped in easing tensions between White Americans

Gender

Imagine suddenly becoming a member of the other sex. What would you have to change? First, you would probably change your appearance—clothing, hairstyle, and any adornments you wear. You would also have to change some of your interpersonal behavior. Contrary to popular belief, men talk more than women, are louder, are more likely to interrupt, and are less likely to recognize others in conversation. Women are more likely to laugh, express hesitance, and be polite. Gender differences also appear in nonverbal communication. Women use less personal space, touch less in impersonal settings (but are touched more), and smile more, even when they are not necessarily happy (Wood 2013). Researchers even find that men and women write email in a different style, women writing less opinionated email than men and using it to maintain rapport and intimacy (Colley and Todd 2002; Sussman and Tyson 2000). Finally, you might have to change many of your attitudes because men and women differ significantly on many, if not most, social and political issues (see ▲ Figure 11.1).

If you are a woman and became a man, perhaps the change would be worth it. You would probably see your income increase (especially if you became a White man). You would have more power in virtually every social setting. You would be far more likely to head a major corporation, run your own business, or be elected to a political office—again, assuming that you are White. Would it be worth it? As a man, you would be far more likely to die a violent death and would probably not live as long as a woman (National Center for Health Statistics 2013).

If you are a man who became a woman, your income would most likely drop significantly. More than fifty years after passage of the Equal Pay Act in 1963, men still earn 22 percent more than women, even comparing those working year-round and full-time (DeNavas-Walt and Proctor 2014). You would probably become resentful of a number of things because poll data indicate that women are more resentful than men about things such as the amount of money available for them to live on, the amount of help they get from their mates around the house, how much men share child care, and how they look. Women also report being more fearful on the

in this chapter, you will learn to:

- Understand gender as a social construction
- Explain the process of gender socialization
- Identify different components of gender stratification
- Compare and contrast different theories of gender stratification
- Relate gender inequality in the United States to that in other nations
- Evaluate the different components of change with regard to gender

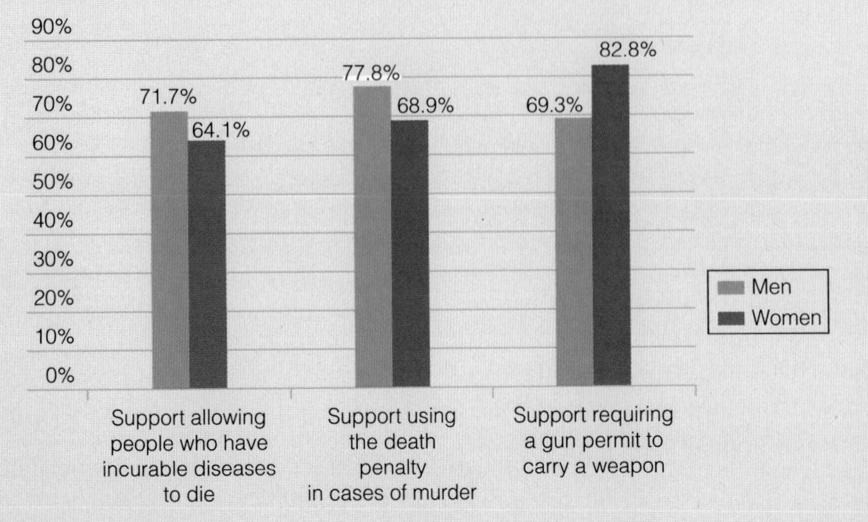

▲ **Figure 11.1** Hot-Button Issues: The Gender Gap in Attitudes When people use the term *gender gap*, they are often referring to pay differences between women and men. As the data here show, there is also a significant gender gap on some of the important issues of the day.

Source: National Opinion Research Center, General Society Survey. Chicago, IL: University of Chicago. **www.norc.org**

streets than men. Women are, however, more satisfied than men with their role as parents and with their friendships outside of marriage.

For both women and men, there are benefits, costs, and consequences stemming from the social definitions associated with gender. As you imagined this experiment, you may have had difficulty trying to picture the essential change in your biological identity: Is this the most significant part of being a man or woman? Nature determines whether you are male or female but society gives significance to this distinction. Sociologists see gender as a social fact, because who we become as men and women is largely shaped by cultural and social expectations.

The Social Construction of Gender

From the moment of birth, gender expectations influence how boys and girls are treated. Now that it is possible to identify the sex of a child in the womb, gender expectations may begin even before birth. Parents and grandparents might select pink clothes and dolls for baby girls, sports clothing and brighter colors for boys. Even if they try to do otherwise, it will be difficult because baby products are so typed by gender. Much research shows how parents and others continue to treat children in stereotypical ways. Girls may be expected to cuddle and be sweet, whereas boys are handled more roughly and given greater independence.

SEE for YOURSELF

Changing Your Gender

Try an experiment based on the example of changing gender that opens this chapter.

1. First, list everything you think you would have to do to change your behavior if you were the opposite gender. Separate your list items according to whether they are related to such factors as appearance, attitude, or behavior.
2. Second, for twenty-four hours, try your best to change any of these things that you are willing to do. Record how others react to you during this period and how the change makes you feel.
3. When your experiment is over, write a report on what your brief experiment tells you about how gender identities are (or are not) supported through social interaction.

Defining Sex and Gender

Sociologists use the terms *sex* and *gender* to distinguish biological sex identity from learned gender roles. **Sex** refers to biological identity, being male or female. For sociologists, the more significant concept is **gender**—the socially learned expectations, identities, and behaviors associated with members of each sex. This distinction emphasizes that behavior associated with gender is culturally learned. Gender is a "system of social practices" (Ridgeway 2011: 9) that creates categories of people—men and women—who are defined in relationship to each other on unequal terms.

The definitions that surround these categories stem from culture—made apparent especially by looking at other cultures. Across different cultures, gender expectations associated with men and women vary considerably. In Western industrialized societies, people tend to think of men and women (and masculinity and femininity) in dichotomous terms, even defined as "opposite sexes." The views from other cultures challenge this assumption. Historically, the *berdaches* (pronounced berdash) in Navajo society were anatomically normal men defined as a third gender between male and female. Berdaches, considered ordinary men, married other men who were not berdaches. Neither the berdaches nor the men they married were considered homosexuals, as they would be considered in other places (Nanda 1998; Lorber 1994).

There are also substantial differences in the construction of gender across social classes and within subcultures in a given culture. Within the United States, there is considerable variation in the experiences of gender among different racial and ethnic groups (Andersen and Collins 2013; Baca Zinn et al. 2011). Differences within a given gender can be greater than differences between men and women. That is, the variation on a given trait, such as aggression or competitiveness, can be as great within a given gender group as the difference across genders. Thus, some women are more aggressive than some men, and some men are less competitive than some women (see ▲ Figure 11.2).

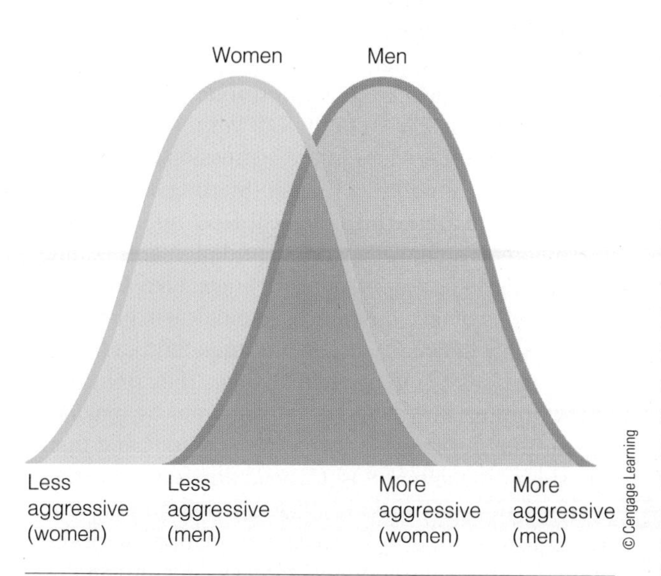

▲ **Figure 11.2** Gender Differences: Aggression
Even when men and women as a whole tend to differ on a given trait, within-gender differences can be just as great as across-gender differences. Some men, for example, are less aggressive than some women.

Sex Differences: Nature or Nurture?

Is gender a matter of nature or nurture? Obviously, there are some biological differences between women and men, but looking at gender sociologically quickly reveals the extraordinary power of social and cultural influences on things often popularly seen as biologically fixed. Let's first examine some basic biological facts about sex and gender.

A person's sex identity is established at the moment of conception. The mother contributes an X chromosome to the embryo; the father, an X or Y. The combination of two X chromosomes makes a female; the combination of an X and a Y makes a male. Under normal conditions, chemical events directed by genes on the sex-linked chromosomes lead to the formation of male or female genitalia.

Irregularities in the process of chromosome formation or fetal differentiation can produce a person with mixed biological sex characteristics—a condition known as **intersexed** (previously referred to as *hermaphroditism*). For example, an intersexed infant may be born with ovaries or testes but with ambiguous or mixed genitals, or may be born a chromosomal male but have an incomplete penis and no urinary canal.

Case studies of intersexed people reveal the extraordinary influence of social factors in shaping a person's identity (Preves 2003). Parents of intersexed children are usually advised to have their child's genitals surgically assigned to either male or female and also to give the child a new name, a different hairstyle, and new clothes—all intended to provide the child with the social signals judged appropriate to a single gender identity. One physician who has worked on such cases tells parents that they "need to go home and do their job as child rearers with it very clear whether it's a boy or a girl" (Kessler 1990: 9).

Physical differences between the sexes do, of course, exist. In addition to differences in anatomy, boys at birth tend to be slightly longer and weigh more than girls. As adults, men tend to have a lower resting heart rate, higher blood pressure, and higher muscle mass and muscle density. These physical differences contribute to the tendency for men to be physically stronger than women, but this can be altered, depending on level of physical activity. The public now routinely sees displays of women's athleticism. Women can achieve a high degree of muscle mass and muscle density through bodybuilding and can win over men in activities that require high levels of endurance, such as the four women who have won the Iditarod—the Alaskan dog sled race considered one of the most grueling competitions in the world. In other words, until men and women really compete equally in activities from which women have historically been excluded,

a sociological eye on the media

Women in the Media: Where Are Women's Voices?

Who reports the news? Even with the increased presence of women as news reporters, anchors, editors, and writers, the news media continue to be dominated by men—White men at that. Women are the vast majority (74 percent) of those graduating from college with majors in journalism and mass communications (Becker et al. 2010), but their voices and words do not narrate the news of the nation. Some facts from recent research include:

- The number of women staffers in newsrooms is largely unchanged from 1999 at 36 percent;
- Men are quoted three and a half times as often as women, a finding based on a detailed three-month analysis of the front page of *The New York Times*;
- Men outnumber women four to one on the editorial pages of the most prestigious national newspapers;
- Men write 82 percent of film reviews;
- Men vastly outnumber women as experts and guests on TV talk shows.

Women are most likely to report on lifestyle, education, culture, and health, not politics, technology, or criminal justice. Women are most underrepresented as reporters of world affairs. Even when the subject matter is of deep concern to women—abortion and birth control—80 percent of those reporting on these issues are men (4th Estate 2014).

Try observing these patterns yourself: Who writes the bylines in Internet news you read? Who is cited as an expert? Who covers what issues? Although sociologists would argue that the mere presence of women does not necessarily change the presentation of ideas, do you think that it matters that men are the ones most likely to narrate our nation's news?

Gender Gap in the News Media

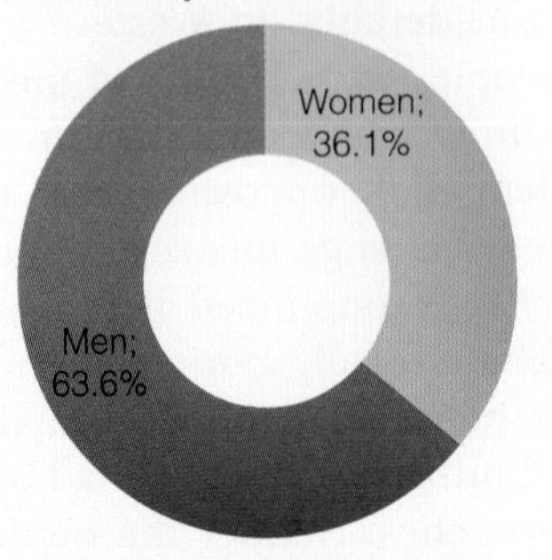

Note: Percentage of women and men with bylines and on-camera appearances in major TV networks, newspapers, news wires, and Internet news services, 2013

Source: Women's Media Center. 2014. *The Status of Women in the U.S. Media 2014*. New York: Women's Media Center. **www.womensmediacenter.com**

we may not know the true extent of physical differences between women and men.

Biological determinism refers to explanations that attribute complex social phenomena to physical characteristics. The argument that men are more aggressive because of hormonal differences (in particular, the presence of testosterone) is a biologically determinist argument. Despite popular belief, studies find only a modest correlation between aggressive behavior and testosterone levels. Furthermore, changes in testosterone levels do not predict changes in men's aggression (such as by "chemical castration," the administration of drugs that eliminate the production or circulation of testosterone). What's more, there are minimal differences in the levels of sex hormones between girls and boys during early childhood, yet researchers find considerable differences in the aggression exhibited by boys and girls as children (Fausto-Sterling 2000, 1992).

Blend Images/Pete Saloutos/The Agency Collection/Getty Images

Title IX, passed in 1972, has opened up athletic opportunities to women that were not available to earlier generations.

Arguments based on biological determinism assume that differences between women and men are "natural" and, presumably, resistant to change. Like biological explanations of race differences, biological explanations of inequality between women and men tend to flourish during periods of rapid social change. They protect the status quo (existing social arrangements) by making it appear that the status of women is "natural" and therefore should remain as it is. If social differences between women and men were biologically determined, there would also be no variation in gender relations across cultures, but extensive differences are well documented.

In sum, we would not exist without our biological makeup, but we would not be who we are without society and culture. As sociologist Cecilia Ridgeway puts it, gender is a "substantial, socially elaborated edifice constructed on a modest biological foundation" (2011: 9).

Culture defines certain behaviors as appropriate (or not) for women and for men; these are learned throughout life, beginning with the earliest practices by which people are raised. Understanding the process of *gender socialization* is then a key part of understanding the formation of gender as a social and cultural phenomenon.

Gender Socialization

As we saw in Chapter 4, socialization is the process by which social expectations are taught and learned. Through **gender socialization**, men and women learn the expectations and identities associated with gender in society. The rules of gender extend to all aspects of society and daily life. Gender socialization affects the self-concepts of women and men, their social and political attitudes, their perceptions about other people, and their feelings about relationships with others. Although not everyone is perfectly socialized to conform to gender expectations, socialization is a powerful force directing the behavior of men and women in gender-typical ways.

Even people who set out to challenge traditional expectations often find themselves yielding to the powerful influence of socialization. Women who consciously reject traditional women's roles may still find themselves inclined to act as hostess or secretary in group settings. Similarly, men may accept equal responsibility for housework, yet they fail to notice when the refrigerator is empty or a child needs a bath—household needs they have been trained to let someone else notice (DeVault 1991). Gender expectations are so pervasive that it is also difficult to change them on an individual basis. If you doubt this, try buying clothing or toys for a young child without purchasing something that is gender-typed, or talk to parents who have tried to raise their children without conforming to gender stereotypes and see what they report about the influence of such things as children's peers and the media.

The Formation of Gender Identity

One result of gender socialization is the formation of **gender identity**, which is one's definition of oneself as a woman or man. Gender identity is basic to our self-concept, shaping our expectations for ourselves, our abilities and interests, and how we interact with others. Gender identity shapes not only how we think about ourselves and others but also influences numerous behaviors, including the likelihood of drug and alcohol abuse, violent behavior, depression, or even how aggressive you are in driving (Andersen 2015).

One area in which gender identity has an especially strong effect is in how people feel about their appearance. Studies find strong effects of gender identity on body image. Concern with body image begins mostly during adolescence. Studies of young children (that is, preschool age) find no gender differences in how boys and girls feel about their bodies (Hendy et al. 2001), but clear differences emerge by early adolescence. At this age, girls report comparing their bodies to others of their sex more often than boys do. By early adolescence, girls report more negativity about their body image than do boys. This type of thinking among girls is related to lower self-esteem (Jones 2001; Polce-Lynch et al. 2001). Among college students, women also are more dissatisfied with their appearance than are men (Hoyt and Kogan 2001). Idealized images of women's bodies in the media, as well as peer pressures, have a huge impact on young girls' and women's gender identity and feelings about their appearance.

Sociologist Debra Gimlin argues that bodies are "the surface on which prevailing rules of a culture are written" (Gimlin 2002: 6). You see this especially with regard to gender. Men and women alike practice elaborate rituals to achieve particular gender ideals, ideals that are established by the dominant culture.

Understanding gender socialization helps you see that gender, like race, is a social construction. We "do" gender—as you will see in the section near the end of this chapter on theories of gender. Understanding gender as a social construction means thinking about the many ways that gender is produced through social interaction instead of seeing it as a fixed attribute of individuals (Connell 2009).

The experiences of transgender people make this point even more obvious. **Transgender** people are those who live as a gender different from that to which they were assigned at birth (Schilt 2011). Transgender individuals experience great pressure to fit within the usual expectations. Because they do not necessarily display the expected routines of gender, they may be ridiculed, shunned, discriminated against, or even violently assaulted. Their experiences show how powerful the social norms are to conform to gender expectations (Westbrook and Schilt 2014; Connell 2010, 2012).

Similarly, those who undergo sex changes as adults report enormous pressure from others to be one sex or the other. Managing such identities can be stressful, largely because of expectations that others have about what are appropriate categories of gender identity. Anyone who crosses these or in any way appears to be different from dominant expectations is frequently subject to exclusion and ridicule, showing just how strong gender expectations are (Stryker and Whittle 2006).

THINKING Sociologically

Look at the products sold to men and women used as part of their daily grooming? How are they packaged? What color are they? What are their names? How do these artifacts of everyday life reflect *norms* about gender?

Sources of Gender Socialization

As with other forms of socialization, there are different agents of gender socialization: family, peers, children's play, schooling, religious training, mass media, and popular culture, to name a few. Gender socialization is reinforced whenever gender-linked behaviors receive approval or disapproval from these multiple influences.

Parents are one of the most important sources of gender socialization. Parents may discourage children from playing with toys that are identified with the other sex, especially when boys play with toys meant for girls. Research finds that parents are more tolerant of girls not conforming to gender roles than they are for boys (Kane 2012). Fathers, especially, discourage sons from violating gender norms (Martin 2005). Although fathers are now more involved than in the past in children's care, they are less likely to provide basic care and more likely to be involved with discipline (LaFlamme et al. 2002).

Expectations about gender are changing, although researchers suggest that the cultural expectations about gender may have changed more than people's behavior. Mothers and fathers now report that fathers should be equally involved in child rearing, but the reality is different. Mothers still spend more time in child-related activities and have more responsibility for children. Furthermore, the gap that mothers perceive between fathers' ideal and actual involvement in child rearing is a significant source of mothers' stress (Milkie et al. 2002).

David Frazier/The Image Works

Changes in gender roles have involved more men in parenting.

Gender socialization patterns also vary within different racial-ethnic families. Latinas, as an example, have generally been thought to be more traditional in their gender roles, although this varies by generation and by the experiences of family members in the labor force. Within families, young women and men learn to formulate identities that stem from their gender, racial, and ethnic expectations.

Peers strongly influence gender socialization—sometimes more so than one's immediate family. Peer relationships shape children's patterns of social interaction. Young people's play also shapes analytical skills, values, and attitudes. Studies find that boys and girls often organize their play in ways that reinforce not only gender but also race and age norms (Moore 2001). Peer relationships often reinforce the gender norms of the culture—norms that are typically even more strictly applied to boys than to girls. Thus boys who engage in behavior that is associated with girls are likely to be ridiculed by friends—more so than are girls who play or act like boys (Sandnabba and Ahlberg 1999). In this way, homophobic attitudes, routinely expressed among peers, reinforce dominant attitudes about what it means to "be a man" (Pascoe 2011). Although girls may be called "tomboys," boys who are called "sissies" are more harshly judged. Note, though, that tomboy behavior among girls beyond a certain age may result in the girl being labeled a "dyke."

THINKING Sociologically

Is your pet gendered? People typically think of the *social construction of gender* in the context of human relations. What about pets? Almost two-thirds of all U.S. households have at least one pet. A recent study of dog owners found that people also gender their dogs, selecting dogs that reflect the owner's gender identity and describing them in gendered terms. Female pets have names with more syllables and that are more likely to have a diminutive ending. You can find lots of advice how to appropriately gender your pet on the Internet. Try to observe this yourself. Apparently, gender norms pervade human relationships with animals as well as with each other!

Sources: Ramirez, Michael. 2006. "'My Dog's Just Like Me': Dog Ownership as a Gender Display." *Symbolic Interaction* 29 (3): 373–391; Abel, Ernest L., and Michael L. Kruger. 2007. "Gender Related Naming Practices: Similarities and Differences between People and Their Dogs." *Sex Roles: A Journal of Research* 57 (1–2): 15–19.

Tony Freeman/PhotoEdit
Golden Pixels LLC/Alamy

Even with social changes in gender roles, boys and girls tend to engage in play activities deemed appropriate for their gender.

Children's play is another source of gender socialization. You might think about what you played as a child and how this influenced your gender roles. Typically, though not in every case, boys are encouraged to play outside more; girls, inside. Boys' toys are more machinelike and frequently promote the development of militaristic values; they tend to encourage aggression, violence, and the stereotyping of enemies—values rarely associated with girls' toys. Children's books in schools also communicate gender expectations. Even with publishers' guidelines that discourage stereotyping, textbooks still depict men as aggressive, argumentative, and competitive. Men and boys are also more likely to be featured in children's books, although interestingly, systematic analysis of children's books shows that fathers are not very present and, when they do appear, they are most often shown as ineffectual (Anderson and Hamilton 2005).

SEE for YOURSELF

Try to purchase a toy for a young child that is not gender-typed. What could you buy? What could you not buy? What does your experiment teach you about gender role socialization? If you take children with you, note what toys they want. What does this tell you about the effectiveness of *gender socialization?*

Schools are particularly strong influences on gender socialization because of the amount of time children spend in them. Teachers often have different expectations for boys and girls. Studies find, for example, that teachers hold gender stereotypes that women are not as capable in math as men (Riegle-Crumb and Humphries 2012). We know from other research that when such stereotypes are present, student performance, such as on standardized tests, is negatively affected by the presence of what is called *stereotype threat* (Steele 2010). Earlier studies have also shown that when teachers respond more to boys in school, even if negatively, they heighten boys' sense of importance (American Association of University Women 1998; Sadker and Sadker 1994). Gender inequality is pervasive in schools and at all levels. Even in college, course-taking patterns, selection of majors, and teachers' interaction with students are shaped by gender (Mullen 2010).

Religion is an often overlooked but significant source of gender socialization. The major Judeo-Christian religions in the United States place strong emphasis on gender differences, with explicit affirmation of the authority of men over women. In Orthodox Judaism, men offer a prayer to thank God for not having created them as a woman or a slave. The patriarchal language of most Western religions as well as the exclusion of women from positions of religious leadership in some faiths signify the lesser status of women in religious institutions. Any religion interpreted in a fundamentalist way can be oppressive to women. Indeed, the most devout believers of any faith tend to hold the most traditional views of women's and men's roles. The influence of religion on gender attitudes cannot be considered separately from other factors, however. For many, religious faith inspires a belief in egalitarian roles for women and men. Christian, Islamic, and Jewish women, along with other religious women, have often organized to resist fundamentalist and sexist practices (Messina-Dysert and Ruether 2014).

The *media* in their various forms (television, film, magazines, music, and so on) communicate strong—some would even say cartoonish—gender stereotypes. Despite some changes in recent years, television and films continue to depict highly stereotyped roles for women and men. In family films, prime-time television, and even children's TV shows, women and girls are shown more so than men and boys wearing sexy attire, with exposed skin, and with thin bodies, reinforcing gendered images of both women and men, even at an early age (Smith et al. 2013). Men are also seen as more formidable, stereotyped in strong, independent roles. Women are more likely now to be portrayed as employed outside of the home and in professional jobs, but it is still more usual to see women depicted as sex objects. In fact, the sexualization of women is so extensive in the media that the American Psychological Association has concluded that there is "massive exposure to portrayals that sexualize women and girls and teach girls that women are sexual objects" (American Psychological Association 2007: 5).

SEE for YOURSELF

Gender and Popular Celebrations

Try to buy a friend a birthday card that does not stereotype women or men. Alternatively, try to find a Father's Day or Mother's Day card without gender stereotypes. How do the gender images in cards overlap with stereotypes about aging, family, and images of beauty? After experiment, ask yourself how products promoted in the media affect ideas about gender.

Popular culture has increasingly sexualized even the youngest of girls.

Do people believe what they see on television? Research with children shows that they identify with the television characters they see. Both boys and girls rate the aggressive toys that they see on television commercials as highly desirable. They also judge them as more appropriate for boys' play, suggesting that something as seemingly innocent as a toy commercial reinforces attitudes about gender and violence (Klinger et al. 2001). Even with adults, researchers find that there is a link between viewing sexist images and having attitudes that support sexual aggression, antifeminism, and more traditional views of women (American Psychological Association 2007).

It is easy to observe the pervasiveness of gendered images by just watching the media with a critical eye. Women in advertisements are routinely shown in poses that would shock people if the characters were male. Consider how often women are displayed in ads dropping their pants, skirts, or bathrobe, or are shown squirming on beds. How often are men shown in such poses? Men are sometimes displayed as sex objects in advertising, but not nearly as often as women. The demeanor of women in advertising—on the ground, in the background, or looking dreamily into space—makes them appear subordinate and available to men.

The Price of Conformity

A high degree of conformity to stereotypical gender expectations takes its toll on both men and women. One of the major ways to see this is in the very high rate of violence against women—both in the United States and worldwide. Too frequently, men's power in society is manifested in physical and emotional violence. Violence takes many forms, including rape, sexual abuse, intimate partner violence, stalking, genital mutilation, and honor killings. Around the world, the United Nations is working to reduce violence against women, including some initiatives to help men examine cultural assumptions about masculinity that promote violence. Violence against women stems from the attitudes of power and control that gender expectations produce and that can lead *some* men to engage in violent behavior—toward women, as well as toward other men.

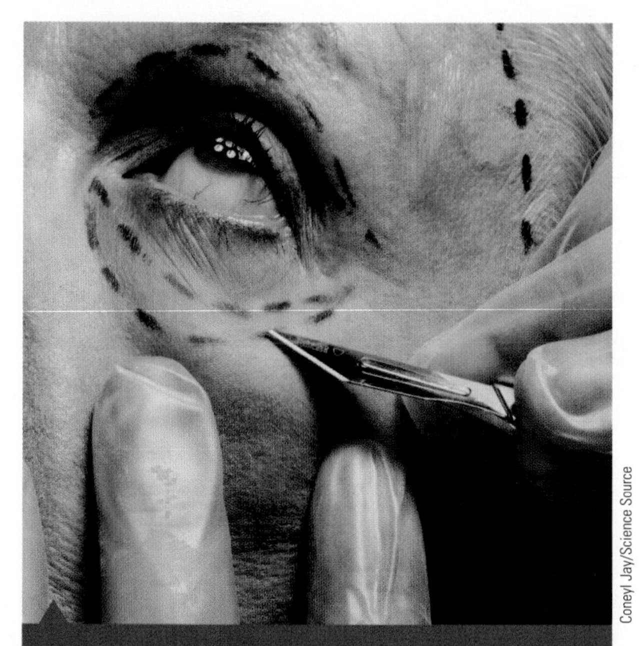

Coneyl Jay/Science Source

Cosmetic surgery is a rapidly growing, and highly profitable, industry. Some do it to try to eliminate signs of aging, others for aesthetic purposes. The pervasive influence of Western images means that some women in other nations pay for procedures to make them look more "Western."

◆ **Table 11.1 Facts about Eating Disorders**

One in five women struggle with an eating disorder or "disordered eating," referring to eating behaviors that are not classified as anorexia or bulimia but that are considered unhealthy.
Ninety percent of those with eating disorders are women between the ages of 12 and 25.
Of those with anorexia, 10 to 15 percent are men; men are less likely to seek treatment for an eating disorder than are women.
Eating disorders have the highest mortality rate of any mental illness.
Over half (58 percent) of women on college campuses report feeling pressure to lose weight; nearly half (44 percent) of them were of normal weight.
The average female fashion model is 5 foot, 11 inches tall and weighs 110 pounds—almost 50 pounds less than what is judged to be the ideal, healthy weight for someone this height.
The diet industry is worth an estimated $50 billion per year.

Sources: Sullivan, Patrick F. 1995. "Mortality in Anorexia Nervosa." *American Journal of Psychiatry*, 152 (July): 1073–1074; Substance Abuse and Mental Health Services Administration (SAMHSA), the Center for Mental Health Services (CMHS), U.S. Department of Health and Human Services; Malinauskas, Brenda, et al. 2006. "Dieting Practices, Weight Perceptions, and Body Composition: A Comparison of Normal Weight, Overweight, and Obese College Females." *Nutrition Journal* 5 (March 31): 5–11; Smolak, L., and M. Levine, eds. 1996. *The Developmental Psychopathology of Eating Disorders: Implications for Research, Prevention, and Treatment*. Hillsdale, NJ: Lawrence Erlbaum Associates Inc.; Spitzer, Brenda L., Katherine A. Henderson, and Marilyn T. Zivian. 1999. "A Comparison of Population and Media Body Sizes for American and Canadian Women." *Sex Roles* 700 (7/8): 545–565.

Violence by men is only one of the harms to women stemming from dominant gender norms. Adhering to gender expectations of thinness for women and strength for men is related to a host of negative health behaviors, including eating disorders, smoking, and for men, steroid abuse. The dominant culture promotes a narrow image of beauty for women—one that leads many women, especially young women, to be disturbed about their body image. Striving to be thin, millions of women engage in constant dieting, fearing they are fat even when they are well within or below healthy weight standards. Many develop eating disorders by purging themselves of food or cycling through various fad diets—behaviors that can have serious health consequences (see ◆ Table 11.1). Many young women develop a distorted image of themselves, thinking they are overweight when they may actually be dangerously thin.

Additionally, despite the known risks of smoking, young women who smoke think that it will keep them thin. Eating disorders can be related to a woman having a history of sexual abuse, but they also come from the promotion of thinness as an ideal beauty standard for women—a standard that can put girls' and women's health in jeopardy (Alexander et al. 2010).

Men also pay the price of conformity if they too thoroughly internalize gender expectations that say they must be independent, self-reliant, and unemotional. Although many men are more likely now than in the past to express intimate feelings, gender socialization discourages intimacy among them, affecting the quality of men's friendships. Conformity to traditional gender roles denies women access to power, influence, and achievement in the public world, but it also robs men of the more nurturing and other-oriented relationships that women have customarily had. Men's physical daring and risk-taking leaves them at greater risk of early death or injury from accidents. On many college campuses, for example, men showing bravado through heavy drinking can result in death or injury—and, as campus leaders know, heavy drinking is also strongly correlated with sexual assault against women on campus (Insight 2013). The strong undercurrent of violence in today's culture of masculinity can in many ways be attributed to the learned gender roles that put men and women at risk.

Gender Socialization and Homophobia

One of the primary ways that gender is produced during socialization is homophobia. **Homophobia** is the fear and hatred of homosexuals. Homophobia plays

doing sociological research

Eating Disorders: Gender, Race, and the Body

Research Question: "A culture of thinness," "the tyranny of slenderness," "the beauty myth": These are terms used to describe the obsession with weight and body image that permeates the dominant culture, especially for girls and women. Just glance at the covers of popular magazines for women and girls and you will very likely find article after article promoting new diet gimmicks, each bundled with a promise that you will lose pounds in a few days if you only have the proper discipline or use the right products. Moreover, the models on the covers of such magazines are likely to be thin, often dangerously so because being too thin causes serious health problems. Do these body ideals affect all women equally?

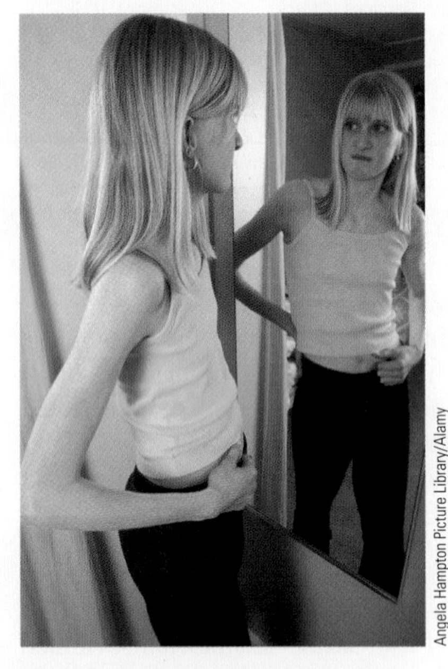

Angela Hampton Picture Library/Alamy

Too much conformity to gender roles can be harmful to your health. Such is the case of anorexic women who starve themselves attempting to meet cultural standards of thinness.

Research Method: Meg Lovejoy wanted to know if the drive for thinness is unique to White women and how gendered images of the body might differ for African American and White women in the United States. Her research is based on reviewing the existing research literature on eating disorders, which has generally concluded that, compared with White women, Black women are less likely to develop eating disorders.

Research Results: Black women are less likely than White women to engage in excessive dieting and are less fearful of fat, although they are more likely to be obese and experience compulsive overeating. White women, on the other hand, tend to be very dissatisfied with their body size and overall appearance, with an increasing number engaging in obsessive dieting. Black and White women also tend to distort their own weight in opposite directions: White women are more likely to overestimate their own weight (that is, saying they are fat when they are not); Black women are more likely to underestimate their weight (saying they are average when they are overweight by medical standards). Why?

Conclusions and Implications: Lovejoy concludes that you cannot understand eating disorders without knowing the different stigmas attached to Black and White women in society. She suggests that Black women develop alternative standards for valuing their appearance as a way of resisting mainstream, Eurocentric standards. Black women who do so are then less susceptible to the controlling and damaging influence of the institutions that promote the ideal of thinness as feminine beauty. On the other hand, the vulnerability that Black women experience in society can foster mental health problems that are manifested in overeating. Eating disorders for Black women can also stem from the traumas that result from racism.

Lovejoy and others who have examined this issue conclude that eating disorders must be understood in the context of social structures—gender, race, class, homophobia, and ethnicity—that affect all women, although in different ways. The cultural meanings associated with bodies differ for different groups in society but are deeply linked to our concepts of ourselves and the basic behaviors—like eating—that we otherwise think of as "natural."

Questions to Consider

1. Pay attention to music and visual images in popular culture. Ask yourself what cultural messages are being sent to different race and gender groups, for example, about appropriate appearance? How do these messages affect people's body image and self-esteem?
2. Lovejoy examines eating disorders in the context of gender, race, class, and ethnicity. What cultural meanings are broadcast with regard to age?
3. Is there a "culture of thinness" among your peers? If so, what impact do you think it has on people's self-concept? If not, are there other cultural meanings associated with weight among people in your social groups?

Source: Lovejoy, Meg. 2001. "Disturbances in the Social Body: Differences in Body Image and Eating Problems among African American and White Women." *Gender & Society* 15 (April): 239–261.

an important role in gender socialization because it encourages stricter conformity to traditional expectations, especially for men and young boys. Slurs directed against individuals who are gay encourage boys to act more masculine as a way of affirming for their peers that they are not gay (Pascoe 2011). Through such ridicule, homophobia discourages so-called feminine traits in men, such as caring, nurturing, empathy,

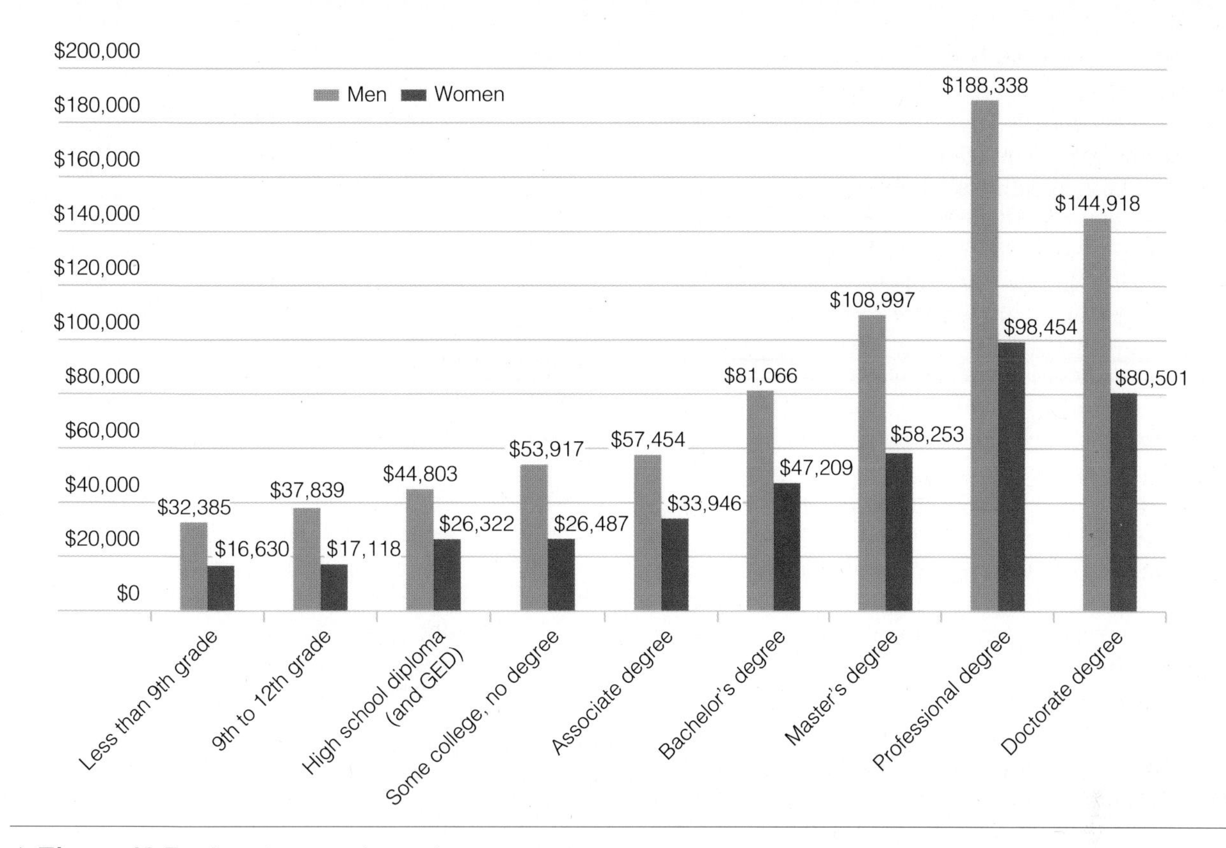

▲ **Figure 11.5** Education, Gender, and Income, 2013 Although the "economic return" on education is high for both men and women at every level of educational attainment, men's income, on average, exceeds that of women. According to this national data, how much education would the average woman have to get to exceed the income of men with a high school diploma? To exceed men with a college degree?

Data: U.S. Census Bureau. 2014. *Detailed Income Tabulations from the Current Population Survey: Selected Characteristics of People 15 Years and Over by Total Money Income in 2013, Work Experience in 2013, Race, Hispanic Origin, and Sex, Table PINC-01*. Washington, DC: U.S. Department of Commerce. **www.census.gov**

44 percent of women were in the labor force, so you can see the rather dramatic change that has occurred (U.S. Bureau of Labor Statistics 2014b). Since 1960, married women with children have also significantly increased their likelihood of employment. Seventy-three percent of mothers are now in the labor force, including more than half of mothers with infants. Current projections indicate that women's labor force participation will continue to rise, and men's will decline slightly (U.S. Bureau of Labor Statistics 2010).

This pattern of women being in the labor market has long been true for women of color but now also characterizes the experience of White women. The labor force participation rates of White women and women of color have, in fact, converged. More women in all racial groups are also now the sole supporters of their families.

Why do women continue to earn less than men, even when laws prohibiting gender discrimination have been in place for more than fifty years? The **Equal Pay Act of 1963** was the first federal law to require that men and women receive equal pay for equal work, an idea that is supported by the majority of Americans. The **Lily Ledbetter Fair Pay Act of 2009** extended the protections of the Civil Rights Bill of 1964. This act states that discrimination claims on the basis of sex, race, national origin, age, religion, and disability accrue with every paycheck, giving workers time to file claims of discrimination and eliminating earlier incentives for employers to cloak discrimination and then claim the employee filed their claims after the fact.

Wage discrimination, however, is rarely overt. Most employers do not even explicitly set out to pay women less than men. Despite good intentions and legislation, however, differences in men's and women's earnings persist. Research reveals four strong explanations for this: human capital theory, dual labor market theory, gender segregation, and overt discrimination, examined as follows.

Human Capital Theory. **Human capital theory** explains gender differences in wages as resulting from the individual characteristics that workers bring to jobs. Human capital theory assumes that the economic system is fair and competitive and that wage discrepancies reflect differences in the resources (or *human capital*) that individuals bring to their jobs. Factors such as age, prior experience, number of hours worked, marital status, and education are human capital variables. Human capital theory asserts that these characteristics will influence people's worth in the labor market. For example, higher job turnover rates or work records interrupted by child rearing and family responsibilities could negatively influence the earning power of women. There is a significant earnings penalty for women because of motherhood (Abendroth et al. 2014).

Education, age, and experience do influence earnings, but when you compare men and women who have the same level of education, previous experience, and number of hours worked per week, women still earn less than men (see Figure 11.5). Although human capital theory explains some of the difference between men's and women's earnings, it does not explain it all. Sociologists look to other factors to complete the explanation of wage inequality.

The Dual Labor Market. **Dual labor market theory** contends that women and men earn different amounts because they tend to work in different segments of the labor market. The dual labor market reflects the devaluation of women's work because women are most concentrated in low-wage jobs. Although it is hard to untangle cause and effect in the relationship between the devaluation of women's work and low wages in certain jobs, once such an earnings structure is established, it is difficult to change. Although equal pay for equal work may hold in principle, it applies to relatively few people because most men and women are not engaged in equal work.

According to dual labor market theory, the labor market is organized in two different sectors: the *primary market* and the *secondary market.* In the primary labor market, jobs are relatively stable, wages are good, opportunities for advancement exist, fringe benefits are likely, and workers are afforded due process. Working for a major corporation in a management job is an example of this. Jobs in the primary labor market are usually in large organizations where there is greater stability, steady profits, benefits for workers, better wages, and a rational system of management. In contrast, the secondary labor market is characterized by high job turnover, low wages, short or nonexistent promotion ladders, few benefits, poor working conditions, arbitrary work rules, and capricious supervision. Many of the jobs students take—such as waiting tables, bartending, or cooking and serving fast food—fall into the secondary labor market. For students, however, these jobs are usually short term, not lifelong.

Within the primary labor market, there are two tiers. The first consists of high-status professional and managerial jobs with potential for upward mobility, room for creativity and initiative, and more autonomy. The second tier comprises working-class jobs, including clerical work, skilled, and semiskilled blue-collar work. Women and minorities in the primary labor market tend to be in the second tier. Although these jobs may be more secure than jobs in the secondary labor market, they are more vulnerable and do not have as much mobility, pay, prestige, or autonomy as jobs in the first tier of the primary labor market.

There is, in addition, an informal sector of the market where there is even greater wage inequality, no benefits, and little, if any, oversight of employment practices. Individuals may hire such workers as private service workers or under-the-table workers who perform a service for a fee (painting, babysitting, car repairs, and any number of services). Although there are no formal data on the informal sector because much of it tends to be in an underground economy, women and minorities form a large segment of this market activity, particularly given their role in care work (Duffy 2011).

Gender Segregation. Dual labor market theory explains wage inequality as a function of the structure of the labor market, not the individual characteristics of workers as suggested by human capital theory. Because of the dual labor market, men and women tend to work in different occupations and, when working in the same occupation, in different jobs. This is referred to as **gender segregation**, a pattern in which different

AP Images/Lincoln Journal Star, Robert Becker

Data show that occupations where women of color predominate also tend to have the lowest wages.

groups of workers are separated into occupational categories based on gender. There is a direct association between the number of women in given occupational categories and the wages paid in those jobs. In other words, the greater the proportion of women in a given occupation, the lower the pay (U.S. Bureau of Labor Statistics 2014b). Segregation in the labor market can also be based on factors such as race, class, age, or any combination thereof.

Despite several decades of legislation prohibiting discrimination against women in the workplace, most women and men still work in gender-segregated occupations. That is, the majority of women work in occupations where most of the other workers are women, and the majority of men work mostly with men. Women also tend to be concentrated in a smaller range of occupations than men. To this day, more than half of all employed women work as clerical workers and sales clerks or in service occupations such as food service workers, maids, health service workers, hairdressers, and child-care workers. Men are dispersed over a much broader array of occupations. Women make up 81 percent of elementary and middle school teachers, 89 percent of secretaries, 90 percent of bookkeepers, 90 percent of registered nurses, and 95 percent of child-care workers—stark evidence of the persistence of gender segregation in the labor force (U.S. Bureau of Labor Statistics 2014b).

Gender segregation also occurs within occupations. Women usually work in different jobs from men, but when they work within the same occupation, they are segregated into particular fields or job types. For example, in sales work, women tend to do noncommissioned sales or to sell products that are of less value than those men sell. Even in fast-food restaurants, where you would expect wages to be routinized, women tend to work in those establishments with lower wages, fewer benefits, and less scheduling control (Haley-Lock and Ewert 2011).

Immigrant women are especially prone to being niched into highly segregated areas of the labor market. Many constraints shape the experience of immigrant women, including their legal status. Immigrant Latinas, for example, routinely find themselves in low-wage jobs regardless of their prior level of education, skill, or experience (Flippen 2014).

Overt Discrimination. A fourth explanation of the gender wage gap is discrimination. **Discrimination** refers to practices that single out some groups for different and unequal treatment. Despite the progress of recent years, overt discrimination continues to afflict women in the workplace. Some argue that men (White men in particular) have an incentive to preserve their advantages in the labor market. They do so by establishing rules that distribute rewards unequally. Women pose a threat to traditional White male privileges, and men may organize to preserve their own power and advantage (Reskin 1988). Historically, White men used labor unions to exclude women and racial minorities from well-paying, unionized jobs, usually in the blue-collar trades. A more contemporary example is seen in the efforts of some groups to dilute legislation that has been developed to assist women and racial–ethnic minorities. These efforts can be seen as an attempt to preserve group power.

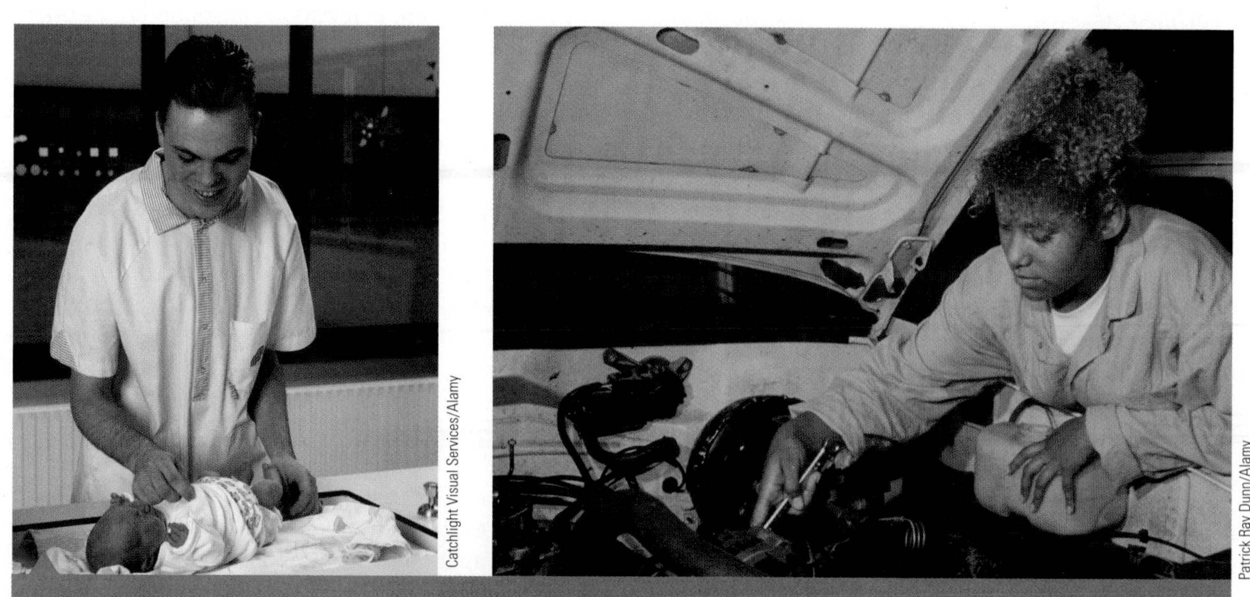
Catchlight Visual Services/Alamy
Patrick Ray Dunn/Alamy

Because gender segregation is so pervasive in the workplace, people may still be surprised when they see women and men in nontraditional occupations.

DEBUNKING Society's Myths

Myth: Black women are taking a lot of jobs away from White men.
Sociological Perspective: Sociological research finds no evidence of this claim. Quite the contrary, women of color work in gender- and race-segregated jobs and only rarely in occupations where they compete with White men in the labor market (Branch 2011).

Another example of overt discrimination is the harassment that women experience at work, including *sexual harassment* and other means of intimidation. Sociologists see such behaviors as ways for men to protect their advantages in the labor force. No wonder that women who enter traditionally male-dominated professions suffer the most sexual harassment. The reverse seldom occurs for men employed in jobs historically filled by women. Although men can be victims of sexual harassment, this is rare. Sexual harassment is a mechanism for preserving men's advantage in the labor force—a device that also buttresses the belief that women are sexual objects for the pleasure of men.

Each of these explanations—human capital theory, dual labor market theory, gender segregation, and overt discrimination—contributes to an understanding of the continuing differences in pay between women and men. Wage inequality between men and women is clearly the result of multiple factors that together operate to place women at a systematic disadvantage in the workplace.

The Devaluation of Women's Work

Across the labor market, women tend to be concentrated in those jobs that are the most devalued, causing some to wonder if the fact that women hold these jobs leads to devaluation of the jobs. Why, for example, is pediatrics considered a less prestigious specialty than cardiology? Why are preschool and kindergarten teachers (98 percent of whom are women) paid less than airplane mechanics (98 percent of whom are men)? The association of preschool teaching with children and its identification as "women's work" lowers its prestige and economic value. Indeed, if measured by the wages attached to an occupation, child care is one of the least prestigious jobs in the nation—paying on average only $418 per week in 2013, which would come out to an income below the federal poverty line if you worked every week of the year. Male highway maintenance workers, roofers, and construction workers all make more (U.S. Bureau of Labor Statistics 2014b).

The representation of women in skilled blue-collar jobs has increased, but it is still a very small fraction (typically less than 2 percent) of those in skilled trades such as plumbers, electricians, and carpenters (U.S. Bureau of Labor Statistics 2014b). Likewise, very few men work in occupations historically considered to be women's work, such as nursing, elementary school teaching, and clerical work. Interestingly, men who work in occupations customarily thought of as women's work tend to be more upwardly mobile within these jobs than are women who enter fields traditionally reserved for men (Budig 2002; Williams 1992).

Gender segregation in the labor market is so prevalent that most jobs can easily be categorized as men's work or women's work. Occupational segregation reinforces the belief that there are significant differences between the sexes. Think of the characteristics of a soldier. Do you imagine someone who is compassionate, gentle, and demure? Similarly, imagine a secretary. Is this someone who is aggressive, independent, and stalwart? The association of each characteristic with a particular gender makes the occupation itself a gendered occupation.

DEBUNKING Society's Myths

Myth: Black and White women's wages are converging as more White women have entered the labor market and Black women have increased their educational attainment.
Sociological Perspective: Although the wage gap between Black and White women decreased in the 1970s as the result of opened opportunities for Black women, the wage gap between Black and White women has increased since the 1980s, especially among young women. Scholars explain this as the result of structural disadvantages that Black women face (Pettit and Ewert 2009).

For all women, perceptions of gender-appropriate behavior influence the likelihood of success within institutions. Even something as simple as wearing makeup is linked to women's success in professional jobs (Dellinger and Williams 1997). When men or women cross the boundaries established by occupational segregation, they are often considered to be gender deviants. Women who work in the skilled trades, for example, are routinely assumed by others to be lesbians, whether or not they are (Denissen and Saguy 2014). Men who are nurses may be stereotyped as effeminate or gay; women marines may be stereotyped as "butch." Social practices like these serve to reassert traditional gender identities, perhaps softening the challenge to traditionally male-dominated institutions that women's entry challenges (Williams 1995).

As a result, many men and women in nontraditional occupations feel pressured to assert gender-appropriate behavior. Men in jobs historically defined as women's work may feel compelled to emphasize their masculinity, or if they are gay, they may feel even

◆ **Table 11.2** **Is It True?***

	True	False
1. Men are more aggressive than women.		
2. Parents have the most influence on children's gender identities.		
3. Most women hold feminist values.		
4. Being a "stay-at-home" mom is the most satisfying lifestyle for women.		
5. In all racial–ethnic groups, women earn less on average than men.		
6. The wage gap between women and men has closed since the 1970s, largely as the result of women being more likely to enter the labor force.		
7. In terms of wages, middle-class women have most benefited from antidiscrimination policies.		

*The answers can be found later in this chapter.

more pressure to keep their sexual orientation secret. Such social disguises can make them seem unfriendly and distant, characteristics that can have a negative effect on performance evaluations. Lesbian, gay, and bisexual workers face the added stress of having to "manage" their identities at work. Whether or not they are comfortable disclosing their identities is related to a number of factors, including the organizational climate and policies related to LGBT employees (King et al. 2014).

Balancing Work and Family

As the participation of women in the labor force has increased, so have the demands of keeping up with work and home life. Research finds that young women and men now want a good balance between work and family life, but they also find that institutions are resistant to accommodating these ideals (Gerson 2010). Men are also more involved in housework and child care than has been true in the past, although the bulk of this work still falls to women—a phenomenon that has been labeled "the second shift" (Hochschild and Machung 1989).

The social speedup that comes from increased hours of employment for both men and women (but especially women), coupled with the demands of maintaining a household, are a source of considerable stress (Jacobs and Gerson 2004). Women continue to provide most of the labor that keeps households running—cleaning, cooking, running errands, driving children around, and managing household affairs. Although more men are engaged in housework and child care, a huge gender gap remains in the amount of such work women and men do. Women are also much more likely to be providing care, not just for children, but also for their older parents. The strains these demands produce have made the home seem more and more like work for many. A large number of women and men report that their days at both work and home are harried and that they find work to be the place where they find emotional gratification and social support. In this contest between home and work, simply finding time can be an enormous challenge (Hochschild 1997). It is not surprising then that women report stress about household finances and family responsibilities more than men (American Psychological Association 2008).

Theories of Gender

Why is there gender inequality? The answer to this question is important, not only because it makes us think about the experiences of women and men, but also because it guides attempts to address the persistence of gender injustice. The major theoretical frameworks in sociology provide some answers, but feminist scholars have developed new theories to address women's lives more directly (◆ Table 11.3).

Functionalist theory traditionally purported that men fill instrumental roles in society whereas women fill expressive roles. Feminist scholars criticize functionalism for interpreting gender as a fixed role and one that is functional for society. Although few contemporary functionalist theorists would argue that now, functionalism does emphasize gender socialization as the major impetus behind gender inequality. Choices women make because of their gender socialization can make them shy away from leadership roles in public institutions, thus also reproducing traditional gender arrangements (Sandberg 2013).

Conflict theorists, in contrast, see women as disadvantaged by power inequities between women and men that are built into the social structure. This includes economic inequality and women's disadvantages in political and social systems. Wage inequality, for example, is produced by men who hold power in

◆ **Table 11.3 Theorizing Gender: Sociological and Feminist Perspectives**

	Functionalism	Conflict Theory	Symbolic Interaction Theory	Doing Gender/ Queer Theory	Liberal Feminism	Radical Feminism	Multiracial Feminism
Gender Identity	Gender roles are learned through socialization.	Conflict theory focuses on social structures, not individual identities.	Identity is constructed through ongoing social interaction and "doing gender."	Identity is constructed as people "perform" expected gender displays.	Except for the differences created by gender socialization, women and men are really no different.	Gender identity is produced by the power of social institutions that are controlled by men.	Gender identity is manifested differently depending on other social factors, such as race, class, and age, among others.
Status of Women	Women's status stems from the social roles of women and men in the family.	Women's status stems from their position as a cheap (or free) source of labor and their relative lack of power compared to men.	Women's status stems from the enactment of gender in social interaction.	Women's status is produced through the constant reenactment of gendered expectations.	Liberal feminism advocates equal rights for women.	Men's power is the basis for women's subordination.	Women of color are caught in a matrix of domination based on the intersection of race, gender, and other social factors.
Status of Men	Men hold instrumental roles; women, expressive roles.	Men hold economic advantages in the labor market and hold power in social institutions.	Masculinity is a learned identity that is created and sustained through social interaction.	Men's status is a performance.	Men's status derives from traditions that are learned and can be changed.	Men use their power to maintain their advantage.	Men's status is not monolithic as there are significant differences in men's power depending on their location in the matrix of domination and the intersection of race and gender.
Social Change	Social change emerges when social institutions become dysfunctional.	Social change comes from transformation of economic institutions and change in power structures that advantage men.	Social change comes when men and women disrupt existing gender displays.	Performances can disrupt existing expectations that redefine gender.	Equal rights is the framework for change most likely to generate equality.	Only the elimination of patriarchy will produce women's liberation.	Social change must come from challenging race, gender, and other forms of oppression simultaneously and as intersecting social systems.

social institutions. Men also benefit as a group from the services that women's labor provides. Conflict theorists have been much more attuned to interactions of race, class, and gender inequality because they see all forms of inequality as stemming from power that dominant groups in society have.

Conflict theory also interprets women's inequality as stemming from capitalism. Influenced by the work of Karl Marx, some feminist scholars argue that women are oppressed because they have historically constituted a cheap supply of labor. Women provide a reserve supply of labor and are pulled into the labor market when underpaid workers are needed. Women also do much of the work that is essential to life, but not paid—that is housework and child care. Conflict theorists understand women's inequality as stemming from economic exploitation and the power that men hold in virtually all social institutions.

Feminist Theory and the Women's Movement

As you will recall, *symbolic interaction theory* focuses on the immediate realm of social interaction. Feminist sociologists have developed an adaptation of symbolic interaction to explain the social construction of gender. An approach known as **doing gender** interprets gender as something accomplished through the ongoing social interactions people have with one another (West and Fenstermaker 1995; West and Zimmerman 1987). Seen from this framework, people produce gender through the interaction they have with one another and through the interpretations they have of certain actions and appearances.

Doing gender sees gender not as a fixed attribute of people, but as constantly made up and reproduced through social interaction. When you act like a man or a woman, you are confirming gender and reproducing existing social order. People act in gender-appropriate ways, displaying gender to others. Others then interpret the gender display that they see and assign people to gender categories, thereby reinforcing the gender order. People can disrupt this taken-for-granted behavior though. From the view of doing gender, gender structures would change if large numbers of people behaved differently. This is one reason the theory has been criticized by those with a more macrosociological point of view. Critics argue that the doing gender perspective ignores the power differences and economic differences that exist based on gender, race, and class. In other words, it does not explain the structural basis of women's oppression (Collins et al. 1995).

Related to doing gender is queer theory, a theoretical perspective that has emerged from gay and lesbian studies. **Queer theory** challenges the idea that sex and gender are binary opposites—that is, either/or categories. Instead, queer theory sees that dichotomous sex and gender categories are enforced by the power of social institutions and those who control them. People who disrupt these categories, such as transgender people, show us how gender is determined through such enforcement practices (Jenness and Fenstermaker 2014; Westbrook and Schilt 2014). Briefly put, queer theory interprets society as forcing people into presumed gender and sexual identities and behaviors. We examine queer theory further in Chapter 12 on sexuality.

Doing gender and queer theory are forms of feminist theory. **Feminist theory** has emerged from the women's movement and refers to analyses that seek to understand the position of women in society for the explicit purpose of improving their position in it.

The feminist movement has fostered widespread changes in and has transformed how people understand women's and men's lives. Simply put, **feminism** refers to beliefs and action that seek a more just society for women. Feminism is not a single way of thinking, as feminists understand the position of women in society in different ways. Different forms of feminism that emerged in the women's movement also provide different theoretical ways of viewing women's status in society.

Liberal feminism emerged from a long tradition that began among British liberals in the nineteenth century. Liberal feminism emphasizes individual rights and equal opportunity as the basis for social justice and reform. From this perspective, inequality for women originates in past and present practices that pose barriers to women's advancement, such as laws that historically excluded women from certain areas of work. From a liberal feminist framework, *discrimination* (that is, the unequal treatment of women) is the major source of women's inequality. Removing discriminatory laws and outlawing discriminatory practices are primary ways to improve the status of women. Calls for equal rights for women are the hallmark of a liberal feminist perspective.

Liberal feminism has broad appeal because it is consistent with American values of equality before the law. More **radical feminists** think that liberal feminism is limited by assuming that social institutions are basically fair were women and men treated the same within them. Radical feminists argue that there cannot be justice for women as long as men hold power in social institutions. As one example, men control the laws that govern women's reproductive lives. In this sense, *patriarchy* (that is, the power of men) is the major source of women's oppression. Patriarchy is found at the institutional level where men have control and at the individual level where men's power is manifested in high rates of violence against women. Indeed, many radical

Is It True ? (Answers)

1. FALSE. Generalizations such as this ignore variation occurring within gender categories; moreover, *aggression* is a broad term that can have multiple meanings.
2. FALSE. There are numerous sources of gender socialization; even parents who try to raise their children not to conform too strictly to gender norms will find that peers, the media, schools, and other socialization agents all push people into the expected behaviors associated with gender (Kane 2012).
3. TRUE. Although many women do not use the label *feminist* to define themselves, surveys show that the majority of women agree with basic feminist principles. Self-identification as a feminist is most likely among well-educated, urban women (McCabe 2005).
4. FALSE. Surveys show that "stay-at-home" moms experience far more depression, worry, anxiety, and anger than employed women (Mendes et al. 2012); half of women say they would prefer to have a job outside the home (Saad 2012a).
5. TRUE. However, the gap in income is not as wide within some groups as it is in others. Among year-round, full-time workers, White women, for example, earn 76 percent of what White men earn, but Hispanic women earn 93 percent of what Hispanic men earn (because both have very low earnings on average). Black women earn 85 percent of Black men's earnings, and Asian women, 75 percent of Asian men's earnings. And White and Asian American women, on average, earn more than Black and Hispanic men (U.S. Census Bureau 2014b).
6. FALSE. The most significant reason for the decline in the wage gap between women and men is the decline in men's wages; a smaller portion of this closing gap is attributed to changes in women's wages (Mishel et al. 2012).
7. FALSE. Although all women do benefit from equal employment legislation, wage data indicate that the group whose wages have increased the most since the 1970s are women in the top 20 percent of earners. Middle- and working-class women have seen far lower gains, and poor women's wages have been relatively flat over this period of time (Mishel et al. 2012).

feminists see violence against women as mechanisms that men use to assert their power in society. Radical feminists think that change cannot come about through the existing system because men control and dominate that system.

Multiracial feminism has also opened new avenues of theory for guiding the study of gender and its relationship to race and class (Andersen and Collins 2013; Baca Zinn and Dill 1996; Collins 1990). Multiracial feminism evolves from studies pointing out that earlier forms of feminist thinking excluded women of color from analysis, which made it impossible for feminists to deliver theories that informed people about the experiences of all women. Multiracial feminism examines the interactive influence of gender, race, and class, showing how they together shape the experiences of all women and men.

From this perspective, gender is not a singular or uniform experience, but rather intersects with race and class in shaping the experience of women and men. Gender is thus manifested differently, depending on the particular location of a given person or group in a system shaped by gender, race, and class, along with other social identities, such as sexual orientation, ability/disability status, age, nationality, and so forth. Also known as *intersectional theory,* analyses that are situated in multiracial feminist thinking have opened up sociological theory to new ways of thinking that include the multiplicity of experiences that people have in a society as diverse as the United States.

Gender in Global Perspective

Increasingly, the economic condition of women and men in the United States is linked to the fortunes of people in other parts of the world. The growth of a global economy and the availability of a cheaper industrial labor force outside the United States mean that U.S. workers have become part of an international division of labor. U.S.-based multinational corporations looking around the world for less expensive labor often turn to the developing nations and find that the cheapest laborers are women or children. The global division of labor is thus acquiring a gendered component, with women workers, usually from the poorest countries, providing a cheap supply of labor for manufacturing products that are distributed in the richer industrial nations.

Worldwide, women work as much as or more than men. It is difficult to find a single place in the world where the workplace is not segregated by gender. On a global scale, women also do most of the work associated with home, children, and the elderly. Although women's paid labor has been increasing, their unpaid labor in virtually every part of the world exceeds that of men. The United Nations estimates that women's unpaid work (both in the home and in the community) is valued as at least $11 trillion (**www.un.org**). Generally speaking, women's status in the world

map 11.1

Viewing Society in Global Perspective: Where's the Best Place to Be a Woman?

This map is based on data measuring a number of factors, including access to education and health care, economic status, and representation in government. Of the 142 countries included, the United States ranks 20th.

Source: World Economic Forum. 2013. *The Global Gender Gap Report*. Geneva, Switzerland.

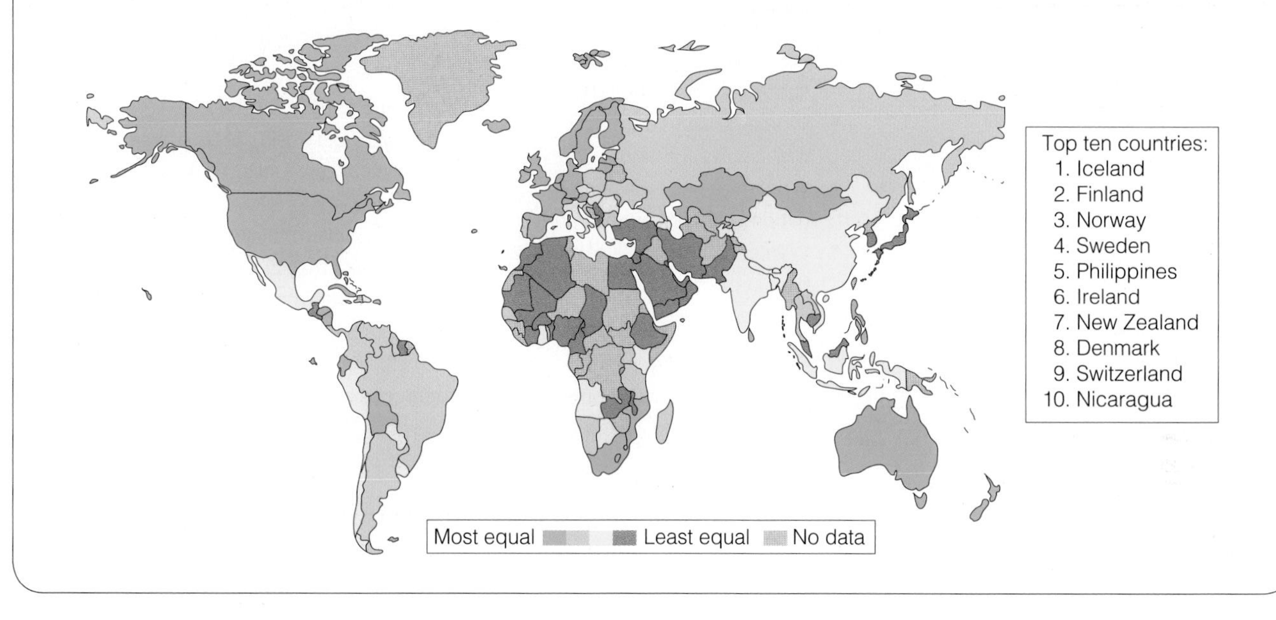

is improving, but slowly and unevenly in different nations (see Map 11.1).

Work is not the only measure by which the status of women throughout the world is inferior to that of men. Women are vastly underrepresented in national parliaments (or other forms of government) everywhere. As late as 2014, only 24 nations had women heads of state or government. Worldwide, women hold only 22 percent of all parliamentary seats. The United States, presumed to be the most democratic nation in the world, ranks 83rd of 145 nations in the representation of women in national parliaments (**www.ipu.org**).

The United Nations has also concluded that violence against women and girls is a global epidemic and one of the most pervasive violations of human rights (United Nations 2012a). Violence against women takes many forms, including rape, domestic violence, infanticide, incest, genital mutilation, and murder (including so-called honor killings, where a woman may be killed to uphold family honor if she has been raped or accused of adultery). Although violence is pervasive, some specific groups of women are more vulnerable than others—namely, minority groups, refugees, women with disabilities, elderly women, poor and migrant women, and women living in countries with armed conflict. Statistics on the extent of violence against women are hard to report with accuracy, both because of the secrecy that surrounds many forms of violence and because of differences in how different nations might report violence. Nonetheless, the United Nations estimates that between 20 and 50 percent of women worldwide have experienced violence from an intimate partner or family member.

As we saw in Chapter 9, many factors put women at risk of violence, including cultural norms, women's economic and social dependence on men, and political practices that either provide inadequate legal protection or provide explicit support for women's subordination.

Gender and Social Change

Few lives have not been touched by the transformations that have occurred in the wake of the feminist movement. The women's movement has opened work opportunities, generated laws that protect women's rights, spawned organizations that lobby for public policies on behalf of women, and changed public attitudes. Many young women and men now take for granted freedoms struggled for by earlier generations. These include access to birth control, equal opportunity legislation,

what would a sociologist say?

The End of Men?

In 2012, journalist Hanna Rosin published a book, *The End of Men: And the Rise of Women,* a book that was widely reviewed and earned Rosin appearances on numerous TV talk shows, news reports, and other media outlets. Her argument was fairly straightforward: that women are replacing men as the heads of households; that women's gender roles make them more adaptable to changes in the economy while men's roles make them more rigid and conforming to past ideals; and, that women are now dominating men in many workplaces and in colleges. For Rosin, "the end of men" signals a time when women are surpassing men in many areas of life, actually leaving men behind who cannot adapt to the new demands of a postindustrial, service-based economy.

Rosin's book points to some major social changes that are affecting both women and men:

- Women are now a majority in the nation's colleges and universities.
- Women's wages have been rising while men's are falling.
- Women can be found among the nation's highest income earners.
- Women find themselves more often than in the past as the major breadwinner in families.

Rosin's book, however, ignored other realities, including that women are still segregated in lower-wage occupations; that women major in fields that are not those that produce top earners; that women have higher rates of poverty than men; and that women remain unequal in every institution in society. Rosin's portrayal of men as resistant to change and women as flexible also rests on overgeneralized gender stereotypes.

Why did Rosin's book strike such a nerve for the public? What changes do you see in your environment that might lead people to think that women have it made and men are falling behind? What specific information would you need to assess the claims that Rosin is making, including information about diverse groups of women and men?

laws protecting against sexual harassment, increased athletic opportunities for women, more presence of women in political life, and greater access to child care, to name a few changes. These impressive changes occurred in a relatively short period of time.

Indeed, many believe that the gender revolution is over and that there is no further need for feminist change. Some say that the nation is *postfeminist,* a term that means different things to different people. For some, it simply means that the women's movement is over because feminism has outlived the need. For others, it means that second-wave feminism (that of the 1970s and 1980s) does not meet the needs of new generations of women. What has the feminist movement accomplished and what remains to be done?

In some regards, women have reached pinnacles of power and influence unprecedented in U.S. history. Women are highly visible as CEOs of major corporations, as Supreme Court justices, as presidential cabinet members, and as extraordinary and highly paid athletes—all of which would have been highly unusual not that many years ago. Women have also risen to positions of political influence, perhaps signaling a new era for women in the political realm. Women have also been especially evident in the new conservative movement, such as in the Tea Party and other organizations.

No doubt there has been substantial progress for women, but the tensions between progressive and conservative politics on women's issues indicate that women's rights are hardly a settled issue. Despite the greater visibility of women, most women still struggle with low wages, managing both work and family, or perhaps struggling alone to support a family. On the conservative side, people feel that the value of women as traditional homemakers is being eroded, perhaps even threatened by women's independence because women in that position need men's economic support. On the progressive side, feminists perceive constant threats to women's reproductive rights and think that the nation has not come far enough in protecting and supporting women's rights. There is currently a complex mix of progressive and conservative politics surrounding gender in the contemporary world. Yet, the changes the feminist movement has inspired have completely transformed many dimensions of women's and men's lives, especially apparent in changed public attitudes.

Contemporary Attitudes

Surveys of public opinion about women's and men's lives are good indicators of the changes we have witnessed as the result of the feminist movement. Now, only a small minority of people disapproves of women being employed while they have young children, and both women and men say it is not fair for men to be the sole decision maker in the household.

Young women and men have different ideals for their future lives than was true for earlier generations. Kathleen Gerson's research finds that most young adults (those she calls "children of the gender revolution") want a lifelong partner and shared responsibilities for work and family. Yet, Gerson also finds a strong gender divide in how men and women imagine what they would do as a backup plan if their ideals were not realized. Men, much more than women, think that if balancing work and family does not work out in their future, they would fall back on traditional arrangements with wives staying home and husbands working. Women disagree, understanding that they have to be self-reliant in the event that their ideals are not met (Gerson 2010).

Gerson also found that, although both men and women want to share work and family life, institutions have not adjusted to this reality. In other words, resistant institutions have not adjusted to the attitudinal changes and desires for new lifestyles that most women and men embrace.

Attitudinal changes are not, however, complete. Many people want more flexible gender arrangements, but traditional gender norms also remain. Beliefs that there are basic differences between women and men and support for traditional gender norms in many aspects of personal life still prevail. In short, change in gender has been uneven—revolutionary in some regards and stagnant in others (Prokos 2011; England 2010).

DEBUNKING Society's Myths

Myth: African American men are less likely than African American women to think that sexism *and* racism are equally important.

Sociological Perspective: African American men and women are equally likely to see racism and sexism as linked together and to see the need to address both (Harnois 2010).

Legislative Change

Attitudes are only one dimension of social change. Some of the most important changes have come from laws that now protect against discrimination—laws that have opened new doors, especially for professional, well-educated women. The *Equal Pay Act of 1963* was one of the first pieces of legislation requiring equal pay for equal work. Following this was the **Civil Rights Act of 1964**, adopted as the result of political pressure from the civil rights movement. You will be interested to learn that adding "sex" to this law actually protects women by accident. A southern congressman added the term, thinking that the idea of giving women these rights was such a joke that the addition of "sex" would prevent passage of the civil rights law.

The Civil Rights Act bans discrimination in hiring, promoting, and firing. This law also created the Equal Employment Opportunity Commission, an arm of the federal government that enforces laws prohibiting discrimination on the basis of race, color, national origin, religion, or sex.

The passage of the Civil Rights Act opened new opportunities to women in employment and education. This was further supported by **Title IX**, adopted as part of the educational amendments of 1972. Title IX forbids gender discrimination in any educational institution receiving federal funds.

Title IX also defines various forms of sexual violence (sexual harassment, sexual assault, and other forms of gender-based violence) as a violation of law. Title IX requires schools and colleges to be free of gender discrimination. The bill also requires that schools and colleges have established procedures for handling complaints of sex discrimination, sexual harassment, and sexual violence. With one in four college women reporting sexual assault while in college, this is clearly a huge problem. Title IX, however, also protects men and gender-nonconforming students, as well as faculty and staff on campus.

Adoption of Title IX, although not perfect in its implementation, has radically altered the opportunities available to female students and has laid the foundation for many coeducational programs that are now an ordinary part of college life. This law has been particularly effective in opening up athletics to women.

Has equality been achieved as the result of Title IX and other laws? In college sports, men still outnumber women athletes by more than two to one, and there is still more scholarship support for male athletes than for women. Title IX allows institutions to spend more money on male athletes if they outnumber women athletes, but the law stipulates that the number of male and female athletes should be closely proportional to their student body representation. Studies of student athletes show that although support for women's athletics has improved since implementation of Title IX, there is still a long way to go toward equality in women's sports. Some argue that Title IX has reduced opportunities

Janine Wiedel Photolibrary/Alamy

Young women are the group most likely to support feminist goals. Ending violence against women is a strong theme in contemporary feminism.

for men in sports. Proponents of maintaining strong enforcement of Title IX counter this by noting that budget reductions in higher education, not Title IX per se, are responsible for any reduction in athletic opportunities for men. Furthermore, they point out that men still greatly predominate in school sports.

In the workplace, a strong legal framework for gender equity is in place, yet equity has not been achieved. The United States has never, as an example, approved the **Equal Rights Amendment**, which would provide a constitutional principle that "equality of rights under the law shall not be denied or abridged by the United States or by any state on the basis of sex" (**www.equalrightsamendment.org**).

It is clear that the passage of antidiscrimination policies is essential, but not a guarantee that discrimination will end. One solution to the problem of gender inequality is to have more women in positions of public power. Is increasing the representation of women in existing situations enough? Without reforming sexism in institutions, change will be limited and may generate benefits only for groups who are already privileged. Feminists advocate restructuring social institutions to meet the needs of all groups, not just those who already have enough power and privilege to make social institutions work for them. The successes of the women's movement demonstrate that change is possible, but change comes only when people are vigilant about their needs.

Chapter Summary

How do sociologists distinguish sex and gender?

Sociologists use the term *sex* to refer to biological identity and *gender* to refer to the socially learned expectations associated with members of each sex. *Biological determinism* refers to explanations that attribute complex social phenomena entirely to physical or natural characteristics.

How is gender identity learned?

Gender socialization is the process by which gender expectations are learned. One result of socialization is the formation of *gender identity.* Overly conforming to gender roles has a number of negative consequences for both women and men, including eating disorders, violence, and poor self-concepts. *Homophobia* plays a role in gender socialization because it encourages strict conformity to gender expectations.

What is a gendered institution?

Gendered institutions are those where the entire institution is patterned by gender. Sociologists analyze gender both as a learned attribute and as an institutional structure.

What is gender stratification?

Gender stratification refers to the hierarchical distribution of social and economic resources according to gender. Most societies have some form of gender stratification, although they differ in the degree and kind. Gender stratification in the United States is obvious in the differences between men's and women's wages.

How do sociologists explain the continuing earnings gap between men and women?

There are multiple ways to explain the pay gap. *Human capital theory* explains wage differences as the result of individual differences between workers. *Dual labor market theory* refers to the tendency for the labor market to be organized in two sectors: the primary and secondary markets. *Gender segregation* persists and results in differential pay and value attached to men's and women's work. *Overt discrimination* against women is another way that men protect their privilege in the labor market.

Are men increasing their efforts in housework and child care?

Many men are now more engaged in housework and child care than was true in the past, although women still provide the vast majority of this labor. Balancing work and family has resulted in social speedup, making time a scarce resource for many women and men.

What is feminist theory?

Different theoretical perspectives help explain the status of women in society. *Functionalist theory* emphasizes how gender roles that differentiate women and men work to the benefit of society. *Conflict theory* interprets gender inequality as stemming from women's status as a supply of cheap labor and men's greater power in social institutions. *Feminist theory,* originating in the women's movement, refers to analyses that seek to understand the position of women in society for the explicit purpose of improving their position in it. *Doing gender* and *queer theory* interpret gender as accomplished in social interaction and enforced through powerful institutions. *Liberal feminism* is anchored in an equal rights framework. *Radical feminism* sees men's power as the primary force that locates women in disadvantaged positions in society. *Multiracial feminism*, or intersectional theory, emphasizes the linkage between gender, race, and class inequality.

When seen in global perspective, what can be observed about gender?

The economic condition of women and men in the United States is increasingly linked to the status of people in other parts of the world. Women provide much of the cheap labor for products made around the world. Worldwide, women work as much or more than men, though they own little of the world's property and are underrepresented in positions of world leadership.

What are the major social changes that have affected women and men in recent years?

Public attitudes about gender relations have changed dramatically in recent years. Women and men are now more egalitarian in their attitudes, although women still perceive high degrees of discrimination in the labor force. A legal framework is in place to protect against discrimination, but legal reform is not enough to create gender equality.

Key Terms

12

Ted Soqui/Ted Soqui Photography/Corbis

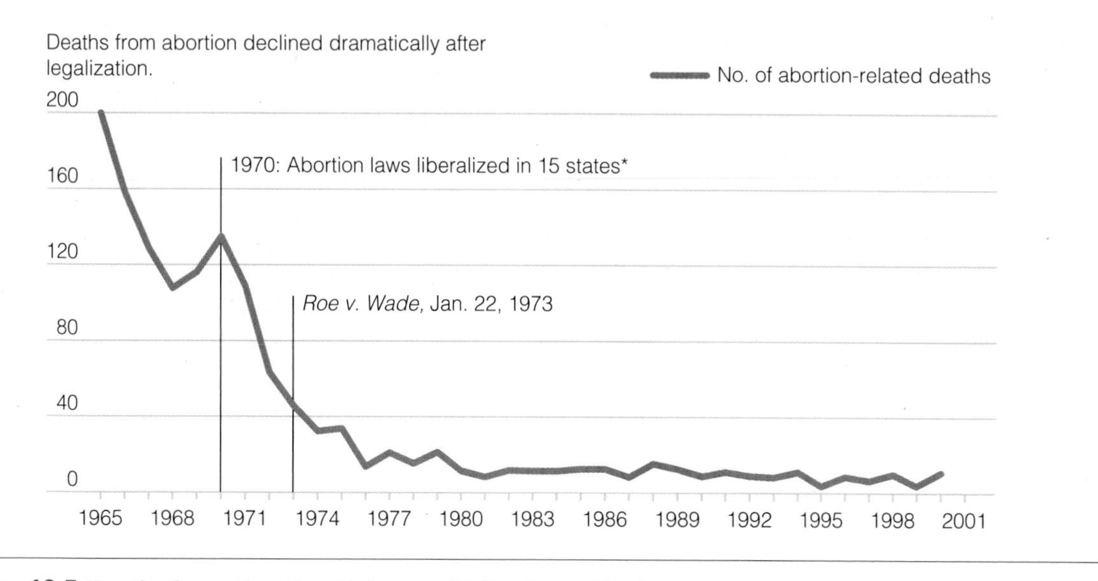

▲ **Figure 12.5** Deaths from Abortion: Before and After Roe v. Wade

*By the end of 1970, four states had repealed their antiabortion laws, and eleven states had reformed them.

Source: Boonstra, Heather, Rachel Benson Gold, Cory L. Richards, and Lawrence B. Finer. 2006. *Abortion in Women's Lives*, New York: Guttmacher Institute, 2006, p. 13.

rate has declined since 1980, from a rate of 25.1 per 1000 women in 1990 to 21.6 in 2010 (among women aged 15 to 44). As you can see in ▲ Figure 12.5, the number of deaths from illegal abortions plummeted in the years following the *Roe v. Wade* decision. Young women (aged 20 to 29) are the most likely group to get abortions; the second most likely group is women under 20. Black and Hispanic women are three times more likely to have abortions than White women; poor women are four times more likely to have abortions than other women. You may be surprised to learn that the majority of women having abortions (80 percent) have already had at least one birth (U.S. Census Bureau 2014b; Boonstra et al. 2006).

The abortion issue provides a good illustration of how sexuality has entered the political realm. Abortion rights activists and antiabortion activists hold very different views about sexuality and the roles of women. Antiabortion activists tend to believe that giving women control over their fertility breaks up the stable relationships in traditional families. They tend to view sex as something that is sacred, and they are disturbed by changes that make sex less restrictive. This belief has been fueled by the activism of the religious right, where strong passions against abortion have driven issues about sexual behavior directly into the political realm. Abortion rights activists, on the other hand, see women's control over reproduction as essential for women's independence. They also tend to see sex as an experience that develops intimacy and communication between people who love each other. The abortion debate can be interpreted as a struggle over the right to terminate a pregnancy as well as a battle over differing sexual values and a referendum on the nature of men's and women's relationships (Luker 1984).

Pornography and the Sexualization of Culture

Little social consensus has emerged about the acceptability and effects of pornography. Part of this debate is about defining what is obscene. The legal definition of obscenity is one that changes over time and in different political contexts. Public agitation over pornography has divided people into those who think it is solidly protected by the First Amendment, those who want it strictly controlled, those who think it should be banned for moral reasons, and those who think it must be banned because it harms women.

An ongoing question in research is whether viewing pornography promotes violence against women. This is a complex question and one that has mixed results in research studies, in part because the connection is difficult to observe directly. Verbal and physical abuse of women in pornography is shockingly common. One analysis of popular pornographic videos found that 88 percent of scenes in such videos include physical aggression against women—hitting, gagging, and slapping, for example. Moreover, these aggressive scenes depict women as either enjoying it or responding neutrally (Bridges et al. 2010).

Many experimental studies have found a correlation between men's viewing of pornography and sexually aggressive behavior. Studies also link consumption of pornography to a greater acceptance of rape myths.

Remember, however, that correlation is not cause. As it turns out, men who are sexually aggressive are more frequent consumers of pornography, meaning that the correlation between viewing pornography and violent attitudes is as much the result of men who are already sexually aggressive being more likely to consume pornography, as it is the use of pornography per se (Malamuth et al. 2011).

The connection between violence against women and consumption of pornography is not settled, but there is a different way to look at the issue—beyond the attitudes and behaviors of those who view it. Instead, think of pornography as an economic industry where, as in other industries, there are owners, bosses, and "employees." Seen in this way, the violence and dehumanization that happens to women in pornography is not just about images and their impact, but is about how women as sex workers are exploited (Voss 2012; Boyle 2011; Weitzer 2009). Women who have limited economic opportunities may turn to work in the pornographic industry as the best possible means of supporting themselves and their dependents, even if the product that results is completely dehumanizing to them.

One thing that is certain about pornography is that it permeates contemporary culture. Once available only in more "underground" places like X-rated movie houses, pornography is now far more public than in the past. Hotel rooms have a huge array of pornographic films available on television; pornographic spam appears regularly in people's email inboxes; casual references to pornography are made in popular shows on prime-time TV; and images that once would have seemed highly pornographic are now commonly found on widely distributed magazines such as *Maxim, Cosmopolitan,* and others.

Pornography has also infiltrated the web. There are now over 4.2 million pornographic websites, a 2.8-billion-dollar industry. Researchers estimate that worldwide, men spend an average of $3000 per second purchasing pornographic material (Johnson 2011; Ropelato 2007). These figures do not account for women's use of pornography, nor do they capture the extent to which pornography is viewed on the web for free.

The United States is both the largest producer and exporter of "hard-core" porn on the web. Studies of Internet porn also find that the most popular website searches are for Black and Asian porn sites; Blacks and Asians have the most porn sites devoted to them, followed by Latinos—a reflection of the sexualized imagery of racism. Sites devoted to "teen porn" also tripled between 2005 and 2013 (Ruvolo 2011).

Pornography clearly reflects the racism and sexism that are endemic in society. Highly sexualized expressions and images are so widely seen throughout society that one commentator has said we are experiencing the *pornification of culture* (Levy 2005). Does this indicate that society has become more sexualized?

According to a major report from the American Psychological Association (APA), the answer is yes. The APA defines *sexualization* as including any one of the following conditions:

- People are judged based only on sexual appeal or behavior, to the exclusion of other characteristics.
- People are held to standards that equate physical attractiveness with being sexy.
- People are sexually objectified—meaning made into a "thing" for others' use. Sexuality is inappropriately imposed on a person (American Psychological Association 2007)

The APA report then details the specific consequences, especially for young girls, of a culture marked by sexualization. This report shows that overly sexualizing young girls harms them in psychological, physical, social, and academic ways. Young girls may spend more time tending to their appearance than to their academic studies; they may engage in eating disorders to achieve an idealized, but unattainable image of beauty; or they may develop attitudes that put them at risk of sexual exploitation. Although the focus of this report is on young girls, one cannot help but wonder what effects the "pornification of culture" also has on young boys, as well as adult women and men.

Despite public concerns about pornography, most people believe that pornography should be protected by the constitutional guarantees of free speech and free press. Yet people also believe that pornography dehumanizes women; women especially think so. Public controversy about pornography is not likely to go away because it taps so many different sexual values among the public.

Teen Pregnancy

Each year about 370,000 teenage girls (under age 19) have babies in the United States (see ■ Map 12.1). The United States has the highest rate of teen pregnancy among developed nations, even though levels of teen sexual activity around the world are roughly comparable. Teen pregnancy has declined since 1990, a decline caused almost entirely from the increased use of birth control (▲ Figure 12.6). Contrary to popular stereotypes, the teen birthrate among African American women has declined more than for White women (U.S. Census Bureau 2014; see also ▲ Figure 12.7). Most teen pregnancies are unplanned, due largely to inconsistent use of birth control (Guttmacher Institute 2014).

Beginning in the early 1980s, the federal government encouraged abstinence policies, putting money behind the belief that encouraging chastity was the best

map 12.1

Teen Births by State

What sociological factors would you say explain these regional differences in the teen birthrate?

Source: Ventura, Stephanie J., Brady E. Hamilton, and T. J. Mathews. 2014. "National and State Patterns of Teen Births in the United States, 1940–2013." *National Vital Statistics Report*. 63 (August). **www.cdc.gov**

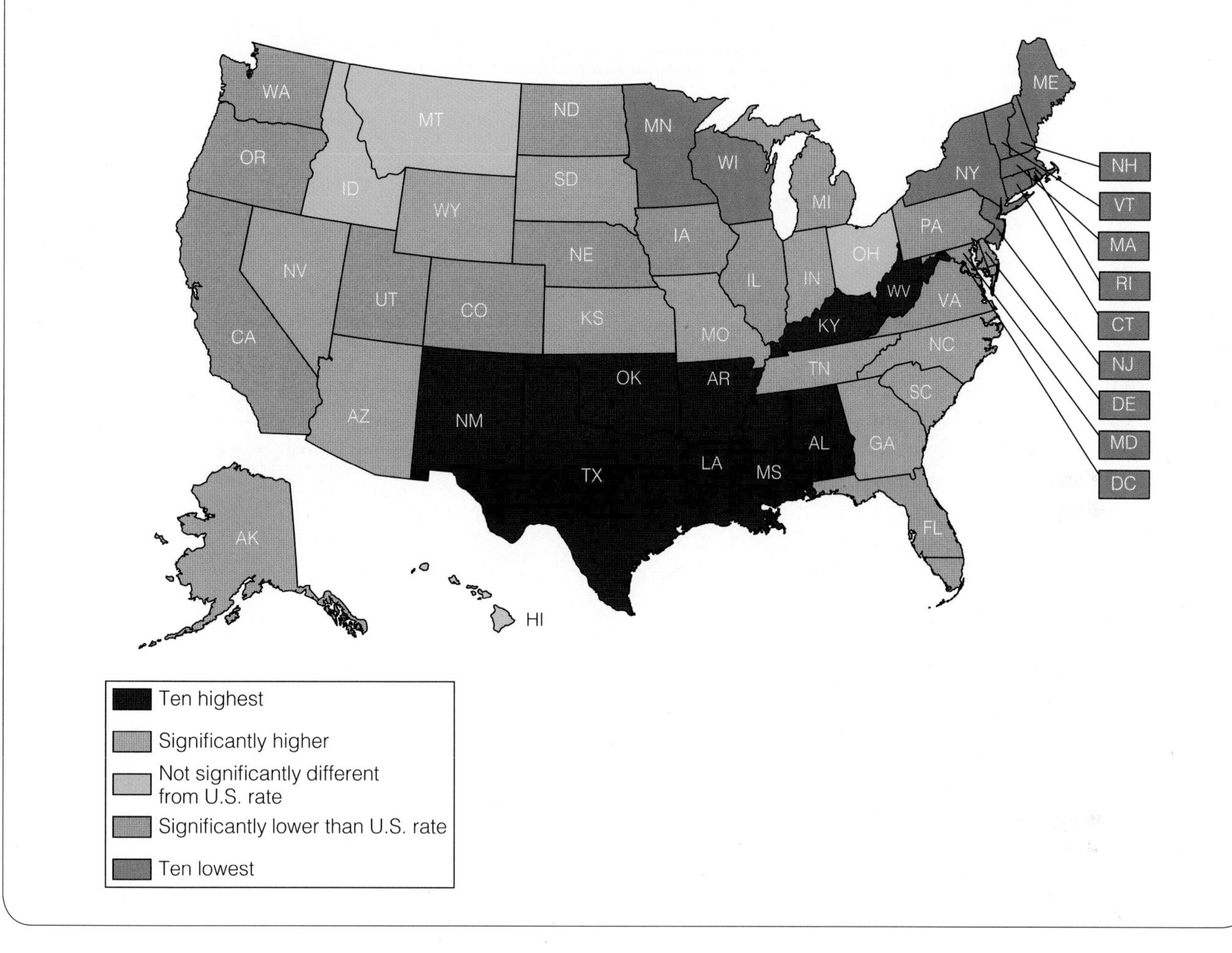

way to reduce teen pregnancy. Under programs that encourage abstinence, young people are encouraged to take "virginity pledges," promising not to have sexual intercourse before marriage. Do such pledges work?

In one of the most comprehensive and carefully controlled studies of abstinence, researchers compared a large sample of teen "virginity pledgers" and "non-pledgers" who were matched on social attitudes such as religiosity and attitudes toward sex and birth control. The study compared the two groups over a five-year period. Results showed that over time, there were *no differences* in the number of times those in each group had sex, the age of first sex, or the practice of oral or anal sex. The main difference between the two groups was that pledgers were less likely to use birth control when they had sex. Also, five years after having taken an abstinence pledge, pledgers denied having done so. The researchers concluded that not only are abstinence pledges ineffective (supporting other research findings) but that taking the pledge makes pledgers less likely to protect themselves from pregnancy and disease when having sex (Rosenbaum 2009). Consistent with these findings are other studies that find that abstinence policies account for a very small portion of the decline in teen pregnancy—probably only about 10 percent of the difference (Santelli et al. 2007; Boonstra et al. 2006).

Although the rate of teen pregnancy has declined, so has the rate of marriage for teens who become pregnant. Now, most babies born to teens will be raised by single mothers—a departure from the past when teen mothers often got married. What concerns people about teen parents is that teens are more likely to be poorer

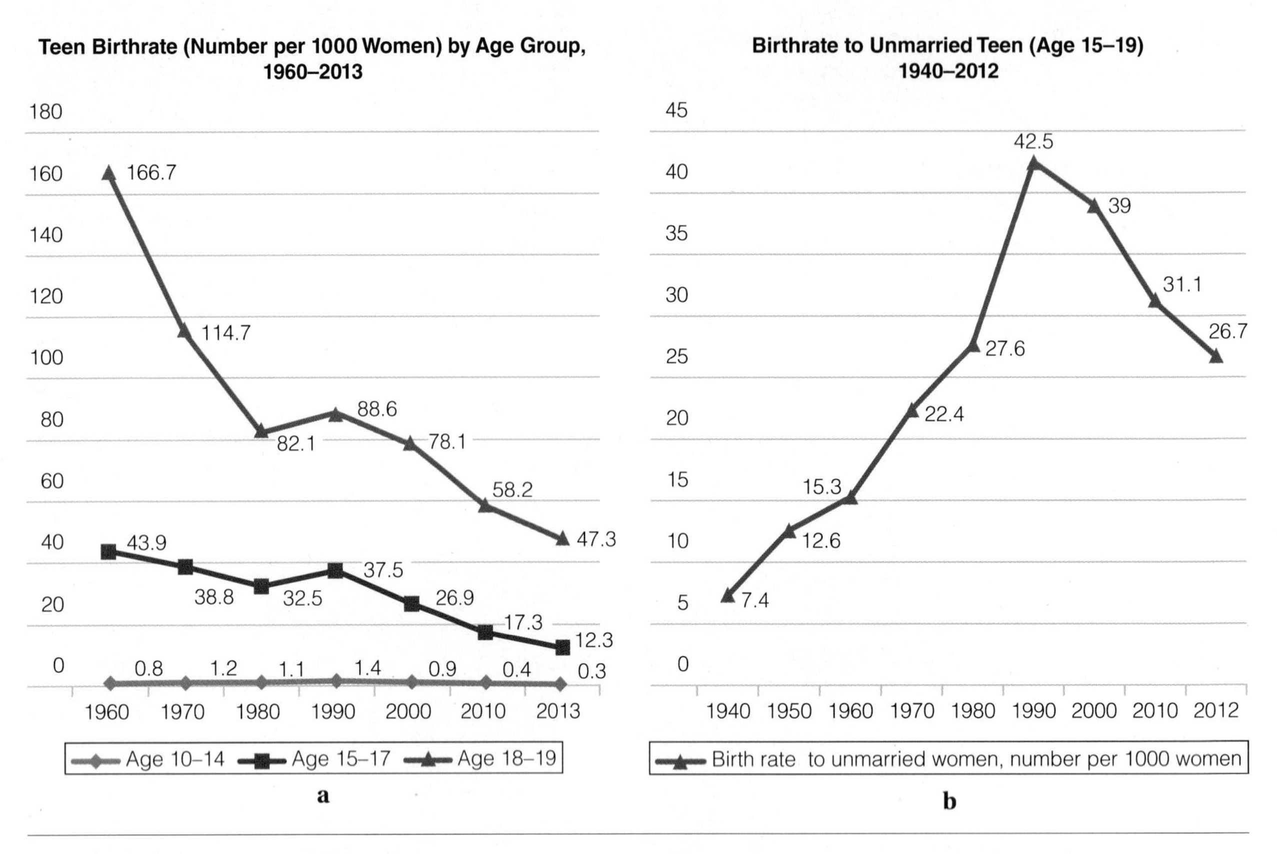

▲ **Figure 12.6** Teen Birthrate (by Age and for Unmarried Teens) Despite public concerns about teen pregnancy, the rate of births to teen women is now lower than at any previous time. But, as you see in Figure 12.6b, births to unmarried teens increased, even while the birthrate in general was declining. How do you explain both facts? In Figure 12.6b, what explains the decline in births to unmarried teens since about 1990? What sociological factors might explain why such behaviors change over time?

Source: Ventura, Stephanie J., Brady E. Hamilton, and T. J. Mathews. 2014. "National and State Patterns of Teen Births in the United States, 1940–2013." *National Vital Statistics Report* 63 (August). **www.cdc.gov**

than other mothers, although sociologists have cautioned that this is because teen mothers are more likely poor *before* getting pregnant (Luker 1996). Teen parents are among the most vulnerable of all social groups.

DEBUNKING Society's Myths

Myth: Providing sex education to teens only encourages them to become sexually active.

Sociological Perspective: Comprehensive sex education actually delays the age of first intercourse; abstinence-only education has not been shown to be effective in delaying intercourse (Risman and Schwartz 2002).

Teenage pregnancy correlates strongly with poverty, lower educational attainment, joblessness, and health problems. Teen mothers have a greater incidence of problem pregnancies and are most likely to deliver low-birth-weight babies, a condition associated with myriad other health problems. Teen parents face chronic unemployment and are less likely to complete high school than those who delay childbearing. Many continue to live with their parents, although this is more likely among Black teens than among Whites.

Although teen mothers feel less pressure to marry now than in the past, if they raise their children alone, they suffer the economic consequences of raising children in female-headed households—the poorest of all income groups. Teen mothers report that they do not marry because they do not think the fathers are ready for marriage. Sometimes their families also counsel them against marrying precipitously. These young women are often doubtful about men's ability to support them. They want men to be committed to them and their child, but they do not expect their hopes to be fulfilled (Edin and Kefalas 2005). Research shows that low-income single mothers are distrustful of men, especially after an unplanned pregnancy. They think they will have greater control of their household if they remain unmarried. Many teen mothers also express fear of domestic violence as a reason for not marrying (Edin 2000).

Why do so many teens become pregnant given the widespread availability of birth control? Teens typically delay the use of contraceptives until several

Education and Health Care

The United States is thought to have the best education system in the world. Compared to underdeveloped countries, Americans are expected to have brighter futures because American schools are presumed to produce better-educated students. In the United States, anyone, regardless of race, class, or gender, can allegedly attend high quality schools. Do all Americans get a quality education? If not, who does and who does not?

These questions ask you to examine the character of education as a social institution—one that, like other social institutions, has a social structure. The social structure of educational institutions means that, for some, schooling provides a path to a good job; for others, schooling provides minimal knowledge with little opportunity for success.

There is much public debate around the issue of education. Teacher unions are under fire. Urban schools in many American cities are facing serious budgetary and performance problems. Higher education is changing in the character of the student body and how educational content is delivered. The high cost of college prevents many from attending and leaves those who do asking if the high cost makes it worthwhile. Sociologists ask: How has the role of formal education in society changed over time? How can we provide quality education to all Americans? How is education linked to America's place in a global economy?

EDUCATION

HEALTH CARE

in this chapter, you will learn to:

- Understand the role of schools in society
- Compare and contrast theoretical perspectives on education
- Explain the connection between education and social mobility
- Outline the race and class inequalities within education
- Summarize the principal ideas regarding educational reform
- Describe the social organization of health care in the United States
- Report data on inequality in health and health care in the United States
- Compare and contrast theoretical perspectives on health care
- Summarize the principal ideas regarding health care reform

Schooling and Society

Education in any society is about the transmission of knowledge. In some societies, such as the United States, education is highly formalized—indeed, even regulated by government (at least for public institutions). In other societies, education may be less formal, perhaps provided solely through the transmission of knowledge by elders or family members (for example, home schooling is the norm in some protected religious communities). In the United States, education teaches formal knowledge, such as reading, writing, and arithmetic, as well as cultural knowledge, such as morals, values, and ethics. Education prepares the young for entry into society and is thus a form of socialization. Sociologists refer to the more formal, institutionalized aspects of education as **schooling**.

In a highly technological society such as the United States, education is increasingly necessary for future opportunities. Why then do some of our schools resemble prisons where entering students are searched and the physical environment is dilapidated and bleak? Other schools look like beautiful campuses, places with modern facilities and sophisticated scientific equipment. To put these inequities in perspective, let us briefly look at how education in the United States has developed over time.

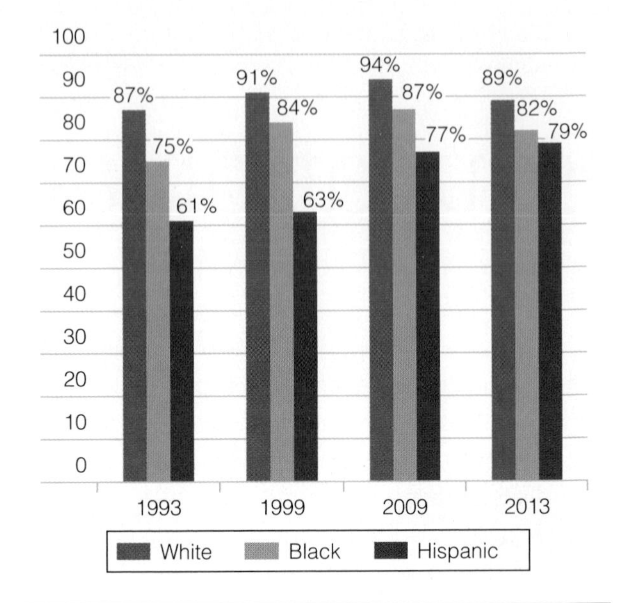

▲ **Figure 14.1** Percentage of 18- to 24-Year-Olds Who Have Completed High School by Race: Recent Trends
The population of high school graduates is evolving, with a greater percentage of Hispanic students graduating each year.
SOURCE: U.S. Census Bureau 2013, Current Population Survey, School Enrollment Supplement. **www.census.gov/hhes/school/data/cps/**

The Structure of Education

During the nineteenth century, education was considered a luxury, available only to White, male children of the upper classes; it was not required for most jobs (Cookson and Persell 1985). In 1900, federal guidelines made education compulsory, yet state laws requiring attendance were generally enforced only for White Americans and then only through eighth grade. ▲ Figure 14.1 shows how high school graduation rates have increased steadily over thirty years, most dramatically for Hispanic students. Inequality still exists, however, with the highest educational attainment for White Americans.

There are traditionally three kinds of education in the United States: public education, private education, and homeschooling. Among public schools, there are now *charter schools*—those that receive public funds, but that are not subject to the same rules and regulations as other public schools. Charter schools are attended by choice, sometimes by a lottery system. Many have a specialized curriculum. Although charter schools are still accountable to local and state school authorities, many think they offer an alternative to failing public schools. The number of students enrolled in charter schools has dramatically increased, from just under a half million in the 2000–2001 school year to over 2.1 million in the 2011–2012 school year (National Center for Education Statistics 2013).

Criticism about charter schools focuses on the illusion of choice. Although most charter schools claim

AP Images/Eric Gay
AP Images/Luke Palmisano

Inequality in education is very apparent in the physical facilities of poor versus wealthy schools. This inequality is further reflected in educational opportunities within schools.

Third, SATs actually do not predict school performance very well for all groups. For example, SAT scores are only modestly accurate predictors of college grades even for White students (Zwick 2004). This fact is not well known. Grade point average in high school (and school class rank as well) is also only a modestly accurate predictor of success in college. High school grades are about as accurate as the SATs in predicting college grades—maybe even a little better (Alon and Tienda 2007).

In general, average scores for tests such as the SAT differ across different groups. Whites score higher on average than minorities, and the higher a person's social class, the higher his or her test score is likely to be. This is where the intelligence debate begins. The segregation of schools discussed previously indicates clear reasons for poor test scores among some students. Lack of resources in schools, inadequate teacher training, and unsafe conditions are environmental factors that likely contribute to below-average academic performance.

Still, occasional claims are made that differences in test scores are somehow genetically inherited. A notorious example was the publication of a book, *The Bell Curve* in 1994. The book caused a major stir, one that is still ongoing among educators, lawmakers, teachers, public officials, policymakers, and the general public. Authors of *The Bell Curve*, Herrnstein and Murray, argued that the distribution of intelligence in the general population closely approximates a bell-shaped curve (called the *normal distribution*). They also asserted that there is one basic, fundamental kind of intelligence and that it is genetically inherited.

Herrnstein and Murray's research was widely criticized for arguing that intelligence is determined primarily by one's genes rather than by one's social and educational environment. Although there is *some* small genetic basis to intelligence, as long as society is marked by the inequalities that we can sociologically observe, then group differences in ability must be seen within that context. Understanding the drastic race and class differences in schools across this country highlights access to educational opportunity as opposed to intellectual ability (Fischer et al. 1996).

Consider the standardized tests commonly used to apply for college admission. On average, students from lower-income families have lower scores on exams such as the Scholastic Assessment Test (SAT) and the American College Testing (ACT) program. As shown in ▲ Figure 14.3, there is a smooth and dramatic increase in average (mean) SAT score as family income increases, for both SAT verbal and math scores. In this sense, a student's SAT score is a proxy, or substitute, measure of that student's social class: Within a certain range, you can guess someone's likely SAT score from knowing only the income and social class of his or her parents! As you can see from Figure 14.3, each additional $10,000 in family income is worth about 10 to 15 points on either the SAT verbal or the SAT math tests.

One possible reason for this is access to test preparation courses. These courses typically cost money and may be inaccessible to some students. Devine-Eller (2012) finds that as household income goes up, the likelihood of participating in test preparation

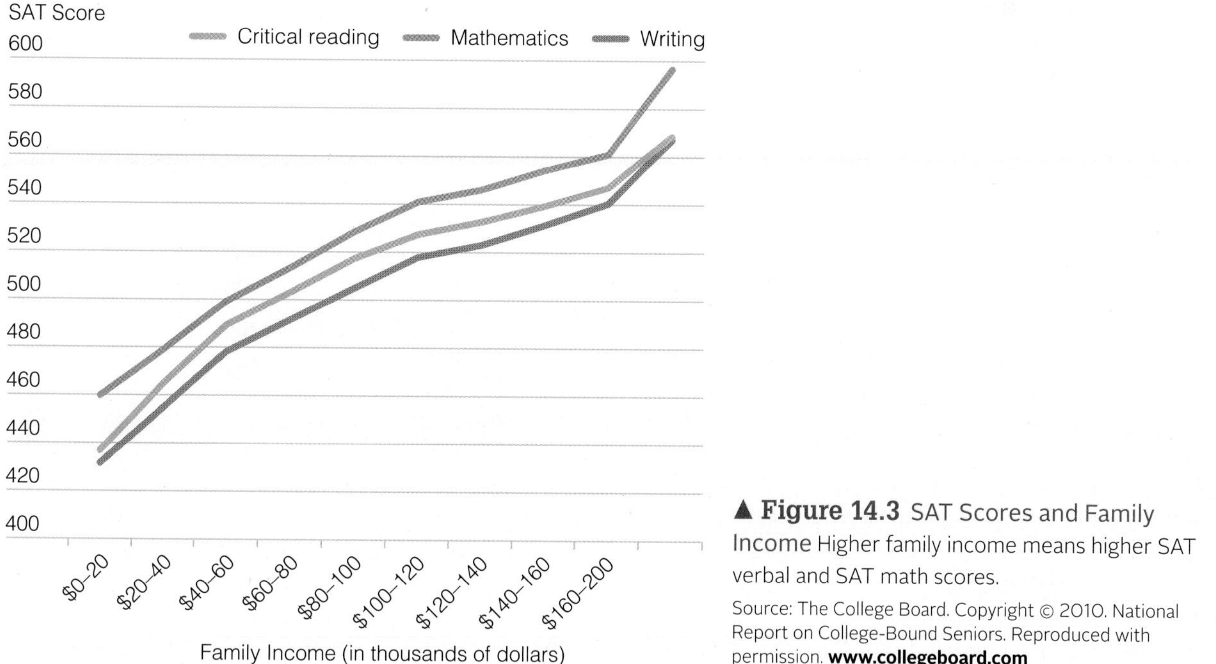

▲ **Figure 14.3** SAT Scores and Family Income Higher family income means higher SAT verbal and SAT math scores.

Source: The College Board. Copyright © 2010. National Report on College-Bound Seniors. Reproduced with permission. **www.collegeboard.com**

courses also goes up for ninth- through eleventh-graders. This research also shows that students are much more likely to participate in a test preparation course if they are active in school activities, have parents who are highly educated themselves, and have parents that are actively involved in school (Devine-Eller 2012). The idea of **cultural capital** in this context suggests that certain types of parents will have access to knowledge and information about preparing their student for college entrance exams. Beyond simply the ability to pay for test preparation courses, parents with knowledge and experience regarding college admissions are able to provide better opportunity for their children.

Less help preparing for standardized tests may diminish a student's chance of getting into the best colleges or universities. The intersection of race and class also contributes to the inequality of educational attainment. Statistics about SAT scores indicate that White students typically score higher in critical reading, mathematics, and writing than Blacks and Latinos (see ◆ Table 14.2). These patterns indicate that college entrance exams continue to stratify student access to educational success.

The educational system in the United States appears to allow for some social mobility as the result of education, but clearly not as much as people believe. A good education is essential for a good job, but the odds of getting such an education are considerably shaped by one's social class of origin, one's race—and, to a lesser extent—one's gender. Too many of the poorest children in the United States lack the very basic necessities for a good education (Kozol 2006). Support and scholarship programs intended to aid those with greater disadvantages in education help, but without such intervention, the forces of social stratification are reproduced in the educational system.

Education and Inequality

Education has reduced many inequalities in society. More high school diplomas are awarded to all race and class groups, and more minorities and women attend and graduate from two- and four-year colleges. Nonetheless, many inequalities still exist in U.S. education. These can be shown in numerous ways. Studies of inequality and poverty, specifically in U.S. urban centers, highlight how schools are organized, often along race and class lines, leaving many of our most marginalized students without quality education (Kozol 2006).

Segregation and Resegregation

In 1954, in a landmark decision, the U.S. Supreme Court ruled in the case of *Brown v. Board of Education* that "separate but equal" in all public facilities, including schools, was unconstitutional. Although it took years before school districts actually began implementing this decision, and only then with substantial pressure from the federal government, some measure of school desegregation followed the *Brown* decision. Although school desegregation was not as great as was hoped, progress was made. By the 1980s, many school districts, particularly in the American South, had made significant gains in integrating schools by race.

Now, however, that historic change is reversing, and the nation is retreating to highly segregated schools. Researchers have found that American schools are now more racially segregated by race and by class than was the case in the 1980s, in every part of the country for both African Americans and Latinos. Communities have reversed the desegregation orders of the 1970s to create neighborhood schools. Residential segregation by race means that schools

◆ **Table 14.2 Average SAT Scores by Ethnicity and Gender**

	Critical Reading		Mathematics		Writing	
	Men	Women	Men	Women	Men	Women
American Indian or Alaska Native	482	480	502	472	456	466
Asian, Asian American, or Pacific Islander	522	520	611	584	522	532
Black or African American	427	433	436	423	408	426
Mexican or Mexican American	453	446	481	450	438	445
Puerto Rican	458	454	468	441	440	449
Other Hispanic, Latino, or Latin American	456	446	480	446	440	445
White	530	525	552	519	508	521
Other	491	493	539	498	483	496

Source: The College Board. Copyright © 2013. Total Group Profile Report. Reproduced with permission. **www.collegeboard.com**

are then also segregated. More than half of African American and Latino students now attend schools that are "majority minority"—that is, more than half of the students in the schools are minority students. This would not be problematic in and of itself were it not for the fact that "majority minority" schools tend to have none of the resources of predominantly white schools (Orfield et al. 2014).

School segregation is problematic on many counts, one of which is the isolation of groups from one another and the resulting loss of friendship, interracial understanding, and comingling. Segregated schools that are heavily minority or poor are also generally of very poor quality—as the *Brown* decision noted. In other words, segregation breeds inequality, and even a cursory look at segregated schools that are predominantly minority and/or poor will reveal this. Unqualified teachers, ill-equipped science labs, a weak curriculum, and a prison-like atmosphere prevail in such schools, thus denying students, who likely are perfectly smart and capable, from achieving the kind of education that will lead to a good job (Kozol 2006). This unequal, subpar education creates a situation in which young African American men are more likely to end up in prison than to graduate from college (National Center for Education Statistics 2013; Carson and Sabol 2012).

The current criticism of the American educational system centers on these disparate conditions between wealthy, predominantly White, suburban schools and poor, predominantly minority, urban schools. Communities with a greater percentage of high-income families simply have better schools. Students raised in low-income communities are denied access to these schools and the strong education provided by them. Despite the perception that the United States is a fair and equitable nation, millions of U.S. children are lacking opportunity to live up to their potential. The educational structure of U.S. schools is not fair and equitable.

DEBUNKING Society's Myths

Myth: Intelligence is mostly determined by genetic inheritance.

Sociological Perspective: Intelligence is a complex concept not easily measured by one thing and is likely shaped as much by environmental factors as by genetic endowment (Zwick 2004).

School Tracking and Individualized Education Plans. **Tracking** (also called *ability grouping*) is the separating of students within schools according to some measure of ability (Oakes 2005). Tracking has taken place for more than seventy years. As early as first grade, children are likely divided into high-track, middle-track, and lower-track groups, or some variation thereof. Perhaps you were assigned to one of these tracks in elementary, junior high, or high school. In high school, the high-track students take college preparatory courses in math and science and read Shakespeare. Middle-track students take courses in business administration and typing. Lower-track students take vocational courses in auto mechanics, masonry, or dental hygiene. Although this kind of tracking is now on the decline in the United States, versions of it still exist.

The original idea behind tracking is that students would get a better education and be better prepared for life after high school if they are grouped early according to ability. Theoretically, students in all tracks learn faster because the curriculum is tailored to their ability level, and the teacher can concentrate on smaller, more homogenous groups.

Advocates of *detracking* give the opposite argument. Detracking is based on the belief that combining students of varying cognitive abilities benefits the students more than tracking, especially by the time students get to junior high and high school. Students of high and low ability can thus learn from each other; the high-ability students are not seen to be "held back" by students with less ability, but are enriched by their presence.

How does tracking work when placing students with learning disabilities or special educational requirements? For years, American students with disabilities were isolated from mainstream students and given an entirely separate curriculum. The Individuals with Disabilities Education Act, however, outlines the federal guidelines for providing quality education for students with disabilities. Since the act was amended in 1997, new trends focus on the need for **individualized education programs** (IEPs), which outline specific types of learning that target specific needs. Not long ago, students with special education needs were taken out of the main classroom. More recently, education research highlights the value of mixed-ability classrooms, including both IEP and mainstream students.

Most researchers and educators who have studied tracking agree that not all students should be mixed together in the same classes. The differences between students can be too great and their needs too dissimilar. Some degree of tracking has always had advocates based on its presumed benefits for all students. This presumption is under attack. One of the most consistent research findings on tracking is that students in the higher tracks receive positive effects, but that the lower-track students suffer negative effects. To begin with, students in the lower tracks learn less because they are, quite simply, taught less. They are asked to read less and do less homework. High-track students are taught more; furthermore, they are consistently rewarded by teachers and administrators for their academic abilities (Oakes 2005). At the elementary level, mixed-ability classrooms are more common. As students progress through middle and high schools, IEPs are developed to address needs for each student and for each subject in school. Structurally, this can be very challenging for teachers and school administrators. Individualized lesson plans could benefit all students, not just those with learning disabilities. Teachers, however, must find ways to teach the same material to a classroom full of students, all of whom learn differently.

THINKING Sociologically

Were you in a *tracked* elementary school? What were the tracks? Did you get the impression that teachers devoted different amounts of actual time to students in different tracks? Did teachers "look down" on those in the lower tracks? What about the students—did they treat some tracks as "better" or "worse" than others (were they perceived as differing in prestige)? Based on your recollections, what does this tell you about tracking and social class?

Who gets assigned to which tracks? Research shows that track assignment is not solely based on the performance in cognitive ability tests. Social class and race are involved. Students with the same test scores often get assigned to different tracks because of differences in their social class and race. Few administrators or teachers consciously and deliberately assign students to tracks based on these criteria, but it occurs nevertheless. Researchers have consistently found that when following two students with identical scores on cognitive ability tests, the student of higher social class is more likely than the student of lower social class status to get assigned to the higher track.

This inequality is at the root of the American education debate. A core American value states that all people are created equal. Through a fair and equitable educational system, all students would have equal access to opportunity and success. Sociological research examines the social institution of education to better understand the consequences for students and how to better improve those consequences.

Educational Reform

There are clearly major challenges facing the educational system in the United States. Calls for reform are many and are coming from parents, communities, teachers, and administrators, as well as presidents and politicians. The No Child Left Behind (NCLB) Act of 2001 was one reform attempt.

The goal of NCLB was, in part, an attempt to narrow the achievement and test the score gap between White students and students of color in U.S. public schools. Of course, this act does not address fundamental problems of public education, including racial segregation and wealthy White students attending private educational institutions. The NCLB Act has meant restructuring education with an emphasis on "accountability"—that is, measured assessments of where and how schools are succeeding or failing. Much of the emphasis in NCLB has then been on *high-stakes testing*. Students cannot graduate or move on to the next grade without reaching certain levels of proficiency, as measured on standardized exams. NCLB also calls for teachers to be replaced based on their students' test scores, and schools that are seen as underachieving are threatened with closure.

Hero Images Inc./Alamy

Greatstock Photographic Library/Alamy

Schools in the United States are rapidly resegregating by race.

Social Class and Health Care

In the United States, social class has a pronounced effect on health and the availability of health services. The lower the social class status of a person or family, the less access available to adequate health care (National Center for Health Statistics 2013). Consequently, the lower one's social class, the less long one will live. People with higher incomes who are asked to rate their own health tend to rate themselves higher than people with lower incomes. The effects of social class are nowhere more evident than in the distribution of health and disease, showing up dramatically in the rates of infant mortality, stillbirths, tuberculosis, heart disease, cancer, arthritis, diabetes, and a variety of other illnesses. The reasons lie partly in personal habits that are themselves partly dependent on one's social class. For example, those with lower socioeconomic status smoke more often, and smoking is the major cause of lung cancer and a significant contributor to cardiovascular disease (Centers for Disease Control 2013).

Social circumstances also have an effect on health. Poor living conditions, elevated levels of pollution in low-income neighborhoods, and lack of access to health care facilities all contribute to the high rate of disease among low-income people. Another contributing factor is the stress caused by financial troubles. Research has consistently shown correlations between psychological stress and physical illness (Taylor 2010). The poor are more subject to psychological stress than the middle and upper classes, and it shows up in their comparatively high level of illness.

Medicaid is the government program that provides medical care in the form of health insurance for the poor, welfare recipients, and the disabled. The program is funded through tax revenues. The costs covered per individual vary from state to state because the state must provide funds to the individual in addition to the funds that are provided by the federal government. Medicaid, Medicare, and now the Affordable Care Act are as close as the United States has come to the ideal of universal health insurance.

Gender and Health Care

Although women live longer on average than men, national health statistics show that hypertension is more common among men than women until age 55, when the pattern reverses. This may reflect differences in the social environment men and women experience, with women finding their situation to be more stressful as they advance toward old age (National Center for Health Statistics 2013).

Health and Disability

The *disability rights movement*, a movement that has defined disabled people as a social group with rights similar to other minority groups in society, has transformed how people think about disability, challenging many preconceived ideas. For example, within a social context, there is a tendency for people to see someone with a disability solely in terms of that social status—what sociologists call a stigma. A **stigma** is a social identity that develops when a person is socially devalued by others because of some identifiable characteristic. When someone is stigmatized, that identity tends to override all other identities, and the person is treated accordingly.

Understanding the social dynamics associated with disabilities has resulted from the efforts of the disability rights movement. The movement has called attention to the social realities of disabilities, even questioning the very language used to identify people with disabilities—for example, using the term *physically challenged* rather than the more negative connotation of *disabled*.

One of the most significant achievements of the disability rights movement is the Americans with Disabilities (ADA) Act, passed by Congress in 1990. This law prohibits discrimination against people with disabilities. The ADA legislates that people with disabilities may not be denied access to public facilities—thus the presence of such things as ramps, wheelchair access on buses and stairways, handicapped parking spaces, and chirping sounds in crosswalk lights for blind pedestrians, all social changes that are now so prevalent that you might even take them for granted. They have resulted, however, from the social mobilization of those who saw a need for social change.

The Americans with Disabilities Act also requires employers and schools to provide "reasonable accommodations" such that those with disabilities are not denied access to employment and education. For

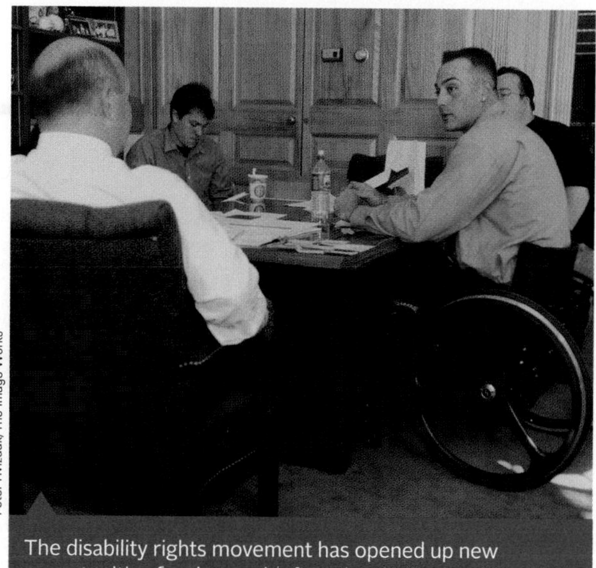
Peter Hvizdak/The Image Works

The disability rights movement has opened up new opportunities for those who face the challenge of disability.

many students with various learning disabilities, this has meant making accommodations for taking tests with extended time or in settings where the test taker is not subject to as much distraction as in a crowded classroom. The increased awareness of disability rights has transformed society in ways that have opened up new opportunities for those who, years ago, would have found themselves with less access to education and jobs and, therefore, more isolated in society.

Age and Health Care

As people age, their health care needs are no doubt likely to increase. Until recently, many of the nation's elderly were also likely to be low income. Although class status varies among the nation's elderly, all older people at this point are beneficiaries of the national Medicare program. Medicare was begun in 1965, under the administration of President Lyndon Johnson. It provides medical insurance, including hospital care, prescription drug plans, and other forms of medical care for all individuals age 65 or older. The Affordable Care Act also aims to strengthen Medicare benefits.

Medicare is partially funded through payroll taxes whereby both employees and employers pay a small percentage of employee wages to cover some of the cost of this large (and costly) federal program. But, with so many people in the population now living longer, and with the now aging baby boomer population being such a large share of the total population, many wonder if Medicare can be sustained in the near future. With the number of workers paying payroll taxes shrinking, the elderly population growing, and the cost of health care rising, there is a looming fear that Medicare simply cannot be financially sustained. Though not the sole basis for the nation's challenges in health care, the health needs of the older population are clearly a major challenge.

Theoretical Perspectives of Health Care

The sociology of health is anchored in the same major theoretical perspectives that we have studied throughout this book: functionalist theory, conflict theory, and symbolic interaction theory (see ◆ Table 14.3).

Functionalist Theory

Functionalism argues that any institution, group, or organization can be interpreted by looking at its positive and negative functions in society. Positive functions contribute to the harmony and stability of society. The positive functions of the health care system are the prevention and treatment of disease. Ideally, this would mean the delivery of health care to the entire population without regard to race, ethnicity, social class, gender, age, or any other characteristic. At the same time, the health care system is notable for a number of negative functions, those that contribute to disharmony and instability of society.

Functionalism also emphasizes the systematic way that various social institutions are related to each other, together forming the relatively stable character of society. You can see this with regard to how the health care system is entangled with government through such things as federal regulation of new drugs and procedures. The government is also deeply involved in health care through scientific institutions such as the National Institutes of Health, a huge government agency that funds new research on various matters of health and health care policy. As a social institution, health care is also one of the nation's largest employers and thus is integrally tied to systems of work and the economy.

◆ Table 14.3 Theoretical Perspectives on the Sociology of Health

	Functionalism	Conflict Theory	Symbolic Interaction
Central point	The health care system has certain functions, both positive and negative.	Health care reflects the inequalities in society.	Illness is partly socially constructed.
Fundamental problem uncovered	The health care system produces some negative functions.	Excessive bureaucratization of the health care system and privatization lead to excess cost.	Patients and health professionals serve specific roles. What is determined as illness is specific to cultural context.
Policy implications	Policy should decrease negative functions of health care system for minority groups, the poor, and women.	Policy should improve access to health care for minority racial–ethnic groups, the poor, and women.	Determining something as disease will make insurance reimbursement more likely.

Conflict Theory

Conflict theory stresses the importance of social structural inequality in society. From the conflict perspective, the inequality inherent in our society is responsible for the unequal access to medical care. Minorities, the lower classes, and the elderly, particularly elderly women, have less access to the health care system in the United States than Whites, the middle and upper classes, and the middle-aged. Restricted access is further exacerbated by the high costs of medical care.

Excessive bureaucratization is another affliction of the health care system that adds to the alienation of patients. The U.S. health care system is burdened by endless forms for both physicians and patients, including paperwork to enter individuals into the system, authorize procedures, dispense medicines, monitor progress, and process payments. Long waits for medical attention are normal, even in the emergency room. Prolonged waits have reached alarming proportions in the emergency rooms of many urban hospitals in the United States and can only deepen the alienation of patients.

Symbolic Interaction Theory

Symbolic interaction theory holds that illness is partly (although obviously not totally) socially constructed (Armstrong 2003). The definitions of illness and wellness are culturally relative—the social context of a condition partly determines whether or not it is sickness. Consider the example of alcoholism and other addictions. During the era of prohibition, people who drank were considered deviants and lacking moral fortitude. Now, however, alcoholism is a diagnosable disease, listed in the *Diagnostic Statistical Manual* as an illness. The medicalization of alcoholism refers to how Americans culturally and socially label abuse of alcohol as a disease that requires treatment. This has profound consequences for how people with alcoholism are treated. People who are *ill* receive more sympathy and more care than those who are labeled *deviant.*

Symbolic interaction also highlights the roles played within the health care institution. There is a hierarchy that puts medical doctors at the top and medical assistants, nursing staff, and orderlies at the bottom. Patients take on the role of a child, with little agency in how treatment is administered. The diagnosis, the treatment plan, and the prognosis are managed with little input from the patient. Insurance companies and pharmaceutical companies play an entirely different role, one that oversees the availability of medical care by determining what procedures or treatments will be financially covered.

The symbolic interaction approach to studying health care institutions focuses on the roles of the patient and medical professionals and on the cultural context within which disease is labeled and treated. Table 14.3 outlines the theoretical perspectives of health care and illness.

Health Care Reform

Currently, the cost of medical care in the United States is approximately 18 percent of our gross domestic product, making health care the nation's third leading industry. The United States tops the list of all countries in per person expenditures for health care (The World Bank 2014a). Other countries spend considerably less money and deliver a level of health care at least as good. For example, Sweden and the United Kingdom spend roughly half as much per capita as the United States, and Turkey spends a bit more than one-third as much.

The Cost of Health Care

One of the challenges of health care is sheer cost. Most health care is provided by a fee-for-service principle in which patients are responsible for paying the fees the health care provider charges. Patients with health insurance are able to pass on health expenses, either in full or partially, to the insurance company, but the cost for health care services is high, in some cases, astronomically expensive. Hospital care can cost thousands, even millions, of dollars for any extended stay. Sophisticated procedures require expensive machinery and technicians, and the nation needs to invest in medical research that allows practitioners to stay abreast of new

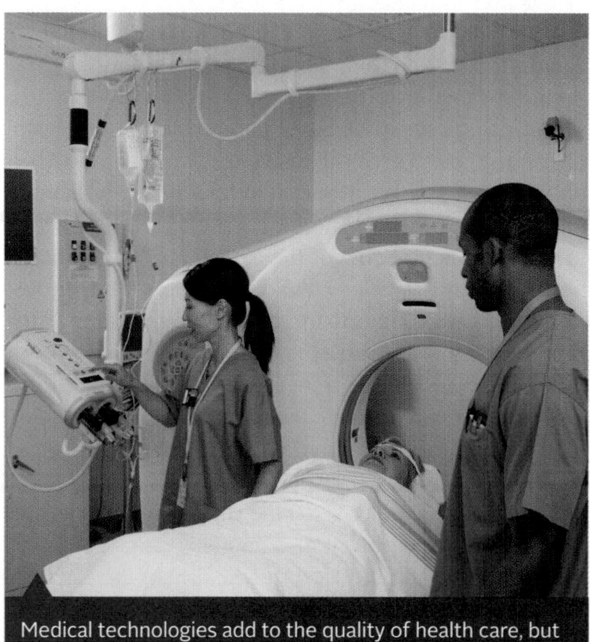

Paul Burns/Shannon Fagan/Photodisc/GettyImages

Medical technologies add to the quality of health care, but also to the cost.

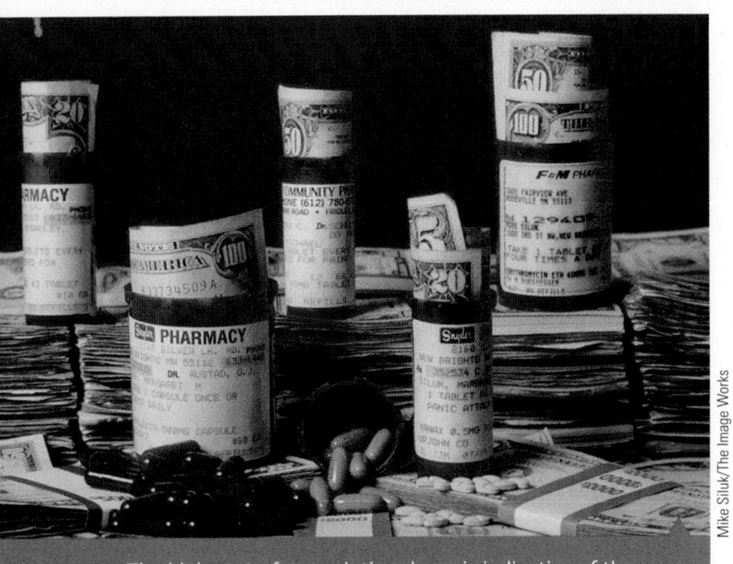

Mike Siluk/The Image Works

The high cost of prescription drugs is indicative of the problems generated by a profit-based health care system.

technologies and new treatments for a wide array of medical conditions.

Most sectors of the health care system (hospitals, pharmaceutical companies, even physician's office practices) are structured as for-profit businesses. Physicians, for example, may have to raise their rates to cover the high cost of malpractice insurance where annual insurance premiums (costs) have skyrocketed. The cost of these insurance premiums is passed along to consumers (patients) and has contributed to the rise in the overall cost of health care.

Adding to the high cost of health care is the role of big pharmaceutical companies. Spending for prescription drugs in the United States has increased from $40 billion in 1990 to a whopping $326 billion in 2013 (Schumock et al. 2014)! There is little sign that this spending will do anything but go further up. Prescription drugs are one of the fastest-growing components of health care costs. The rise in spending on drugs is partially attributed to increased use, but other factors include the actual cost of the drugs, the availability of new drugs for various maladies, and, without question, the cost of advertising directly to the public. The money spent on advertising directly to consumers has doubled since 1999 (Kaiser Family Foundation 2010). You can see this yourself as hardly an hour goes by on television without an advertisement for some kind of prescription drug.

The health care crisis in the United States is largely a question of cost, but it also entails a debate over the nation's responsibility for the health of its citizens. Who should pay for the soaring costs of health care? Who receives the benefits of such sophisticated medicine? Should there be universal health care for all, like we are seeing through the Affordable Care Act? These questions are at the heart of the current national debate about health care reform.

Health Care for All?

Despite the success of the Affordable Care Act in getting many more Americans health insurance, there is strong opposition to the program. Many in Congress are working to repeal the Affordable Care Act. Why are so many in the United States resistant to providing health care to its citizens in line with other Western nations? Sociologists offer several explanations. First, there is an antigovernment attitude among many in the United States that fuels resistance to a national health care system. The argument in Congress is that the government should not *force* people to spend money on health insurance. Second, analysts argue that, unlike in other Western nations, there is a relatively weak labor movement in the United States, resulting in more limited state-based benefits for workers. Third, racial politics have also shaped the nation's health care system; federal social welfare programs are associated in many people's minds with racial groups, and this, too, fuels the politics of health care reform. Finally, the health care system in this country is fundamentally structured on private, for-profit interests (Quadagno 2005).

Without the Affordable Care Act, millions more Americans will be uninsured. This creates vulnerability for people, especially people in poor communities, when they get sick. For Americans without insurance, the main source for medical care is a hospital emergency room, often called the "doctor's office of the poor." This is a very expensive way to deliver routine health care—and there is rarely any follow-up care or comprehensive and preventative treatment.

MICHAEL STRAVATO/AP Images

Many uninsured people wait in long lines to sign up for government-run health and medical plans.

B Christopher/Alamy

The Affordable Care Act was finally passed into law in June 2012, after the Supreme Court ruled it did not violate the Constitution. The debate over whether or not to appeal "Obamacare" continues.

SEE for YOURSELF

Youth and Health insurance

Identify a group of young people you know and ask them if they are covered by health insurance. If they are insured, where does their insurance come from? Who pays? Did they use the Affordable Care Act marketplace to find insurance? If they are not insured, ask them why not and whether they think this is important. Do they support a national health insurance program?

Having conducted your interviews, ask yourself how social factors such as the age, race, ethnicity, gender, and educational/occupational status of those you interviewed might have affected what people say about their insurance. Do you think any or all of these social characteristics are related to the likelihood that people are covered by health insurance and whether these characteristics are related to their attitudes about coverage? What are the implications of your results for public support for new health care policies?

Chapter Summary

What is the importance of the education institution?

Education is the social institution that is concerned with the formal transmission of society's knowledge. It is therefore part of the socialization process. Although the U.S. education system has long produced students at the top of the world's educational achievements, the United States is falling behind other nations on standardized test scores.

How does sociological theory inform our understanding of education?

Functionalism interprets education as having various purposes for society, such as socialization, occupational training, and social control. *Conflict theory* emphasizes the power relationships within educational institutions, as well as how education serves the powerful interests in society. *Symbolic interaction theory* focuses on the subjective meanings that people hold. These meanings influence educational outcomes.

How does education link to social mobility?

The number of years of formal education for individuals has important effects on their ultimate occupation and income. Social class origin affects the extent of educational attainment (the higher the social class origins, the more education is ultimately attained), as well as occupation and income (higher social class origin likely means a more prestigious occupation and more income).

Does the educational system perpetuate or reduce inequality?

Although the education system in the United States has traditionally been a major means for reducing racial, gender, and class inequalities among people, the education institution has perpetuated these inequalities. Segregation of schools and communities keep minority and poor children in schools that lack resources for success.

What current reforms are guiding education?

The No Child Left Behind Act program emphasized accountability in the schools, largely through testing. Current educational reforms focus on achieving educational standards, assessing school progress, and developing strong measures of student and teacher success. Free community college is also an educational reform idea.

How does the United States compare to other nations in the area of health care?

The United States is only recently providing universal health care for its citizens, through the Affordable Care Act. Despite disagreement with this program, more Americans now have health care insurance. The health care system is organized according to social patterns, including that disease itself is influenced by social facts, such as race, gender, and social class.

How does sociological theory inform our understanding of health and health care?
Functionalism interprets the health care system in terms of the systematic way that health care institutions are related to each other. Conflict theory addresses the inequalities that occur within the health care system. Symbolic interaction analyzes the interpretations that can affect people's health care, such as the tendency to place patients in a sick role and label some ailments as disease and others not.

What is the health care crisis in the United States?
High costs and questions about universal health care have created a policy crisis today in the U.S. health care system. The Affordable Care Act addresses some of the problems on universal health care, but the policy remains controversial.

Key Terms

achievement test **343**
Affordable Care Act **352**
Brown v. Board of Education **341**
cultural capital **348**
individualized education programs **349**
Medicaid **353**
Medicare **352**
schooling **340**
self-fulfilling prophecy **344**
stigma **355**
teacher expectancy effect **342**
tracking **349**

15

ZUMA Press, Inc/Alamy

Economy and Politics

Because you are reading this book, chances are that you are in school, probably seeking a college degree that will help you find a good job. Your hopes and dreams probably rest on getting a decent education, discovering what you plan to do in a job, and gaining the skills you need to succeed. You might be wondering how your education will prepare you for work and how you will match your interests to the current job market.

On one level, these are individual issues that involve your interests and talents, your motivation, and how well prepared you are to study and work. But behind these individual matters lie other social structures—structures that shape the options you face, the resources you have to pursue your dreams, and whether good jobs are available to you. The background structures that frame your individual life are the result of social institutions, and how those institutions are structured by social stratification. Finding work situates you within economic institutions—that is, the economy, which like other social institutions, has a particular social structure that includes an organized system of social roles, norms, and values. Social institutions extend beyond us, but they shape day-to-day life. Typically, people do not think about such structures and may not even be aware of the institutional structures that shape their lives.

To illustrate, think about finding a job. The likelihood of doing so depends not only on your individual attributes (level of education, skill, region where you live, and so forth) but also on the economic conditions of the time. For some, these

in this chapter, you will learn to:

- Compare different types of economic systems
- Identify the components of change in the contemporary global economy
- Identify the different organizational components of the workplace
- Explain the conditions affecting diverse groups in the workplace
- Compare and contrast sociological theories of work
- Compare different types of political systems
- Define different forms of power and authority and the organization of bureaucracies
- Analyze patterns of political participation
- Comprehend the organization of the military as a social institution

conditions may mean long-term unemployment. Young people entering the labor market for the first time, even as college graduates, may find themselves facing possible joblessness—through no fault of their own—merely because they exited school at a time when broader social forces were shaping their life opportunities. Although understanding the institutional forces at work at any given time does not reduce the suffering that people may be experiencing, such an understanding can help you think about social changes that can transform people's experience in society. In this chapter, we look at the sociological analysis of the economy and work in society.

Economy and Society

All societies are organized around an economic base. The **economy** of a society is the system by which goods and services are produced, distributed, and consumed. The historic transformation from an agriculturally based society to an industrial society and now a post-industrial society is the foundation that will guide an understanding of the economy and work.

The Industrial Revolution

In Chapter 5, we discussed the evolution of different types of societies. Recall that one of the most significant of these changes was, first, the development of agricultural societies and, later, the far-ranging impact of the Industrial Revolution. Now, the Industrial Revolution is giving way to a postindustrial revolution—a development with far-reaching consequences for how work in society is organized.

The Industrial Revolution is usually pinpointed as beginning in mid-eighteenth-century Europe, soon thereafter spreading throughout other parts of the world. The Industrial Revolution led to numerous social changes because Western economies became organized around the mass production of goods. The Industrial Revolution created factories, separating work and family by relocating the place where most people were employed.

We still live in a society that is largely industrial, but that is quickly giving way to a new kind of social organization: the postindustrial society. Whereas industrial societies are primarily organized around the production of goods, **postindustrial societies** are organized around the provision of information and services. The United States has thus moved from being a manufacturing-based economy to an economy centered on the provision of services. *Service* is a broad term meant to encompass a wide range of economic activities now common in the labor market. It includes banking and finance, retail sales, hotel and restaurant work, and health care. It also includes parts of the information technology industry—not electronics assembly, but areas such as software design and the exchange of information (through the Internet, publishing, video production, and the like).

Comparing Economic Systems

Economic systems are also characterized by different forms; the three major forms are *capitalism, socialism,* and *communism.* These are not totally distinct, that is, many societies have a mix of these economic systems. **Capitalism** is an economic system based on the principles of market competition, private property, and the pursuit of profit. Within capitalist societies, stockholders own corporations—or a share of the corporations' wealth. Under capitalism, owners keep a surplus of what is generated by the economy; this is their *profit,* which may be in the form of money, financial assets, and other commodities.

Socialism is an economic institution characterized by state (government) ownership and management of the basic industries; that is, the means of production are the property of the state, not of individuals. Modern socialism emerged from the writings of Karl Marx, who predicted that capitalism would give way to egalitarian, state-dominated socialism, followed by a transition to stateless, classless communism. Many European nations, for example, have strong elements of socialism that mix with the global forces of capitalism. Sweden supports an extensive array of state-run social services, such as health care, education, and social welfare programs, but Swedish industry is capitalist. Other world nations are more strongly socialist, although they are not immune from the penetrating influence of capitalism. China was formerly a strongly socialist society that is currently undergoing great transformation to capitalist principles, including state encouragement of a market-based economy, the introduction of privately owned industries, and increased engagement in the international capitalist economy.

Communism is sometimes described as socialism in its purest form. In pure communism, industry is not the private property of owners. Instead, the state is the sole owner of the systems of production. Communist philosophy argues that capitalism is fundamentally unjust because powerful owners take more from laborers (and society) than they give, and use their power to maintain the inequalities between the worker and owner classes. Communist theorists in the nineteenth century declared that capitalism would inevitably be overthrown as workers worldwide united against owners and the system that exploited them. Class divisions were supposed to be erased at that time, along with private property and all forms of inequality. History has not borne out these predictions.

The Changing Global Economy

One of the most significant developments of modern times is the creation of a global economy, affecting work in the United States and worldwide. The concept of the **global economy** acknowledges that all dimensions of the economy now cross national borders, including investment, production, management, markets, labor, information, and technology. Economic events in one nation now can have major reverberations throughout the world. When the economies of any major nation are unstable, the effects are felt worldwide.

Multinational corporations—those that draw a large share of their revenues from foreign investments and conduct business across national borders—have become increasingly powerful, spreading their influence around the globe. The global economy links the lives of millions of Americans to the experiences of other people throughout the world. You can see the internationalization of the economy in everyday life: Status symbols such as high-priced sneakers are manufactured for just a few cents in China. The Barbie dolls that young girls accumulate are inexpensive by U.S. standards, yet it would require one month's wages for an Indonesian or Chinese worker who makes the doll to buy it for her child.

In the global economy, the most developed countries control research and management, and assembly-line work is performed in nations with less privileged positions in the global economy. A single product, such as an automobile, may be assembled from parts made all over the world—the engine assembled in Mexico, tires manufactured in Malaysia, and electronic parts constructed in China. The relocation of manufacturing to wherever labor is cheap has led to the emergence of the **global assembly line**, a new international division of labor in which research and development is conducted in the United States, Japan, Germany, and other major world powers, and the assembly of goods is done primarily in underdeveloped and poor nations—mostly by women and children.

Related to the global assembly line is the phenomenon known as outsourcing. **Outsourcing** is the transfer of a specialized task from one organization to another that occurs for cost saving; often, the work is transferred to a different nation, as you have likely witnessed when calling someone for help with your computer. The person who answers may well be working in India or somewhere else and is part of an economy that is deeply entangled with that in the United States.

Within the United States, the development of a global economy has also created anxieties about foreign workers, particularly among the working class. Because it is easier to blame foreign workers for unemployment in the United States than it is to understand the complex processes that have produced this phenomenon, U.S. workers have been prone to **xenophobia**, the fear and hatred of foreigners. Campaigns to "buy American" reflect this trend, although the concept of buying American is becoming increasingly antiquated in a global economy.

When buying a product from a U.S. company, it is likely that the parts, if not the product itself, were built overseas. In a global economy, distinctions between U.S. and foreign businesses blur. Moreover, the label "made in U.S.A." does not necessarily mean that the product was made by well-paid workers in the United States. In the garment industry, sweatshop workers—many of whom are recent immigrants and primarily women—are likely to have stitched the clothing that bears such a label. Moreover, these workers are likely to be working under exploitative conditions.

The development of a global economy is part of the broad process of **economic restructuring**, which refers to the contemporary transformations in the basic structure of work that are permanently altering the workplace. This process includes the changing composition of the workplace, deindustrialization, and use of enhanced technology. Some changes are *demographic*—that is, resulting from changes in the population. The labor force is becoming more diverse, with women and people of color becoming the majority of those employed. Other changes are driven by *technological developments.* For example, the economy is based less on its earlier manufacturing base and more on service industries.

THINKING Sociologically

Identify a job you once held (or currently hold) and make a list of all the ways that workers in this segment of the labor market are being affected by the various dimensions of *economic restructuring*: demographic changes, globalization of the economy, and technological change. What does your list tell you about how social structure shapes people's individual work experiences?

A More Diverse Workplace

A more diverse workplace is becoming a common result of economic restructuring. Today's workforce is both older, includes more women, and is more racially and ethnically diverse than ever before. People in the older age group (age 55 and older) have lower labor force participation rates than middle-age people, yet they comprise, at least for the time being, a somewhat larger percentage of workers than was true in the past—due in large part to the size of the baby boomer population.

As the U.S. population becomes more diverse, diversity in the labor market will continue. In fact, the White, non-Hispanic share of the labor market has fallen to 68 percent of the labor force, compared to 76 percent as recently as 1990. White, non-Hispanics in the labor market are expected to fall to 62 percent by the year 2020. Meanwhile, Asian and Hispanic workers—and all women—are expected to become an increased proportion of the labor market. The workforce is also growing more slowly than in the past, even though the U.S. population is growing (Toossi 2012). These basic facts about population change in the workforce will shape the experience of generations to come.

THINKING Sociologically

Think about the labor market in the region where you live. What racial and ethnic groups have historically worked in various segments of this labor market, and how would you now describe the racial and ethnic *division of labor*?

These changes in the social organization of work and the economy are creating a more diverse labor force, but much of the growth in the economy is projected to be in service industries, where, for the better jobs, education and training are required. People without these skills will not be well positioned for success. Manufacturing industries, where racial minorities have in the past maintained a foothold on employment, are now in decline. New technologies and corporate layoffs have reduced the number of entry-level corporate jobs that recent college graduates have always used as a starting point for career mobility. Many college graduates are employed in jobs that do not require a college degree. College graduates, however, do still have higher earnings than those with less education.

Deindustrialization

Deindustrialization refers to the transition from a predominantly goods-producing economy to one based on the provision of services. This does not mean that goods are no longer produced, but that fewer workers in the United States are required to produce goods. Machines can do the work people once did, and many goods-producing jobs have moved overseas.

Deindustrialization is most easily observed by looking at the decline in the number of jobs in the manufacturing sector of the U.S. economy since the Second World War. The manufacturing sector includes workers who actually produce goods. At the end of the war in 1945, the majority of workers (51 percent) in the United States were employed in manufacturing-based jobs. Now, manufacturing accounts for only about 10 percent of the total labor force (U.S. Department of Labor 2014).

The *service sector* employs the other 90 percent, including two segments: the actual delivery of services (such as food preparation, cleaning, or child care) and the transmission and processing of information (such as banking and finance, computer operation, clerical work, and even education workers, such as teachers). Parts of the service sector are higher-wage and prestigious jobs, such as physicians, lawyers, financial professionals, and so forth, but huge parts of the service sector—and those with the largest occupational growth—are low-wage, semi-skilled, and unskilled jobs. This lower end of the service sector employs many women, people of color, and immigrants.

The human cost of deindustrialization can be severe. Deindustrialization has led to job displacement, the permanent loss of certain job types that occurs when employment patterns shift. When a manufacturing plant shuts down, many people may lose their jobs at the same time, and whole communities can be affected. Communities that were heavily dependent on a single industry, such as steel towns or automobile-manufacturing cities such as Detroit, are among the areas hardest hit by deindustrialization. Rural communities, too, can be harmed by deindustrialization because they often have only one major employer, such as a textile plant.

Job displacement hits people in both rural and inner cities hard because emerging new industries tend to be located in suburban, not urban or rural, areas—a phenomenon called *spatial mismatch*. It is then no surprise that poverty is highest in central cities and rural America. Unless young people have the educational and technical skills for employment in a new economy, or if they live in areas hard hit by job displacement, they have little opportunity for getting a good start in the now global economy. You see this in the extremely high unemployment rates for Black and Hispanic teens (see Figure 15.3 later in this chapter; Wilson 2009, 1996).

Technological Change

Coupled with deindustrialization, rapidly changing and developing technologies are bringing major changes in work, including how it is organized, who does it, and how much it pays. One of the most influential technological developments of the twentieth century has been the invention of the semiconductor. Computer technology has made possible workplace transactions that would have seemed like science fiction not that many years ago. Electronic information can be transferred around the world in less than a second. Employees can provide work for corporations located on another continent. A woman in Southeast Asia or the Caribbean can download and produce a book manuscript for a publishing house in New York. Some argue that the computer chip

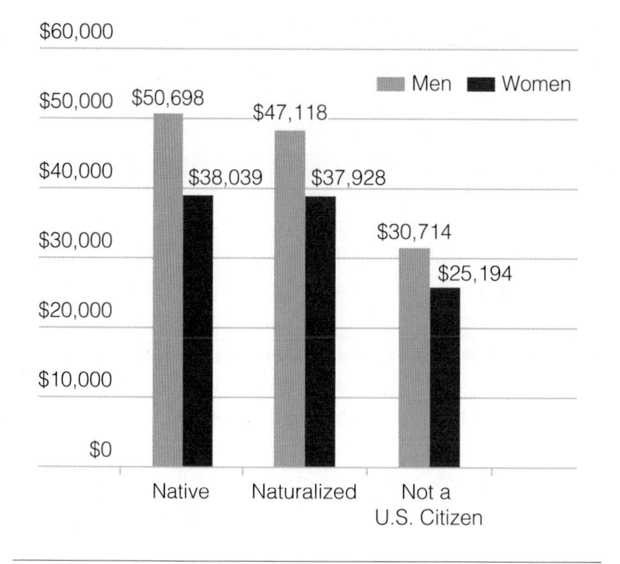

▲ **Figure 15.2** The Income Gap: U.S.-Born, Naturalized, and Non–U.S. Citizen Workers, 2012

These median income figures are for full-time workers only. Does anything surprise you about what you see in the graph? What social factors do you think influence these earnings differences?

Data: U.S. Census Bureau. 2013. *Total Earnings of Full-Time, Year-Round Workers 15 Years and Over with Earnings by Sex, Nativity, and U.S. Citizenship Status: 2011*. Washington, DC: U.S. Department of Commerce. **www.census.gov**

Earnings. Sociologists have extensively documented that earnings from work are highly dependent on race, gender, and class, as shown in ▲ Figure 15.2. White men earn the most, with a gap between men's and women's earnings among all groups. African American women and Hispanic men and women earn the least. Occupations in which White men are the numeric majority tend to pay more than occupations in which women and minorities are a majority of the workers (Levanon et al. 2009).

Diverse Groups/Diverse Work Experiences

Data on characteristics of the U.S. labor force typically are drawn from official statistics reported by the U.S. Department of Labor. The labor force now includes approximately 155 million people (U.S. Bureau of Labor Statistics 2014a). Who works, where, and how varies considerably for different groups in the population, however.

One of the most dramatic changes in the labor force since the Second World War has been the increase in the number of women employed. Since 1948, the employment of women has increased from 35 to 57 percent of all women. Women now constitute almost half (47 percent) of all workers. Other changes in the labor force include that racial minorities (Hispanics, Asians, and African Americans) are the fastest-growing segment of the labor force, although White, non-Hispanic workers are still 70 percent of the workforce. Hispanics and Asians have the fastest growth in the labor force, largely because of immigration (U.S. Bureau of Labor Statistics 2014a).

These trends, however, do not mean—as popularly believed—that minorities and women are routinely taking jobs from White men. Jobs where White men have predominated are precisely those that have been declining because of economic restructuring, such as in the manufacturing sector. Jobs in areas of the labor market that are race- and gender-segregated are increasing, such as fast-food work and other low-wage service jobs. The conflicts that exist about work—such as the belief that immigrant and foreign workers are taking U.S. jobs or that women and minorities are taking away White men's jobs—stem from social structural transformations in the economy, not the individual behaviors of people affected by these changes (see ◆ Table 15.1).

◆ **Table 15.1** Occupational Status by Nativity and Citizenship Status

	Native	Naturalized	Not a U.S. Citizen
Management, business, financial	16.5%	14.6%	8.1%
Professional	22.9%	23.5%	13.3%
Service	16.4%	20.7%	30.9%
Sales	24.9%	20.3%	13.8%
Farming, fishing, forestry	0.5%	0.6%	2.5%
Construction, extraction, maintenance	7.6%	6.4%	14.1%
Production, transportation, material moving	11.2%	13.9%	17.3%

Source: U.S. Census Bureau. 2012. "Occupation of Employed Civilian Workers 16 Years and Over by Sex, Nativity, and U.S. Citizenship Status: 2012." Washington, DC: U.S. Department of Commerce. **www.census.gov**

Halfpoint/Shutterstock.com

Without work, people can become detached from society, producing depression and stress, but also producing social problems such as crime and homelessness.

Unemployment and Joblessness

The U.S. Department of Labor regularly reports the **unemployment rate**, defined as the percentage of those not working but officially defined as looking for work. As of 2013, the unemployment rate was 7.6—that is, 7.6 percent of the labor force. Full employment is considered to be around 4 percent. During the recession that hit the nation in 2010, unemployment was as high as 9.6 (U.S. Bureau of Labor Statistics 2014a). Although there has been some recovery from this very difficult time, many say it is a jobless recovery because so many people remain jobless.

How can there be a decline in the unemployment rate while so many people are still out of work? The official unemployment rate does not include all people who are jobless. It includes only those who meet the official definition of unemployment—those who do not have a job and who have looked for work in the period being reported. Thus the official measure excludes people who earned money at any job during the time prior to the data being collected; it excludes people who have given up looking for full-time work (so-called *discouraged workers*); those who have settled for part-time work; people who are ill or disabled; those who cannot afford the child care, transportation, or other necessities for getting to work; and those who work only a few hours a week even though their economic position may differ little from that of someone with no work at all.

DEBUNKING Society's Myths

Myth: The unemployment rate is a good measure of the nation's joblessness.

Sociological Perspective: Unemployment is not the same as joblessness. The unemployment rate only counts those who are actively seeking work and excludes those who are working part-time, those who are so discouraged they have quit working, and those who are underemployed, among others. The unemployment rate is one indicator of the state of the economy, but joblessness is typically higher.

Migrant workers and other transient populations are also undercounted in the official statistics. Workers on strike are also counted as employed, even if they receive no income while they are on strike. Because so many people are excluded from the official definition of unemployment, the official unemployment rate seriously underestimates actual joblessness. The people most likely to be left out of the unemployment rate are those for whom unemployment runs the highest—the youngest and oldest workers, women, and racial minority groups. These groups are also those most likely to have left jobs that do not qualify them for unemployment insurance, because to be eligible for unemployment you have to have worked a certain period of time and earned enough wages to qualify for a claim.

Official unemployment rates also ignore **underemployment**—the condition of being employed at a skill level below what would be expected given a person's training, experience, or education. This condition can also include working fewer hours than desired. A laid-off autoworker flipping hamburgers at a fast-food restaurant is underemployed as is a person with a college degree cleaning houses for a living.

Even given the problems in measuring unemployment, the highest rates of unemployment are among Black men, followed by Hispanic men (see ▲ Figure 15.3). When the government reports the national unemployment rate, it is a safe bet that unemployment among African Americans will be at least twice the national rate—a pattern that has persisted over time. Although not regularly reported by the Department of Labor, unemployment among Native Americans is also staggeringly high—11 percent in 2013. In several areas, unemployment among Native Americans is even higher—16.8 percent in the Midwest; 15 percent in the High Plains and Southwest (Austin 2013). The unemployment rate that is found among Native Americans, African Americans, and Hispanics (with the exception of Cuban Americans) exceeds the unemployment rate that was considered a national crisis during the Great Depression of the 1930s.

People often attribute unemployment to the individual failings of workers, claiming that unemployed people do not try hard enough to find jobs or prefer a welfare check to hard work and a paycheck. This leads some to attribute unemployment to the "laziness" of unemployed individuals rather than to actual, factual structural conditions in society. This viewpoint reflects the common myth that *anyone* who works hard enough and puts forward sufficient effort can succeed.

Digital Storm/Shutterstock.com

The influence of money in politics raises serious concerns about how representative the U.S. political system really is.

these organizations and the interests of the government. These interests naturally overlap and reinforce each other.

Members of the upper class do not need to occupy high office themselves to exert their will, as long as they are in a position to influence people who are in power (Domhoff 2013). The majority of the power elite are White men, which means that the interests and outlooks of White men dominate the national agenda.

The Autonomous State Model

A third view of power developed by sociologists, the **autonomous state model**, interprets the state as its own major constituent. From this perspective, the state develops interests of its own, which it seeks to promote independently of other interests and the public that it allegedly serves. The state does not reflect the needs of the dominant groups, as Marx and power elite theorists would contend. It is an administrative organization with its own needs, such as maintenance of its complex bureaucracies and protection of its special privileges (Rueschmeyer and Skocpol 1996; Skocpol 1992).

The huge government apparatus now in place in the United States is a good illustration of autonomous state theory. The government provides a huge array of social support programs, including Social Security, unemployment benefits, agricultural subsidies, public assistance, and other economic interventions intended to protect citizens from the vagaries of a capitalist market system. The purpose of these programs is to serve people in need. Autonomous state theory argues that the government has grown into a massive, elaborate bureaucracy, run by bureaucrats more absorbed in their own interests than in meeting the needs of the people. As a consequence, government can become paralyzed in conflicts between revenue-seeking state bureaucrats and those who must fund them. You can imagine autonomous state theory as well explaining what many now perceive as a completely stalled federal government.

DEBUNKING Society's Myths

Myth: Congress cannot get anything done because it is filled with obstructionist individuals.

Sociological Perspective: It is true that bipartisanship in Congress has been very difficult to achieve in recent years. Beyond individual behavior, however, autonomous state theory suggests that the Congress takes on an organizational life of its own, operating to protect the status quo and its own interests. It thus becomes a very stalled bureaucracy.

Feminist Theories of the State

Feminist theorists diverge from the preceding theoretical models by seeing men as having the most power in society. The pluralists see power as widely dispersed through the class system, power elite theorists see political power directly linked to upper-class interests, and autonomous state theorists see the state as relatively independent of class interests.

Some feminist theorists argue that all state institutions reflect men's interests; they see the state as fundamentally patriarchal, its organization embodying the fixed principle that men are more powerful than women. Feminist theories of the state conclude that despite the presence of a few powerful women, the state is devoted primarily to men's interests. Moreover, the actions of the state will tend to support gender inequality (Haney 1996; Blankenship 1993). One historical example would be laws denying women the right to own property once they married. Such laws protected men's interests at the expense of women.

Evidence that "the state is male" (MacKinnon 2006, 1983) is easy to observe by looking at powerful political circles. Despite the inclusion of more women in powerful circles and the presence of some notable women as major national figures, most of the powerful are men. Both the U.S. Senate and House of Representatives, despite recent gains for women, are still 80 percent men. Groups that exercise state power, such as the police and military, are predominantly men. Moreover, these institutions are structured by values and systems that can be described as culturally masculine—that is, based on hierarchical relationships, aggression, and force. Feminist theory begins with the premise that an understanding of power cannot be sound without a strong analysis of gender (Haney 1996).

Government: Power and Politics in a Diverse Society

The terms *government* and *state* are often used interchangeably. More precisely, the government is one of several institutions that make up the state. The **government** includes those institutions that represent the population, making rules that govern the society. The government of the United States is a *democracy*; therefore, it is based on the principle of representing all people through the right to vote.

The actual makeup of the government, however, is far from representative of society. Not all people participate equally in the workings of government, neither as elected officials nor as voters. Women, the poor and working class, and racial-ethnic minorities are less likely to be represented by government than are White middle- and upper-class men. Sociological research on political power has concentrated on inequality in government affairs and has demonstrated large, persistent differences in the political participation and representation of various groups in society.

Diverse Patterns of Political Participation

One would hope that all people in a democratic society would be equally eager to exercise their right to vote and be heard. That is far from the case. Among democratic nations, the United States has one of the *lowest* voter turnouts (see ▲ Figure 15.4). In the 2012 presidential election, the percentage of eligible voters who went to the polls was only 57 percent of the population, less than the all-time high of 62 percent in the 2008

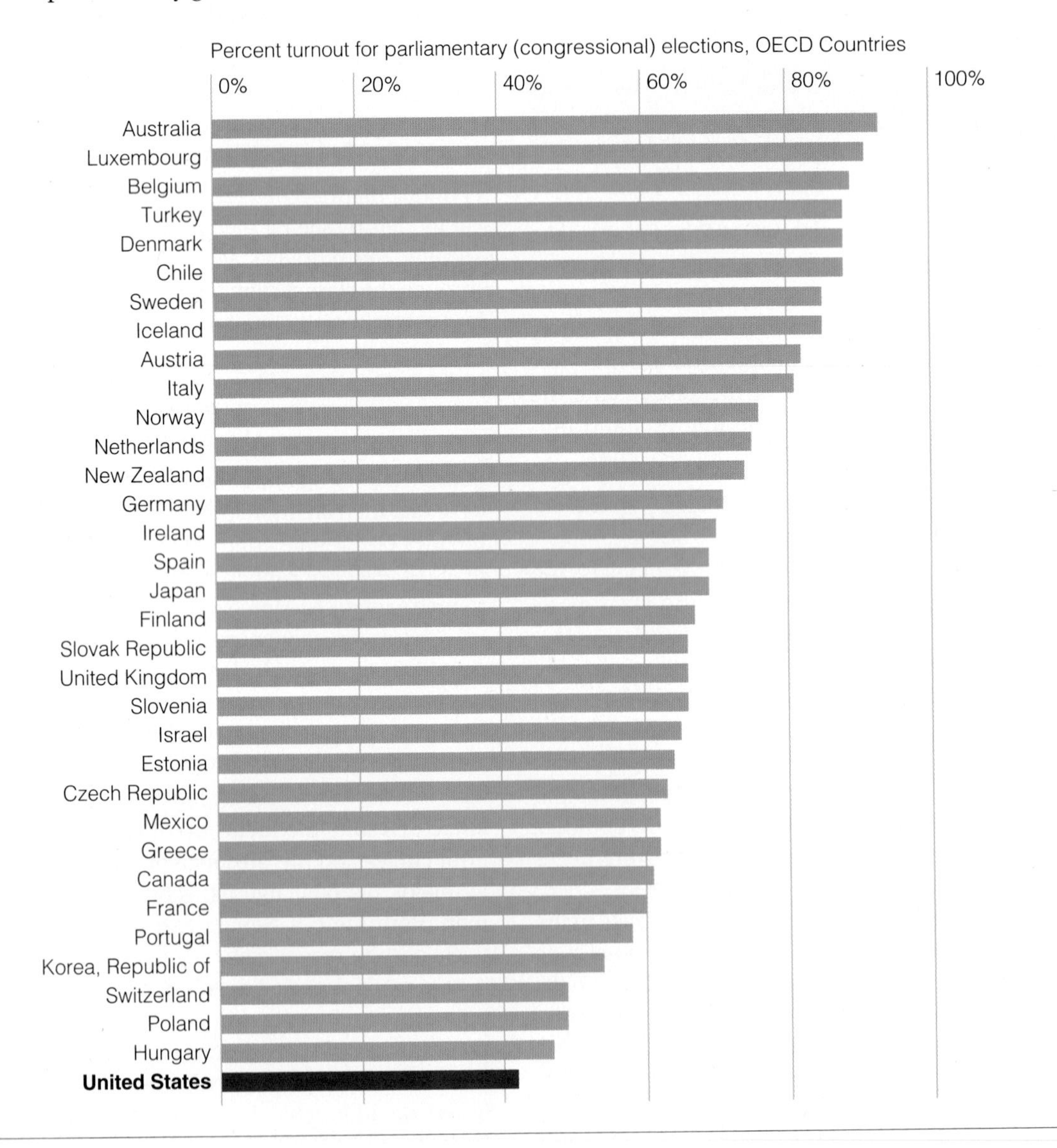

▲ **Figure 15.4** International Voter Turnout, 2009–2011 As you can see, the United States has lower voter turnout in national elections than other industrialized nations.

Source: Institute for Democracy and Electoral Assistance. 2012. "Voter Turnout." Stockholm: Sweden. **www.idea.int**

understanding **diversity**

Diversity in the Power Elite

As society has become more diverse, has it made a difference in the makeup of the power elite? Various groups—women, racial–ethnic groups, lesbians, and gays—have vied for more representation in the halls of power, but have their efforts succeeded? If they make it to power, does this change the corporations, military, or government—the major institutions composing the power elite?

Sociologists Richard L. Zweigenhaft and G. William Domhoff examined these questions by analyzing the composition of boards of directors and chief executive officers (CEOs) of the largest banks and corporations in the United States, as well as analyzing Congress, presidential cabinets, and the generals and admirals who form the military elite. In addition, they examined the political party preferences and the political positions of people found among the power elite. Do women and minorities bring new values into power, thereby changing society as they move into powerful positions, or do their values match those of the traditional power elite or become absorbed by a system more powerful than they are? Zweigenhaft and Domhoff's study looks as well at whether those who do make it into the power elite are within the innermost circles or whether they are marginalized.

They find that women, Jews, gays, lesbians, Black Americans, and Hispanics have become more numerous within the power elite, but only to a small degree. The power elite is still overwhelmingly White, wealthy, Christian, and male. Women and other minorities who make it into the power elite also tend to come from already privileged backgrounds, as measured by their social class and education. Among African Americans and Latinos, skin color continues to make a difference, with darker-skinned Blacks and Hispanics less likely to achieve prominence compared with lighter-skinned people. Furthermore, Zweigenhaft and Domhoff find that the perspectives and values of women and minorities who rise to the top do not differ substantially from their White male counterparts. Some of this is explained by the common class origins of those in the power elite. The researchers also attribute the managing of one's identity to avoid challenging the system as a sorting factor that perpetuates the dominant worldview and practices of the most powerful.

The authors of this study conclude that "the irony of diversity" is that greater diversity may have strengthened the position of the power elite because its members appear to be more legitimate through their inclusion of those previously left out. But, by including only those who share the perspectives and values of those already in power, little is actually changed.

Sources: Zweigenhaft, Richard L, and G. William Domhoff. 2011. *The New CEOs: Women, African Americans, Latino, and Asian American Leaders of Fortune 500 Companies*. Lanham, MD: Rowman and Littlefield; Zweigenhaft, Richard L., and G. William Domhoff. 2006. *Diversity in the Power Elite: Have Women and Minorities Reached the Top?* New Haven: Yale University Press; Domhoff, G. William. 2002. *Who Rules America?* New York: McGraw-Hill.

are now in the U.S. Senate (out of 100) and 83 in the House of Representatives (out of 435). In the 114th Congress, there are 49 African Americans (one in the Senate); 32 Latinos (three in the Senate); 9 Asian Americans; and six openly gay or bisexual members, including the first openly lesbian member of the Senate.

There has been an increase in religious diversity in Congress, but the overwhelming majority of senators and representatives are Protestant and Catholic. To a limited extent, the Congress includes people of different faiths, including those who are Jewish (5 percent) and a very small number (about 3 percent) of those from diverse religious backgrounds, such as Muslims, Hindus, Buddhist, and one self-identified atheist, among others (Wolf 2014). There is a long way to go before Congress truly represents the diversity in the population.

Researchers offer several explanations for why women and racial–ethnic minorities continue to be underrepresented in government. Certainly, prejudice plays a role. It was not long ago, in the 1960 Kennedy-Nixon election, that Kennedy became the first Catholic president elected. Joseph Lieberman was the first Jewish candidate to appear on a major national ticket. Mitt Romney in 2012 was the first Mormon candidate to appear on a national ballot, but 18 percent of the public now say they would not vote for a Mormon as president, even if qualified (Newport 2012a). We have also witnessed much prejudice directed toward President Obama with false accusations that he is "Muslim" and not a U.S. citizen, even though he was born in Hawaii.

Gender and racial prejudice run deep in the public mind. Although 71 percent of Americans say it would not matter if a presidential candidate were a woman, a significant number do not say so (Pew Research Center for the People and the Press 2014). Prejudice does not, however, fully account for the lack of a more representative government. Societal causes are a major factor in the successful elections of women and people of color. Women and minority candidates receive a great deal of political support from local groups, but at the national level, they do not fare as well. The power of incumbents, most of whom are White men, is also a disadvantage to any new office seeker.

The Military as a Social Institution

Social institutions are stable systems of norms and values that fulfill certain functions in society. The military is a social institution whose function is to defend the nation against external (and sometimes internal) threats. A strong military is often considered an essential tool for maintaining peace. The military arm of the state is among the most powerful and influential social institutions in almost all societies. In the United States, the military is the largest single employer. Approximately 2.4 million men and women serve in the U.S. military, 1.3 million on active duty, and the rest in the reserves. This does not include the many hundreds of thousands who are employed in industries that support the military, nor does it include civilians who work for the Department of Defense and other military-affiliated agencies (U.S. Census Bureau 2012a).

The military is one of the most hierarchical social institutions, and its hierarchy is extremely formalized. People who join the military are explicitly labeled with rank, and if promoted, they pass through a series of additional well-defined levels (ranks), each with clearly demarcated sets of rights and responsibilities. An explicit line exists between officers (lieutenants and higher ranks) and enlisted personnel, and officers have many privileges that others do not. Higher ranks are also entitled to absolute obedience from the ranks below them, with elaborate rituals created to remind both dominants and subordinates of their status.

As in other social institutions, military enlistees are carefully socialized to learn the norms of the culture they have joined. Military socialization places a high premium on conformity and eliminates individuality. All new recruits are issued identical uniforms and are allowed to retain very few of their personal possessions. They are endlessly harangued by the infamous "D.I." (drill instructor). They must quickly learn new, strictly enforced codes of behavior.

Most of the military is a part of the institution of government, but there has also been *privatization of the military*—meaning that an increasing number of military functions have been paid on a contract basis to private, for-profit employers. Under this development, the military becomes like a business, with people and corporations reaping profits on activities that once would have been not for profit. The privatization of the military can include companies that provide specific services (such as security), as well as engineering and building contracts. Critics of this trend warn that it will sacrifice safety and national security for the sake of corporate and individual profit and could lure the brightest people away from traditional military service if they see economic gains from private military service (Singer 2007).

Race and the Military

The greatest change in the military as a social institution is the representation of racial minority groups and women within the armed forces. Picture a U.S. soldier. Whom do you see? At one time, you would have almost certainly pictured a young White male, possibly wearing army green camouflage and carrying a weapon. Today, the image of the military is much more diverse. Drawing on the cultural images you have stored in your mind, you are just as likely to picture a young African American man in a military dress uniform with a stiffly starched shirt and a neat and trim appearance or perhaps a woman wearing a flight helmet in the cockpit of a fighter plane.

African Americans have served in the military for almost as long as the U.S. armed forces have been in existence. Except for the Marines, which desegregated in 1942, the armed forces were officially segregated until 1948, when President Harry Truman signed an executive order banning discrimination in the armed services. Although much segregation continued after this order, the desegregation of the armed forces is often credited with promoting more positive interracial relationships and increased awareness among Black Americans of their right to equal opportunities than has been the case in society at large. Until that time, the widespread opinion among Whites was that to allow Black and White soldiers to serve side by side would destroy soldiers' morale.

Currently, 18 percent of active military personnel are African American and 11 percent are Hispanic (who fall into various other racial-ethnic categories). Almost 3.6 percent are Asian Americans; 1.7 percent, American Indian or Alaskan Native; 3.2 percent multiracial; and 1.1 percent native Hawaiian or Pacific Islander (U.S. Department of Defense 2012). Enlistees have many reasons to join the military, but the desire for education and job training is certainly among the strongest motivators, along with wanting to serve one's country.

Within the military today, there is a policy of equal pay for equal rank. African Americans and Latinos, however, are overrepresented in lower-ranking support positions. Often, they are excluded from the higher-status, technologically based positions—those most likely to bring advancement and higher earnings both in the military and beyond. Most minorities remain in positions with little supervisory responsibility, such as

JIM WATSON/Getty Images

Women are an increasing presence in the U.S. military.

service and supply jobs. Although the number of racial minorities in officers' positions has been increasing, they are still underrepresented and are less likely to get there via the route of military academies, as is the case for White officers (Segal and Segal 2004). Still, for both Whites and racial minorities, serving in the military leads to higher earnings relative to one's nonmilitary peers.

Women in the Military

The military academies did not open their enrollment to women until 1976. Since then, the armed services have profoundly changed their admission policies, and in 1996, the Supreme Court ruled (in *United States v. Virginia*) that women cannot be excluded from state-supported military academies such as the Citadel and the Virginia Military Institute (VMI). This was a landmark decision that opened new opportunities for women who want the rigorous physical and academic training that military academies provide (Kimmel 2000).

The involvement of women in the military has reached an all-time high in recent years. Women are 14 percent of enlisted military personnel. Now, almost 204,000 women are on active duty in the United States. The Army has the highest proportion of women, followed by the Air Force, the Navy, and the Marines. Women now comprise 17 percent of military officers and 21 percent of military academy cadets and midshipmen (U.S. Department of Defense 2014).

The former exclusion of women from military service was rationalized by the popular conviction that women should not serve in combat. Despite this attitude, women have been fighting in active combat to defend the nation and were officially made eligible for combat role in 2013.

The presence of women in the military has transformed the armed forces, but it also has raised new issues for military personnel. Slightly more than half (56 percent) of military personnel are married, and 6 percent of active-duty members are in dual-military marriages. Family separations, frequent moves (on average every three years), risk of injury or death, and living in a foreign country are only some of the challenges that military personnel face in trying to manage their lives (U.S. Department of Defense 2012; Segal and Segal 2004).

For women in the military, the highly gendered organization of which they are a part is also a challenge. Indeed, recent reports have documented an alarmingly high rate of sexual assault and sexual harassment against women in the military—by other military personnel. Reports from the Pentagon have found that one-third of women in the military (and 6 percent of men) experienced sexual harassment, including unwanted crude and offensive behavior, unwanted sexual attention, and sexual coercion; the same report found that 5 percent of women in the military experienced some form of unwanted sexual contact, such as rape, unwanted sodomy, or indecent assault (U.S. Department of Defense 2012). Periodic scandals involving rape, sexual harassment, and other forms of intimidation against women in the military (including the military academies) reveal that, although certainly not all military men engage in these behaviors, institutions organized around such masculine characteristics as aggression, domination, and hierarchy put women at risk.

Wavebreak Media ltd/Alamy

The men and women who serve in the armed forces, such as this young woman returning from Iraq, are often separated from families and loved ones for long periods of time.

Gays and Lesbians in the Military

Gays and lesbians have long served in military duty, despite the policies that have attempted to exclude them. The military has admitted that there always have been gays and lesbians in all branches of the U.S. armed forces, but homophobia is a pervasive part of military culture (Becker 2000; Myers 2000).

The Obama administration ended the "don't ask, don't tell" policy by which recruiting officers could not ask about sexual preference. It remains unclear whether gays and lesbians will be permitted to live openly as gay while also pursuing careers in the armed services. Supporters of the ban on gays in the military often use arguments similar to the arguments used before 1948 to defend the racial segregation of fighting units. As in 1948, detractors claim that the morale of soldiers will drop if forced to serve alongside gay men and women, national security will be threatened, and known homosexuals serving in the military will upset the status quo and destroy the fighting spirit of military units. Seeing these arguments in historical perspective helps you see through some of the myths perpetuated by such attitudes.

Military Veterans

Now, almost two million veterans have returned from the wars in Iraq and Afghanistan. Add to that the veterans of the Gulf War, Vietnam War, Korean War, and the living veterans of World War II, and it totals 22 million veterans living in the United States (U.S. Census Bureau 2012a). For all veterans, the return home—though joyful—also has risks, risks that result from social, as well as physical, needs for recovery and adaptation.

The changed nature of combat in the two Iraq and Afghanistan wars has meant that returning veterans have more complex forms of physical and emotional injury. Exposure to repeated blasts of IEDs (improvised explosive devices) has resulted in more traumatic brain injuries. Having an all-volunteer army has also produced more frequent redeployment, resulting in longer-term exposure to war trauma, as well as greater exposure to blasts and other forms of violence. Changes in military and medical technology have also increased survival rates from injuries that would have killed military personnel in the past.

All of these factors have meant increased risks for returning veterans, including not only difficult recoveries from physical injuries, but also high rates of mental health disorders, a high risk of suicide, depression, and/or drug and alcohol addiction. In addition to these social problems, veterans face various adaptation challenges as they transition back into the civilian workforce—that is, if they find work. Veterans and their families and partners also have to adapt to new family roles, perhaps even including a readjustment as parents because their children will have matured in their absence. Adding to this complexity is the fact that there may be a significant readjustment to a new physical or mental disability. Managing the health problems that may have developed during deployment produces new forms of stress on preexisting relationships (Institute of Medicine of the National Academies 2010).

When veterans return home, often the social supports they need are not strong. One consequence has been an increase in the number of homeless veterans—a figure that has doubled since 2010 (Zoroya 2012). African American veterans, for whom the military has been a path for social mobility, may face the additional fact that social institutions fail them again in the form of unemployment and persistent racism (Finkel 2014; Fleury-Steiner 2012).

The situation for U.S. veterans shows how critical social institutions are in the lives of these men and women. For all members of society, the support—or lack thereof—provided by social institutions is a critical backdrop to the character of everyday life.

Chapter Summary

How are societies economically organized?

Societies are organized around an economic base. The *economy* is the system on which the production, distribution, and consumption of goods and services are based. *Capitalism* is an economic system based on the pursuit of profit, market competition, and private property. *Socialism* is characterized by state ownership of industry; *communism* is the purest form of socialism.

How has the global economy changed?

As capitalism has spread throughout the world, *multinational corporations* conduct business across national borders. A number of countries have undergone *deindustrialization*, or changeover from a goods-producing economy to a services-producing one. This has caused many heavy-industry jobs in U.S. cities to vanish, thus increasing the unemployment rate in those cities. Changes in information technology, plus increased *automation*, have resulted in the further elimination of jobs in both the United States and abroad.

What is the social organization of work?

Sociologists define *work* as human activity that produces something of value. Some work is judged to be

more valuable than other work. *Emotional labor* is work that is intended to produce a desired state of mind in a client. The *division of labor* is the differentiation of work roles in a social system. In the United States, there is a class, gender, and racial division of labor. The labor market in the United States is described as a *dual labor market.* Jobs in the primary sector of the labor market carry better wages and working conditions, whereas those in the secondary labor market pay less and have fewer job benefits. Women and minorities are disproportionately employed in the secondary labor market. Patterns of occupational distribution also show tremendous segregation by race and gender in the labor market. Race and gender also affect the occupational prestige, as well as the earnings, of given jobs.

How is diversity reflected in the workplace?

The workplace is becoming more diverse with greater numbers of racial–ethnic groups, women, and an older workforce. Official *unemployment rates* underestimate the actual extent of joblessness. Women and minorities often encounter the *glass ceiling*—a term used to describe the limited mobility of women and minority workers in male-dominated organizations. In addition, women more often than men face *sexual harassment* at work—defined as the unequal imposition of sexual requirements in the context of a power relationship. Homophobia in the workplace also negatively affects the working experience of gays and lesbians. New protections are in place for disabled workers through the *Americans with Disabilities Act* (ADA).

What is the state?

The *state* is the organized system of power and authority in society. It comprises different institutions, including the government, the military, the police, the law and the courts, and the prison system. The state is supposed to protect its citizens and preserve society, but it often protects the status quo, sometimes to the disadvantage of less powerful groups in the society. States can also be organized as *democracies*, as *authoritarian*, or as *totalitarian.*

How do sociologists define power and authority?

Power is the ability of a person or group to influence another. *Authority* is power perceived to be legitimate and formal. There are three kinds of authority: *traditional authority,* based on long-established patterns; *charismatic authority,* based on an individual's personal appeal or charm; and *rational–legal authority,* based on the authority of rules and regulations (such as law).

What theories explain how power operates in the state?

Sociologists have developed four theories of power. The *pluralist model* sees power as operating through the influence of diverse interest groups in society. The *power elite model* sees power as based on the interconnections between the state, industry, and the military. *Autonomous state theory* sees the state as an entity in itself that operates to protect its own interests. *Feminist theorists* argue that the state is patriarchal, representing primarily men's interests.

How well does the government represent the diversity of the U.S. population?

An ideal democratic government would reflect and equally represent all members of society. The makeup of the U.S. government does not reflect the diversity of the general population. African Americans, Latinos, Native Americans, Asians, and women are underrepresented within the government. Political participation also varies by a number of social factors, including income, education, race, gender, and age. African Americans and Latinos, however, are overrepresented in the military, in part because of the opportunity the military purports to offer groups otherwise disadvantaged in education and the labor market; however, both are underrepresented at the levels of high-level commissioned officers. There is an increased presence of women in the military; however, prejudice and discrimination continue against lesbians and gays in the military.

Key Terms

alienation **375**
Americans with Disabilities Act (ADA) **374**
authoritarian **377**
authority **378**
automation **367**
autonomous state model **381**
bureaucracy **378**
capitalism **364**
charismatic authority **378**
communism **364**
contingent worker **368**
deindustrialization **366**
democracy **377**
division of labor **369**
dual labor market **370**
economic restructuring **365**
economy **364**
emotional labor **368**
glass ceiling **369**
global assembly line **365**
global economy **365**
government **382**
interest group **379**
interlocking directorate **380**
multinational corporations **365**
nationalism **377**
occupational segregation **370**
outsourcing **365**
pluralist model **379**
political action committees (PAC) **379**
postindustrial society **364**
power **377**
power elite model **380**
propaganda **377**
rational–legal authority **378**
sexual harassment **373**
socialism **364**
state **376**
totalitarian **377**
traditional authority **378**
underemployment **372**
unemployment rate **372**
work **368**
xenophobia **365**

16

Stuwdamdorp/Alamy

Environment, Population, and Social Change

Can we preserve the Earth's resources as we know them? Scientists and others are warning us that the polar ice cap is melting, causing ocean levels to rise. Climate patterns are changing. Although the specific effects of climate change are being debated, people worry that more severe storms, extreme heat, or perhaps bone-chilling cold will become more common. In some parts of the world, population growth outpaces the ability to feed people. In other places, including in the United States, water can no longer be assumed to be available or safe to drink. Air pollution is so bad in some parts of the world that people routinely wear masks. Is our current lifestyle sustainable?

Sustainability, in fact, has become an organizing cry—a cry for new social policies that will protect the Earth's environment and the people who live within it. New movements have developed—movements to eat local food, to support the creation of urban community gardens, and to recycle used products by transforming them into something else. These and other developments signal the public's concern with the environment and the related phenomena of population, pollution, and social change.

You might think that studying such things as sustainable energy and environmental pollution is solely the work of scientists and engineers. No doubt these are critical scientific problems, yet as scientists and engineers will tell you, social issues are just as important in understanding how we can preserve the Earth's resources. What lifestyles consume the highest

THE ENVIRONMENT

in this chapter, you will learn to:

- Identify the social dimensions of environmental change
- Explain how inequality affects environmental quality for different groups
- Understand the basic processes of population change
- List the changes that affect population diversity in the United States
- Explain theories of population growth
- Describe the different components and sources of social change
- Compare and contrast sociological theories of social change
- Analyze the social implications of globalization and modernization

amounts of energy? What social and cultural changes are needed to protect our environment and not deplete the Earth's natural resources? How is social inequality related to the degradation of the environment? Are there just too many people for the world to sustain human society as we know it?

These and other questions drive the substance of this chapter—a chapter that looks at the sociological issues that come from studying population and the environment. *Environmental sociology* now has particular urgency as people become more attuned to the potential crises that our planet faces.

A Climate in Crisis: Environmental Sociology

Human beings, animals, and plants all depend on one another and on the physical environment for their survival. **Environmental sociology** is the scientific study of the interdependencies that exist between humans and our physical environment. A *human ecosystem* is any system of interdependent parts that involves human beings in interaction with one another and the physical environment. As examples, a city is a human ecosystem; so is a rural farmland community. In fact, the entire world is a human ecosystem.

The examination of ecosystems has demonstrated two things:

1. The supply of many natural resources is finite, and
2. If one element of an ecosystem is disturbed, the entire system is affected.

For much of the history of humankind, the natural resources of the Earth were abundant. Compared with the amounts humans used, they may as well have been infinite. No more. Some resources, such as certain fossil fuels, are simply nonrenewable and may be gone soon. Other resources, such as timber or seafood, are renewable only if we do not plunder the sources of supply so recklessly that they disappear. Some natural resources are so abundant that they still seem infinite, such as the planet's stock of air and water. But, at this stage of our societal development, we are learning that without more vigilance, we can destroy even the near-infinite resources (Gore 2006). Understanding social behavior as it affects the environment is critical to thinking about and solving our environmental problems.

Society at Risk: Air, Water, and Energy

An engineer might invent a new way to heat our homes and power our cars, but without understanding the social dimensions of issues like energy, pollution, water usage, and other environmental behaviors, we cannot make progress in maintaining and improving our environmental sustainability. The challenges we face in protecting the environment are many. Gaseous wastes are gnawing away at the ozone layer. Buried chemical wastes are trickling into the water table, creating underground pools of poison. Pollution has damaged the Earth's surface water so badly that worldwide underground water reserves are being mined faster than nature can replenish them.

China is the world's leading producer of carbon-based emissions.

The skies of all major cities around the world are stained with pollution hazes. In cities that rest within geological basins, such as Mexico City and Los Angeles, the concentrations of pollutants can rise so high that pollution-sensitive individuals cannot leave their homes or must wear masks when they go outdoors. For many, alerts about unsafe air quality have become a routine phenomenon.

Rather small changes in the average temperature of the Earth can have dramatic consequences (see ■ Map 16.1). A few degrees of difference can cause greater melting in the Arctic regions, which raises the level of the sea, affecting water, land, and weather systems worldwide. Today, we see images of polar bears drowning because of the breakup of ice floes, that is, the melting of the polar cap. Sea levels are rising because of the melting of the polar ice cap, threatening to drown major urban areas.

Human ingenuity has, no doubt, produced inventions that have vastly improved the quality of life. Sometimes, though, these inventions have unintended and possibly dangerous consequences. To illustrate, the

map 16.1

Global Warming: Viewing the Earth's Temperature

Scientists at the National Aeronautics and Space Administration (NASA) can measure and map changes in the Earth's temperature. The average temperature on Earth has increased by 1.4 degrees Fahrenheit since 1980 (0.8 degrees Celsius). Two-thirds of this change has occurred since 1975. A change of just one degree has massive consequences. Earth was plunged into a "little ice age" with only a one- to two-degree drop in the seventeenth century. Twenty thousand years ago, much of North America was buried under a towering mass of ice from a five-degree drop. You can almost see why people want to face the reality of climate change. What must be done to protect us from this pending disaster? How are social policies and social behaviors involved in such changes?

Source: Carlowicz, Michael. 2014. "Global Temperatures." National Aeronautics and Space Administration. **www.earthobservatory.nasa.gov**

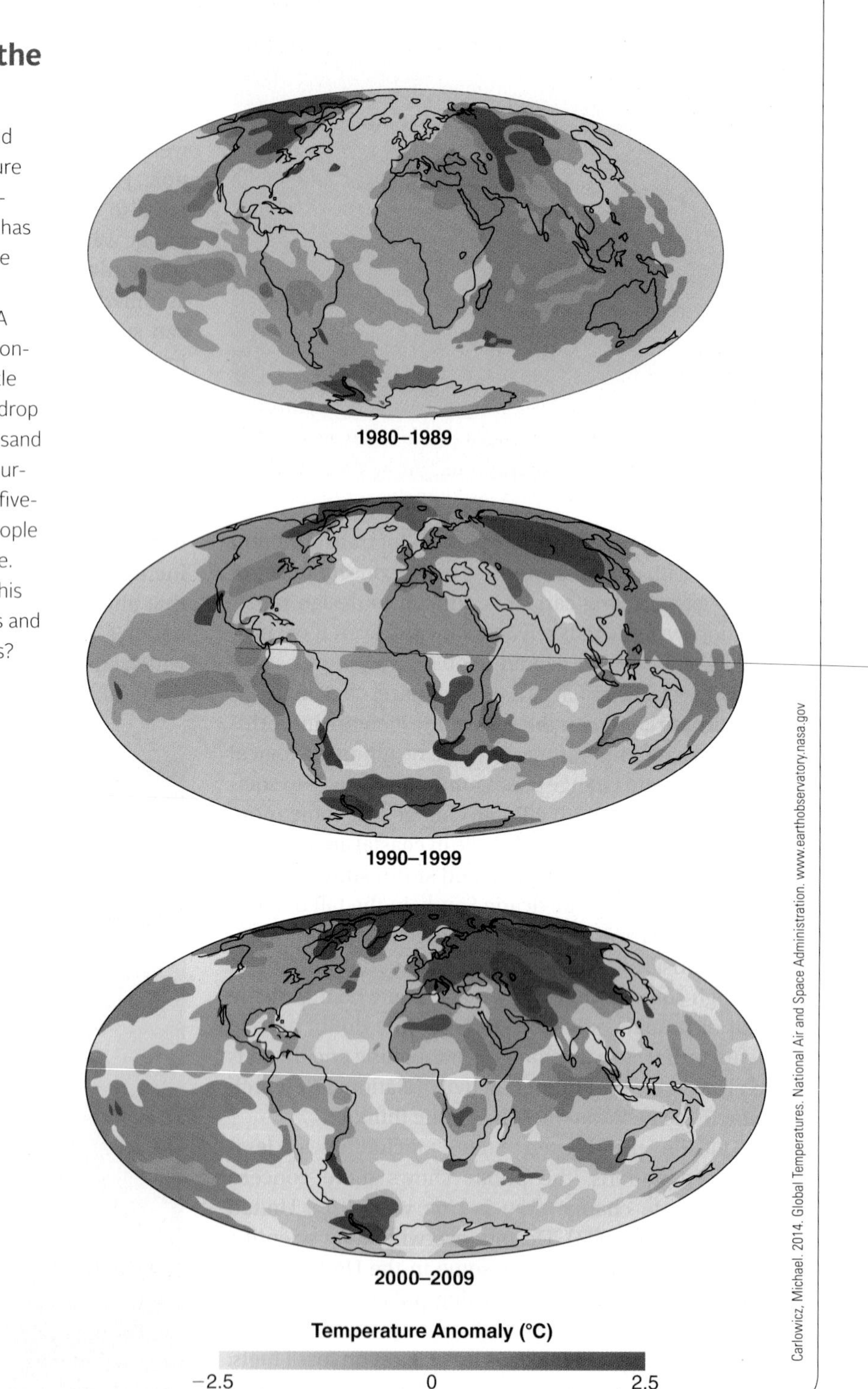

Carlowicz, Michael. 2014. Global Temperatures. National Air and Space Administration. www.earthobservatory.nasa.gov

invention of the automobile has transformed society. Today, many cannot imagine getting around without a car. We have designed many of our cities and, especially, our suburbs, in ways that require people to drive. But, a huge portion of the pollutants released into the air comes from the exhaust pipes of motor vehicles. The major component of this exhaust is carbon monoxide, a highly toxic substance. Also found in exhaust fumes are nitrogen oxides, the substances that give smog its brownish-yellow color. Sunlight causes these oxides to combine with hydrocarbons, also emitted from exhausts, forming a host of health-threatening substances. We have

iStockphoto.com/Ekvals

Miners used to use canaries, which have a fragile respiratory system, to signal gas leaks in the mines. If the canary died, it was a signal of danger to the miners. Polar bears may be the new "canaries" in that the melting of their habitat is a warning of the rise in temperature of polar water—an indication of global warming.

become so dependent on automobiles for transportation that even with greater awareness of the consequences of driving gas-guzzling cars, it is difficult to design transportation systems that rely less on cars. Even if we produced such a design, would people give up their cars? That is a social as much as a technological challenge.

Climate change is the systematic increase in worldwide surface temperatures and the resulting ecological change. Climate change poses numerous threats to society as we know it. With climate change will come more extreme weather patterns. People in coastal areas will be prone to rising coastal waters and storm surges, all too vividly seen during Hurricane Sandy in the fall of 2012. Some will have too much water; others, too little.

Scientists see climate change as a serious problem. Although some people deny that climate change is the result of human behavior, there is little doubt among scientists that climate change is happening and that it is largely the result of human activity (National Research Council 2012). What does the public think?

Outside the United States, including in less-developed nations, the public shows more concern about climate change than is true within the United States (Brechin 2003). Despite overwhelming scientific evidence of climate change, some in the United States deny its existence, thus thwarting policy changes that could address its causes and consequences, such as programs that would make us less reliant on fossil fuels. Denial about climate change is only part of the problem, though. Even when people have information about the potential effects of climate change, they often ignore taking action. Why? The *social organization of denial*, according to sociologist Kari Norgaard, comes from people holding unpleasant emotions—such as the fear of flooding or devastation—at a distance (Norgaard 2006). Although some may deny climate change for purely political reasons or for lack of information, for many, the sheer unpleasantness of facing such catastrophic change is more than people can willingly admit or face.

What, for example, would we do without water? Most Americans have come to think of water as abundant, free, and safe, but the safety and availability of water is now threatened. Thousands of rural water wells have had to be abandoned due to contamination. Households served by municipal water systems are also endangered; fully 20 percent of the country's public water systems do not meet the minimum toxicity standards set by the government. Although many have assumed that the nation's water is plentiful, whether that can still be assumed is questionable (Fishman 2011). In the western and southwestern United States, the groundwater supply is being depleted at a rapid pace, making water one of the causes of political conflict between different states in the region (Espeland 1998).

Threats to our water supply also spill into other issues. As people have become concerned about water quality, drinking water from plastic bottles has increased. Where do the bottles go? Estimates are that about 40 million plastic water bottles go into the nation's trash *every day*—only about 23 percent of which are recycled (Fishman 2011). From production to disposal, water bottles reveal that technical know-how

Joel Sartore/National Geographic Image Creative

This woman was told by a gas company in Powder River Basin, Wyoming, that the drilling for gas near her house "would never cause you to lose your water." Shortly after drilling began near her home, her well water turned into a muddy methane slurry, which she unhappily holds in a glass in this photo.

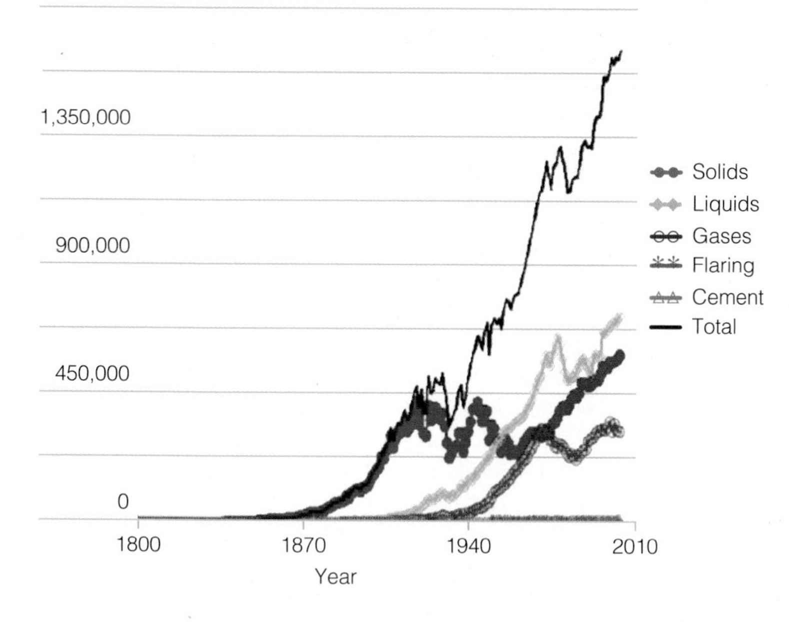

▲ **Figure 16.1** U.S. Carbon Dioxide (CO_2) Emissions, 1800–2010 You can vividly see in this graph the increased emission from fossil fuels and other substances. What historical events and changes are reflected in the ups and downs that you see here? What changes would be needed in society to produce a decrease in the future?

Source: Carbon Dioxide Information Analysis Center. 2007. **http://cdiac.ornl.gov/**

merges with human behavior, creating a complex system of environmental challenges.

The nation's water is also threatened by the chemical pollutants that industries discharge into rivers, lakes, and the oceans, including solid wastes, sewage, nondegradable by-products, synthetic materials, toxic chemicals, and radioactive substances. Add the polluting effects of sewage systems of towns and large cities, detergents, oil spills, pesticide runoff, and runoff from mines, and the enormity of the problem is clear.

Federal and state statutes now prohibit industry from polluting the nation's water, but the pollution continues. Why? The answer is economic, political, and sociological. Industries that contribute to a vigorous economy have traditionally met with little interference from the government. Public awareness and outrage can force the government to crack down on major polluters.

We are racing through our nonrenewable natural resources and destroying much that could be renewable. Addressing this problem also requires looking at some of the inequalities that are revealed when we examine such things as energy usage. On a global scale, the use of natural resources is not evenly shared around the world. The United States, which is a little under 5 percent of the world's population, consumes 20 percent of the world's energy and emits about 20 percent of the carbon dioxide emissions from fossil fuels. China now exceeds the United States in the release of carbon-based emissions, sending 8,547 million metric tons of carbon dioxide emissions into the atmosphere; the United States ranks second in the world, spewing 5,270 million metric tons of carbon dioxide emissions into the atmosphere per year (U.S. Energy Information Administration 2012; see ▲ Figure 16.1). How much is a metric ton? The average car now weighs about two tons, so 5 million metric tons would be the weight of about two and a half million cars sent into the atmosphere, if that were even possible!

SEE for YOURSELF

The Wasteful Society

For just one day (a full twenty-four-hour period), make a list of everything that you use up or discard. Include everything that you throw away, including garbage, waste from cooking and eating, gasoline in your car, and so forth. At the end of the day, list the things you discarded. Indicate

Huguette Roe/shutterstock.com

Bottled water (not counting other plastic containers) produces 1.5 million tons of waste every year, only a small percentage of which is recycled.

whether there were any alternatives to discarding these things. How might one reduce the amount of waste produced in society generally?

Disasters: At the Interface of Social and Physical Life

Even while the normal practices of everyday life threaten the Earth, periodic hazards also come from natural disasters. Floods, earthquakes, tsunamis, hurricanes, tornadoes—these are just a few of the disasters that disrupt communities, families, and public health. Many disasters are forces of nature; some are predictable, some are not. Other disasters, such as chemical spills, explosions, and huge forest fires, are more directly attributable to human actions. But, either way, disasters are not solely the result of physical or natural factors.

According to sociologists who study them, disasters juxtapose physical events, such as floods, hurricanes, earthquakes, and the like with vulnerable populations (Tierney 2007). And, although people think of disasters as "nature's wrath," the impact of disasters is often the result of social behavior too. Hurricane Katrina is a good example. Although the hurricane itself emerged from nature, neglect of an inadequate levy system made communities in low-lying areas more vulnerable than others. Likewise, when Hurricane Sandy hit the northeastern United States in 2012, it affected many, but it had a disproportionate impact on poor and lower-income communities. Even when disasters affect hundreds of thousands of people, it is generally true that poor people and people of color are most negatively impacted. They are more vulnerable even before a disaster, but then have the greatest difficulties during recovery from poor access to health care, insurance, housing, and other social services (Tierney 2012).

Craig Ruttle/AP Images

Researchers find that, in the aftermath of disasters or other unexpected events, people tend to come together to provide assistance to each other, as happened in the aftermath of Hurricane Sandy.

Time and time again, human behavior is implicated in the impact of natural disasters. Overdevelopment can destroy natural environments, such as barrier islands, that mitigate the effects of a natural disaster. During the 1930s, the devastation wrecked by the dust bowl resulted from agricultural practices (the overproduction of wheat) that had stripped the prairies of natural grasses in the southern plains. Although drought brought on the devastating dust storms, had humans not destroyed the grasslands, it is doubtful that the consequences would have been so dire (Egan 2006).

Social systems are also disrupted in the aftermath of disasters. The 2011 nuclear power plant leakage in Japan that occurred following the major earthquake and tsunami (the resulting massive ocean wave) showed how vulnerable social systems can be to natural disasters. In the United States, the massive Gulf Oil spill in 2010 spewed thousands of gallons of crude oil into the Gulf. The spill was so large and disastrous that large numbers of shrimpers and fishermen were forced out of business. The spill rapidly polluted major marshes surrounding the Gulf, killing off much flora and fauna, including birds of several species and all varieties of fish, shrimp, and mussels.

Who is most vulnerable during disasters is also shaped by social factors (Tierney et al. 2001). The poor and the elderly are often the most vulnerable. During the infamous Chicago heat wave of 1993, temperatures soared above 105 degrees and over 700 people died. Research by sociologist Eric Klinenberg (2002) has found that social factors influenced these very high mortality rates. The isolation of elderly people, retrenchment of social services, and little institutional support in poor neighborhoods meant that those most likely to die from this disaster were the poor, the elderly, African Americans (because of their concentration in poor neighborhoods), and women (because they are more of the older population).

DEBUNKING Society's Myths

Myth: In the aftermath of a natural disaster, people live in chaos and with a breakdown of the social order.

Sociological Perspective: Even when the ordinary course of life is disrupted following a catastrophic event or disaster, people rely on what disaster researchers call "pro-social behavior"—that is, assisting each other, developing support networks, and organizing informal systems of social control (Tierney et al. 2006).

Government responses to disasters also show the consequences of human behavior for understanding the social dimensions of otherwise natural disasters. The slow work of the federal bureaucracy, as well as partisan politics, both play a role in social responses to disasters. Social stereotypes also figure in how victims of

▲ **Figure 16.2** Environmental racism refers to the pattern whereby people living in predominantly minority communities are more likely exposed to toxic dumping and other forms of pollution. Nuclear waste and testing in the American Southwest, for example, have been located in areas predominantly inhabited by Native Americans. In other areas, African Americans and Latinos are exposed to the effects of industrial waste.

disasters are portrayed. Following Hurricane Katrina in New Orleans, African Americans were depicted in media coverage as wild looters and thieves—an image not seen so much when the predominantly White, working- and middle-class communities in New York and New Jersey were so profoundly disrupted following Hurricane Sandy.

Now, given climate change, people even wonder if disasters will be more frequent. The warming of the Earth's oceans could produce more frequent and more damaging hurricanes. Heat waves could become more frequent, overpowering power supplies as people try to cool their homes. Drought could intensify, fueling political struggles over who controls water supplies in the driest areas of our nation. Some populations will be more vulnerable than others, showing once again how factors such as race, social class, age, and gender shape the impacts and portrayals of so-called natural disasters.

Environmental Inequality and Environmental Justice

Many argue that of all environmental problems facing the United States today, the most urgent is the dumping of hazardous wastes, if only for the sheer noxiousness of the materials being dumped. Since 1970, the production of toxic wastes has increased ninefold (Weeks 2012). Of course, any degradation in the Earth's well-being affects everyone, but who is most vulnerable to pollutants and toxic waste dumping reveals patterns of social inequality.

Environmental racism is the pattern whereby toxic wastes and other pollutants are disproportionately found in minority and poor neighborhoods, a pattern with clear health consequences (Brulle and Pellow 2006). Research has determined that it is virtually impossible that dumps are being placed so often in communities of minority and lower socioeconomic status by chance alone (Bullard and Wright 2009). Is class or race to blame? Race and class both influence toxic waste disposal (Mohai and Saha 2007). Wealthier communities are better able to resist dumping in their neighborhoods, and housing discrimination and other race-related disparities are strongly linked to toxic waste being more present in minority areas, as illustrated in ▲ Figure 16.2.

Studies find that Native American, Hispanic, and particularly African American populations reside disproportionately closer to toxic sources than do Whites. Such patterns are *not* explainable by social class differences alone. That is, when communities of the same socioeconomic characteristics but different racial-ethnic compositions are compared, Native Americans, Hispanics, and African Americans of a given socioeconomic level live closer to toxic dumps than do Whites of the *same* socioeconomic level (Bullard and Wright 2010; Mohai and Saha 2007).

Take a look around your own neighborhood. Are there industrial waste sites nearby? Where are toxic products being disposed? For that matter, is there recycling available and to whom? You are likely to find patterns of waste disposal that are significantly linked to the class and racial composition of your neighborhood.

Within minority and poor neighborhoods, many groups have mobilized to protest and stop dumping in their communities. The *environmental justice movement* is the broad term used to refer to the social action that communities have taken to ensure that toxic waste dumping and other forms of pollution do not fall disproportionately on groups because of their race, class, or gender (Pellow 2004).

Environmental justice encompasses a wide array of programs for change. Developing more organic methods of growing food—indeed, encouraging community gardens where people can grow their own food—and other "green" programs are important trends to promote social change for a more sustainable society. But social change is hard to accomplish, in part because it takes more than individual effort. As we will see, social change requires individual action, but it is also collective, that is, a fundamentally *social* process.

DEBUNKING Society's Myths

Myth: Environmental pollution is in fact more common in or near economically poor areas; social class is a more important reason for this than race.

Sociological Perspective: Even when comparing areas of the same low economic status but different racial compositions, the areas with a higher percentage of minorities are on average closer to polluted areas than those with a lower percentage of minorities (Mohai and Saha 2007).

Counting People: Population Studies

Studies of the environment raise fundamental questions about how human societies relate to the physical and natural world. Population growth and density are responsible for some of the challenges we face with our environment: urban overcrowding and sprawl, traffic jams, pollution, and the threat of diminishing or tarnished Earth resources. Can we sustain the current way of living, given the size of the national and world populations? Are there simply too many people for our planet to support?

There are seven billion people living in this world. What do we know about how population is shaped? When a baby is born, what are his or her odds of survival beyond the first year? How many others will be born the following year? Will the population of people born in a given year influence the future of society simply because of the size of this age group?

These questions can be studied through the sociology of population. The scientific study of population is called **demography**. Demography includes studying the size, distribution, and composition of human populations as well as studying population changes over time, both those of the past and those predicted for the future.

Basic population facts drive many of the experiences and attitudes of some people. Young people may feel insecure about their future; decisions about having or not having children may loom; young people will likely have to care for older people; and hotly debated topics like immigration are likely to continue to shape national politics. The decisions people make—both personal and national—will ripple forward for years to come. Will there be a need for more senior centers? Will minority children get a good education? If there is a decline in the number of middle-aged people, who will take care of the old? Will environmental resources hold up in such a way as to maintain current lifestyles?

Counting People: Demographic Processes

Demography draws on huge bodies of data generated by a variety of sources. One major source is the U.S. Census Bureau. A **census** is a head count of the entire population of a country, usually done at regular intervals. The U.S. census is conducted every ten years, as required by the U.S. Constitution; the latest was conducted in 2010. The census attempts to enumerate every individual and to obtain information such as gender, race, ethnicity, age, education, occupation, and other social factors. The census is updated annually through a much smaller sample of the population that can then be used to track changes more frequently, although in less detail, than the decennial census (conducted every ten years).

The current population of the United States is more than 316 million—a milestone when the 300-million mark was passed in 2011. By 2060, the U.S. population is not only predicted to be larger (416 million) but also older and more diverse. White Americans are expected to decline as a percentage of the population; Hispanics and Asians are expected to double in number; African Americans will also increase, though not as much as Hispanics and Asians. As soon as 2030, one in five Americans will be over 65. The older population is expected to almost double in size by mid-century (U.S. Census Bureau 2014).

Even with this detail, however, it is known that the census undercounts a small percentage of the country's population. It is simply impossible to have every single person complete the census form that is distributed. Who would be most likely undercounted? You can probably guess. Those most likely to have been undercounted by the census are the homeless, immigrants, minorities living in poor neighborhoods, and others of low social status. In general, the lower your overall social status (such as by income, occupation, race-ethnicity, gender, immigrant status, or other measures), the less likely you are to be counted in the U.S. census.

The constitutional requirement for a census was included to ensure fair apportionment of representatives in the federal government. Undercounting specific groups of people leaves them underrepresented in government. The Census Bureau itself estimated that Whites were overcounted in the 2010 census by close to one percent. African Americans were undercounted by 2 percent; Hispanics, by 1.5 percent. Native Americans living on reservations are estimated to have been undercounted by about 5 percent. Even whether you own or rent your residence affects the likelihood of being

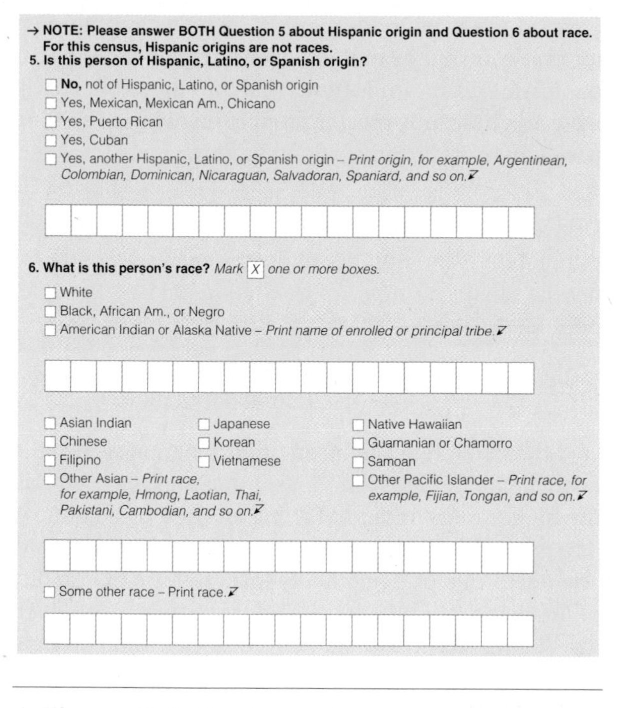

→ **NOTE: Please answer BOTH Question 5 about Hispanic origin and Question 6 about race. For this census, Hispanic origins are not races.**

5. Is this person of Hispanic, Latino, or Spanish origin?

☐ **No,** not of Hispanic, Latino, or Spanish origin
☐ Yes, Mexican, Mexican Am., Chicano
☐ Yes, Puerto Rican
☐ Yes, Cuban
☐ Yes, another Hispanic, Latino, or Spanish origin – *Print origin, for example, Argentinean, Colombian, Dominican, Nicaraguan, Salvadoran, Spaniard, and so on.*

6. What is this person's race? *Mark* ☒ *one or more boxes.*

☐ White
☐ Black, African Am., or Negro
☐ American Indian or Alaska Native – *Print name of enrolled or principal tribe.*

☐ Asian Indian
☐ Chinese
☐ Filipino
☐ Other Asian – *Print race, for example, Hmong, Laotian, Thai, Pakistani, Cambodian, and so on.*
☐ Japanese
☐ Korean
☐ Vietnamese
☐ Native Hawaiian
☐ Guamanian or Chamorro
☐ Samoan
☐ Other Pacific Islander – *Print race, for example, Fijian, Tongan, and so on.*

☐ Some other race – Print race.

▲ **Figure 16.3** The Census Counts Race This is the form that the U.S. Census Bureau uses in its decennial census to tally the racial–ethnic composition of the U.S. population. How would you answer? Does this adequately measure your racial identity? If so, why? If not, how might you revise it, and would that be correct for others?

Source: U.S. Census Bureau. 2010. "Overview of Race and Hispanic Origin: 2010." Washington, DC: U.S. Department of Commerce. **www.census.gov**

counted—renters being more likely to be undercounted; homeowners, overcounted (U.S. Census Bureau 2012c).

Although counting people may seem tedious and dry to you, it can actually be a fiercely debated topic. Now (and beginning in the 2000 census), people are allowed to select multiracial (or "mixed race") as a response regarding their racial and ethnic identity on the census questionnaire (see ▲ Figure 16.3). If you are, for example, Hispanic and Black or Black and White, how should the census "count" you in a racial or ethnic category—a category that will later be used to determine such facts as you have seen in this book, such as income distribution by race or the ethnic makeup of neighborhoods? The use of the multiracial response option gives individuals an opportunity to define themselves as mixed race. One argument against this option is that it subtracts from the number of people who would have otherwise indicated only one category, thus further undercounting African Americans, Hispanics, and Native Americans. Currently, only 2.5 percent of people responding to the census indicate a multiracial response, but this is expected to increase substantially again in 2020, as well as into the future. Of course, how people identify can change according to the social and political climate of the time, but population experts use these projections to anticipate changes that are likely to occur (U.S. Census Bureau 2014).

The world population is currently seven billion people, and it is expected to grow to nine billion by 2050. Most of the growth will be in developing areas of the world, not the already industrialized nations, as you can also see in ▲ Figure 16.4. Many of the nations with the highest rates of population growth also have high rates of poverty (United Nations 2012b). Barring some major catastrophe, such as a health epidemic, these nations will have a higher **population density**, defined as the number of people per unit of area, usually per square mile.

The total number of people in a society at any given moment is determined by only three variables: births, deaths, and migrations. These three variables show different patterns for different racial and ethnic groups, different social strata, and both genders. Births add to the total population, and deaths subtract from it. Migration into a society from outside, called **immigration**, adds to the population, whereas

▲ **Figure 16.4** World Population Growth, 1750–2050 As you can see from this graph, population in the most developed parts of the world is expected to remain somewhat flat or even decline, while population in the less-developed areas will increase dramatically. What implications does this have for feeding the world and protecting people's health?

Data: United Nations Population Division. 2012. "World Population Trends, 2012 Revision." **www.prb.org**

emigration, the departure of people from a society (also called *out-migration*), subtracts from the population.

Birthrate. The **crude birthrate** (or **birthrate**) of a population is the number of babies born each year for every 1000 members of the population or, alternatively, the number of births divided by the total population, multiplied by 1000. It is labeled crude because it does not take into account age or sex differences:

$$\text{Crude birth rate (CBR)} = \frac{\text{number of births}}{\text{total population}} \times 1000$$

Nations vary considerably in their birthrates, with the highest being Uganda, with 47.6 births per 1000 people, and the lowest, Japan, with only 7.4 births per 1000 people (U.S. Census Bureau 2012c). The birthrate for the United States is approximately 14 births per 1000 people now, lower than at any other time in U.S. history. By way of comparison, the birthrate in 1910 was 30.1, but has declined rather steadily since then, with the exception of the years just after World War II when the population now called baby boomers was born (National Center for Health Statistics 2012).

The effects of birthrates are somewhat cumulative. For example, minorities tend to be overrepresented at the lower end of the socioeconomic scale, compounding the likelihood of a high birthrate. Similarly, religious and cultural differences affect the birthrate. Catholics, for example, have a higher birthrate than non-Catholics of the same socioeconomic status. Hispanic Americans have a high likelihood of being Catholic, another factor that contributes to the higher birthrate among Hispanic Americans. Projections that the United States will have a significantly greater proportion of minorities are based on births, deaths, and migration rates.

Death Rate. The **crude death rate** (or **death rate**) of a population is the number of deaths each year per 1000 people, or the number of deaths divided by the total population, times 1000:

$$\text{Crude death rate (CDR)} = \frac{\text{number of deaths}}{\text{total population}} \times 1000$$

The death rate can be an important measure of the overall standard of living for a population. In general, the higher the standard of living enjoyed by a country, or a group within the country, the lower the death rate. The death rate of a population also reflects the quality of medicine and health care. Poor medical care, which goes along with a low standard of living, will correlate with a high death rate. The death rate can be an important indicator of a population's overall standard of living. In general, the higher the standard of living, the lower the death rate.

In nations with a poor standard of living, infant mortality is typically high. The **infant mortality rate** is measured by the number of deaths per year of infants less than one year old for every 1000 live births. In the United States, the overall infant mortality rate is generally low (6.1 in 2010), although not compared to other industrialized nations, as you can see in ▲ Figure 16.5.

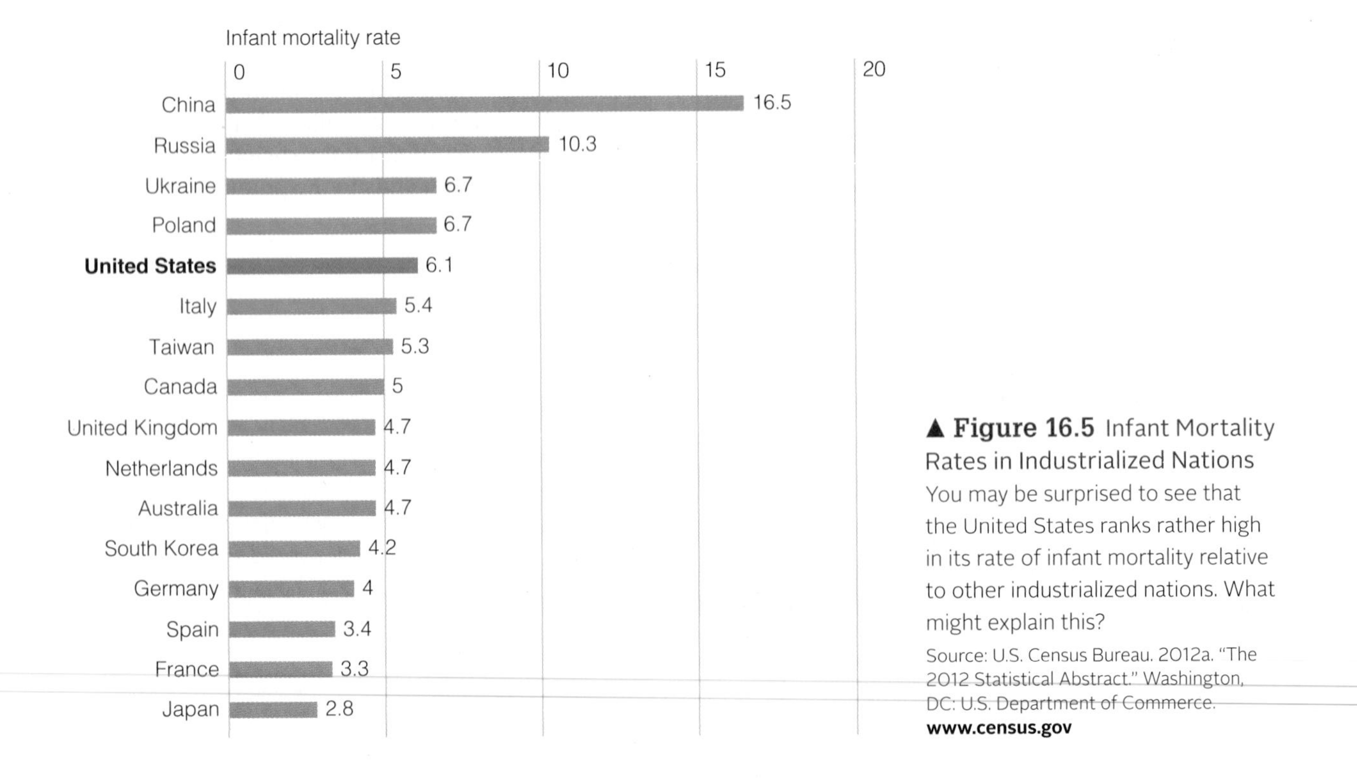

▲ Figure 16.5 Infant Mortality Rates in Industrialized Nations
You may be surprised to see that the United States ranks rather high in its rate of infant mortality relative to other industrialized nations. What might explain this?
Source: U.S. Census Bureau. 2012a. "The 2012 Statistical Abstract." Washington, DC: U.S. Department of Commerce. **www.census.gov**

Infant mortality rates, a measure of the chances of the very survival of members of the population, are important to compare across racial–ethnic groups and across social class strata. The relatively higher rate of infant mortality in the United States stems in large part from the poverty and inequality that exist, especially among racial and ethnic minorities and also the White poor. Infant mortality is a good indicator of the overall quality of life, as well as the survival chances for members of that racial or class group. There are also many other causes of higher infant mortality, such as presence of toxic wastes, malnutrition of the mother, inadequate food, and outright starvation.

Migration. Joining the birthrate and death rate as factors in determining the size of a population is the migration of people into and out of the country. We see the impact of migration in current policy debates about immigration. Who should be allowed into a country? Should those who have immigrated illegally be given amnesty and allowed to stay? Should children who came to the United States at a very young age, but now have never known another country, be allowed to attend college by paying in-state tuition? These questions stem from population changes that are rather dramatically shaping the nation's future.

Migration affects society in many ways. Immigration has ebbed and flowed over the years, but some waves of immigration have, at certain times, had a huge impact on society. Of course, the United States has always been a land of immigrants. Only American Indians and Mexicans, settled in what is now the American Southwest, are indigenous people to this land. Of course, one could hardly call African Americans who came here as slaves "immigrants," as their entry was forced. Still, our nation has become a diverse mix of peoples, given the different origins of our population.

In 1924, National Origins Quota Act encouraged immigration from northern and western Europe (England, France, Germany, Switzerland, and the Scandinavian countries), but discouraged immigration from eastern and southern Europe (Greece, Italy, Poland, Turkey, and eastern European Jews generally, among others). Despite this openly discriminatory law, millions of eastern Europeans successfully made the journey to Ellis Island and then the U.S. mainland, only to face prejudice, discrimination, and the accusation that they were taking jobs that would have otherwise gone to the already-present White majority.

Unless there is a change in immigration law as the result of the current national debate about immigration, immigration to the United States is governed by the Hart Celler Act of 1965. This law abolished the national origins quotas that had been mandated since 1920. This meant that the doors were open for immigrants from Asia, Africa, and Latin and Central America—places that had been excluded from the prior policies that favored those from northern and western Europe. Neighborhoods are now invigorated and culturally enriched by mosques or Buddhist temples; by whole neighborhoods of Vietnamese, Koreans, or Asian Indians; or by war refugees from Somalia and Bosnia. Whereas immigrants once settled almost entirely in a small number of cities, now immigration is affecting communities throughout the country (Hirschman and Massey 2008). The simple change in law brought by the Hart Celler Act has had an enormous impact on population diversity, the effects of which we see today.

Diversity and Population Change

The composition of a society's population can reveal a tremendous amount about the society's past, present, and future. To begin with, many nations, including the United States, have a striking imbalance in the number of men and women, with many fewer men than would be expected. The **sex ratio** is the number of males per 100 females, or the number of males divided by the number of females, times 100.

$$\text{Sex ratio} = \frac{\text{number of males}}{\text{number of females}} \times 100$$

A sex ratio above 100 indicates there are more males than females in the population; below 100 indicates there are more females than males. A ratio of exactly 100 indicates the number of males equals the number of females.

In almost all societies, there are more boys born than girls, but because males have a higher infant mortality rate and a higher death rate after infancy, there are usually more females in the overall population. In the United States, approximately 105 males are born for every 100 females, thus giving a sex ratio for live births of 105. After factoring in male mortality, the sex ratio for all ages for the entire country ends up being 94; there are 94 males for every 100 females.

The *age composition* of the U.S. population is presently undergoing major changes. More and more people are entering the sixty-five and older age bracket. This trend is known as the *graying of America*. The elderly are now the largest population category in our society. Whereas those over age 65 were 8 percent of the population in 1950, they are now 13 percent of the population and are expected to be 20 percent of the population by 2050 (U.S. Census Bureau 2014). As our society gets grayer, older people have more influence on national policy and a greater say in matters such as health care and housing.

Population pyramids are graphic depictions of the age and sex distribution of a given population at

what would a sociologist say?

The End of the White Majority?

As the U.S. population is becoming more diverse, many are saying that White people will no longer be the majority. Even after the 2012 presidential election, when the votes of racial–ethnic minorities helped to reelect President Obama (along with 39 percent of the White vote), many pundits pontificated about whether White people were losing their historic hold on national power. Conservative commentator Bill O'Reilly even declared, "The white establishment is the new minority" (Fox News, November 6, 2012). What's true here?

First, even with the population projections indicating that Whites will become a smaller share of the U.S. population, Whites will still constitute the largest racial–ethnic group in the nation.

Second, and perhaps more importantly, from a sociological perspective, the terms *majority* and *minority* do not refer to numbers alone, as you learned in Chapter 10. Sociologically speaking, *majority* refers to a group that holds political, social, and economic power over others—and that group can, indeed, be a small percentage of the population (as we saw in apartheid South Africa when Whites, who had total rule over the whole population, were a mere 10 percent of the population). It remains true, even with the increased presence of people of color in social, political, cultural, and economic institutions, Whites are still the dominant group—in terms of power, privilege, and prestige.

Third, White is not a monolithic category. White people are diverse by many social-demographic characteristics—including age, gender, social class, region of residence—and all of these social facts affect the degree to which White people actually hold any power at all! So the next time you hear someone saying that White people are the new minority, you should have the sociological tools to challenge that assertion.

a point in time (see ▲ Figure 16.6). The bulge in the pyramid for those in their late fifties and sixties represents the baby boom generation. As the baby boomers age, the bulge will continue rising toward the top of the pyramid, to be replaced underneath by whatever birth trends occur in the coming years. You can also see that there is another bulge for those who are in their twenties, baby boomlets—that is, children of baby boomers. You might ask yourself how these generational structures are likely to affect society and its institutions over time as these "bulges" in population move upward.

The bulges you are seeing in Figure 16.6 are *birth cohorts*. A **cohort** consists of all the people born within a given period. A cohort can include all people born within the same year, decade, or other time period. Over time, cohorts either stay the same size or get smaller owing to deaths, but can never grow larger. If we have knowledge of the death rates for this population, we can predict quite accurately the size of the cohort as it passes through the stages of life from infancy to old age. This enables us to predict things such as how many people will enter the first grade in a given period, how many are likely to enroll in college, and how many will arrive at retirement decades down the road. Administrators of social entities such as schools and pension funds can make preparations on the basis of cohort predictions.

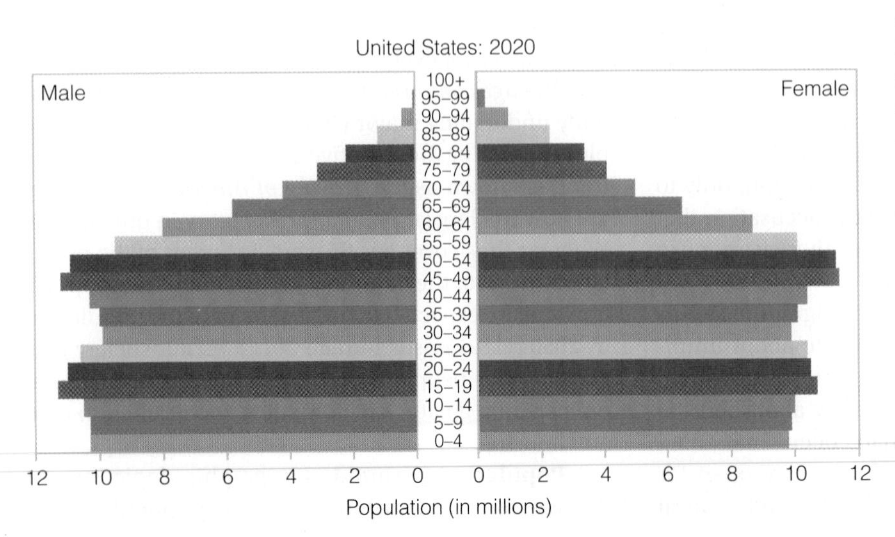

▲ **Figure 16.6** The Age–Sex Pyramids A graphic depiction, such as this population pyramid, can capture social realities—such as generational differences—that might drive important issues for public policy. Here, you can see two primary "bulges" in the U.S. population. How do you think these two populations might have a different stake in national debates about issues such as Medicare, taxes, Social Security, and the like?

Source: U.S. Census Bureau. 2012. "Annual Estimates of the Resident Population by Sex and Five-Year Age Group for the United States." **www.census.gov**

immigration (*versus emigration*) the migration of people into a society from outside it (also called in-migration).

implicit bias nonconscious form of racism where individuals unconsciously associate negative characteristics with racial-ethnic groups.

impression management a process by which people attempt to control how others perceive them.

income the amount of money brought into a household from various sources during a given year (wages, investment income, dividends, etc.).

independent variable a variable that is the presumed cause of a particular result (*see* dependent variable).

indicator something that points to or reflects an abstract concept.

individualized education plan (IEP) programs and services that provide options for students with learning disabilities and physical disabilities.

inductive reasoning the process of arriving at general conclusions from specific observations.

infant mortality rate the number of deaths per year of infants under age 1 for every 1000 live births.

informant in covert participant observation research, a single group member who provides "inside" information about the group being studied.

informed consent a formal acknowledgment by research subjects (respondents) that they understand the purpose of the research and agree to be studied.

institutional racism racism involving notions of racial or ethnic inferiority that have become ingrained into society's institutions.

instrumental needs emotionally neutral, task-oriented (goal-oriented) needs.

interest group a constituency in society organized to promote its own agenda.

interlocking directorate organizational linkages created when the same people sit on the boards of directors of a number of different corporations.

internalization a process by which a part of culture becomes incorporated into the personality.

international division of labor system of labor whereby products are produced globally, while profits accrue only to a few.

intersection perspective analytical framework that interprets race, class, and gender as simultaneously overlapping social factors.

intersexed person a person born with the physical characteristics of both sexes.

issues problems that affect large numbers of people and have their origins in the institutional arrangements and history of a society.

kinship system the pattern of relationships that defines people's family relationships to one another.

labeling theory a theory that interprets the responses of others as most significant in understanding deviant behavior.

labor force participation rate the percentage of those in a given category who are employed.

laissez-faire racism maintaining the status quo of racial groups by persistent stereotyping and blaming of minorities for achievement and socioeconomic gaps between groups.

laws the written set of guidelines that define what is right and wrong in society.

liberal feminism a feminist theoretical perspective asserting that the origin of women's inequality is in traditions of the past that pose barriers to women's advancement.

life chances the opportunities that people have in common by virtue of belonging to a particular class.

life course the connection between people's personal attributes, the roles they occupy, the life events they experience, and the social and historical context of these events.

life expectancy the average number of years individuals and particular groups can expect to live.

Lily Ledbetter Fair Pay Act law stating that discrimination claims on the basis of sex, race, national origin, age, religion, and disability accrue with every paycheck.

looking-glass self the idea that people's conception of self arises through reflection about their relationship to others.

macroanalysis analysis of the whole of society, how it is organized and how it changes.

mass media channels of communication that are available to very wide segments of the population.

master status some characteristic of a person that overrides all other features of the person's identity.

material culture the objects created in a given society.

matriarchy a society or group in which women have power over men.

matrilineal kinship kinship systems in which family lineage (or ancestry) is traced through the mother.

matrilocal a pattern of family residence in which married couples reside with the family of the wife.

McDonaldization the increasing and ubiquitous presence of the fast-food model in vast numbers of organizations.

mean the sum of a set of values divided by the number of cases from which the values are obtained; an average.

mechanical solidarity unity based on similarity, not difference, of roles.

median the midpoint in a series of values that are arranged in numerical order.

median income the midpoint of all household incomes.

Medicaid a governmental assistance program that provides health care assistance for the poor, including the elderly.

medicalization of deviance explanations of deviant behavior that interpret deviance as the result of individual pathology or sickness.

Medicare a governmental assistance program established in the 1960s to provide health services for older Americans.

meritocracy a system in which one's status is based on merit or accomplishments.

microanalysis analysis of the smallest, most immediately visible parts of social life, such as people interacting.

minority group any distinct group in society that shares common group characteristics and is forced to occupy low status in society because of prejudice and discrimination.

mode the most frequently appearing score among a set of scores.

modernization a process of social and cultural change that is initiated by industrialization and followed by increased social differentiation and division of labor.

modernization theory a view of globalization in which global development is a worldwide process affecting nearly all societies that have been touched by technological change.

monogamy the marriage practice of a sexually exclusive marriage with one spouse at a time.

monotheism the worship of a single god.

mores strict norms that control moral and ethical behavior.

multidimensional poverty index measure of poverty that accounts for health, education, and the standard of living.

multinational corporations corporations that conduct business across national borders.

multiracial feminism form of feminist theory noting the exclusion of women of color from other forms of theory and centering its analysis in the experiences of all women.

nationalism the strong identity associated with an extreme sense of allegiance to one's culture or nation.

neocolonialism a form of control of poor countries by rich countries, but without direct political or military involvement.

net worth the value of one's financial assets minus debt.

nonmaterial culture the norms, laws, customs, ideas, and beliefs of a group of people.

nonverbal communication communication by means other than speech, as by touch, gestures, use of distance, eye movements, and so on.

normative organization an organization having a voluntary membership and that pursues goals; examples are the PTA or a political party.

norms the specific cultural expectations for how to act in a given situation.

nuclear family family in which a married couple resides together with their children.

occupational prestige the subjective evaluation people give to jobs as better or worse than others.

occupational segregation a pattern in which different groups of workers are separated into different occupations.

organic solidarity unity based on role differentiation, not similarity.

organic metaphor refers to the similarity early sociologists saw between society and other organic systems.

organizational culture the collective norms and values that shape the behavior of people within an organization.

organizational ritualism a situation in which rules become ends in themselves rather than means to an end.

organized crime crime committed by organized groups, typically involving the illegal provision of goods and services to others.

outsourcing transferring a specialized task or job from one organization to a different organization, usually in another country, as a cost-saving device.

overt participant observation the form of participant observation wherein the observed individuals are told that they are being studied.

participant observation a method whereby the sociologist becomes both a participant in the group being studied and a scientific observer of the group.

patriarchy a society or group where men have power over women.

patrilineal kinship a kinship system that traces descent through the father.

patrilocal a pattern of family residence in which married couples reside with the family of the husband.

peers those of similar status.

percentage the number of parts per hundred.

peripheral countries (nations) poor countries, largely agricultural, having little power or influence in the world system.

personal crimes violent or nonviolent crimes directed against people.

personality the cluster of needs, drives, attitudes, predispositions, feelings, and beliefs that characterize a given person.

play stage the stage in childhood when children begin to take on the roles of significant people in their environment.

pluralism pattern whereby groups maintain their distinctive culture and history.

pluralist model a theoretical model of power in society as coming from the representation of diverse interests of different groups in society.

political action committees (PACs) groups of people who organize to support candidates they feel will represent their views.

polygamy a marriage practice in which either men or women can have multiple marriage partners.

polytheism the worship of more than one deity.

popular culture the beliefs, practices, and objects that are part of everyday traditions.

population a relatively large collection of people (or other unit) that a researcher studies and about which generalizations are made.

population density the number of people per square mile.

population pyramids graphic depictions of the age and sex distribution of a given population at a point in time.

positivism a system of thought that regards scientific observation to be the highest form of knowledge.

postindustrial society a society economically dependent upon the production and distribution of services, information, and knowledge.

poverty line the figure established by the government to indicate the amount of money needed to support the basic needs of a household.

power a person or group's ability to exercise influence and control over others.

power elite model a theoretical model of power positing a strong link between government and business.

preindustrial society one that directly uses, modifies, and/or tills the land as a major means of survival.

prejudice the negative evaluation of a social group, and individuals within that group, based upon conceptions about that social group that are held despite facts that contradict it.

prestige the value with which different groups of people are judged.

primary group a group characterized by intimate, face-to-face interaction and relatively long-lasting relationships.

profane that which is of the everyday, secular world and is specifically not religious.

propaganda information disseminated by a group or organization (such as the state) intended to justify its own power.

Protestant ethic belief that hard work and self-denial lead to salvation.

psychoanalytic theory a theory of socialization positing that the unconscious mind shapes human behavior.

qualitative research research that is somewhat less structured than quantitative research but that allows more depth of interpretation and nuance in what people say and do.

quantitative research research that uses numerical analysis.

queer theory a theoretical perspective that recognizes the socially constructed nature of sexual identity.

race a social category, or social construction, that we treat as distinct on the basis of certain characteristics, some biological, that have been assigned social importance in the society.

racial formation process by which groups come to be defined as a "race" through social institutions such as the law and the schools.

racial profiling the use of race alone as a criterion for deciding whether to stop and detain someone on suspicion of having committed a crime.

racialization a process whereby some social category, such as a social class or nationality, is assigned what are perceived to be race characteristics.

racism the perception and treatment of a racial or ethnic group, or member of that group, as intellectually, socially, and culturally inferior to one's own group.

radical feminists feminist theory that locates the source of women's inequality in the power that men hold in society.

random sample a sample that gives everyone in the population an equal chance of being selected.

rate parts per some number (for example, per 10,000; per 100,000).

rational-legal authority authority stemming from rules and regulations, typically written down as laws, procedures, or codes of conduct.

reference group any group (to which one may or may not belong) used by the individual as a standard for evaluating her or his attitudes, values, and behaviors.

reflection hypothesis the idea that the mass media reflect the values of the general population.

relative poverty a definition of poverty that is set in comparison to a set standard.

reliability the likelihood that a particular measure would produce the same results if the measure were repeated.

religion an institutionalized system of symbols, beliefs, values, and practices by which a group of people interprets and responds to what they feel is sacred and that provides answers to questions of ultimate meaning.

religiosity the intensity and consistency of practice of a person's (or group's) faith.

replication study research that is repeated exactly, but on a different group of people at a different point in time.

research design the overall logic and strategy underlying a research project.

resegregation the process by which once integrated schools become more racially segregated.

residential segregation the spatial separation of racial and ethnic groups in different residential areas.

resocialization the process by which existing social roles are radically altered or replaced.

revolution the overthrow of a state or the total transformation of central state institutions.

risky shift (also polarization shift) the tendency for group members, after discussion and interaction, to engage in riskier behavior than they would while alone.

rite of passage ceremony or ritual that symbolizes the passage of an individual from one role to another.

ritual a symbolic activity that expresses a group's spiritual convictions.

role behavior others expect from a person associated with a particular status.

role conflict two or more roles associated with contradictory expectations.

role modeling imitation of the behavior of an admired other.

role set all roles occupied by a person at a given time.

role strain conflicting expectations within the same role.

sacred that which is set apart from ordinary activity, seen as holy, and protected by special rites and rituals.

salience principle categorizing people on the basis of what initially appears prominent about them.

sample any subset of units from a population that a researcher studies.

Sapir–Whorf hypothesis a theory that language determines other aspects of culture because language provides the categories through which social reality is defined and perceived.

schooling socialization that involves formal and institutionalized aspects of education.

scientific method the steps in a research process, including observation, hypothesis testing, analysis of data, and generalization.

secondary analysis any analysis of data, qualitative or quantitative, which has been already gathered and analyzed by another researcher.

secondary group a group that is relatively large in number and not as intimate or long in duration as a primary group.

sect group that has broken off from an established church.

secular the ordinary beliefs of daily life that are specifically not religious.

segregation the spatial and social separation of racial and ethnic groups.

self our concept of who we are, as formed in relationship to others.

self-concept a person's image and evaluation of important aspects of oneself.

self-fulfilling prophecy the process by which merely applying a label changes behavior and thus tends to justify the label.

semiperipheral countries semi-industrialized countries that represent a kind of middle class within the world system.

serendipity unanticipated, yet informative, results of a research study.

sex used to refer to biological identity as male or female.

sex ratio (gender ratio) the number of males per 100 females.

sex tourism practice whereby people travel to engage in commercial sexual activity.

sex trafficking refers to the practice whereby women, usually very young women, are forced by fraud or coercion into commercial sex acts.

sexual harassment unwanted physical or verbal sexual behavior that occurs in the context of a relationship of unequal power and that is experienced as a threat to the victim's job or educational activities.

sexual identity the definition of oneself that is formed around one's sexual relationships.

sexual orientation the attraction that people feel for people of the same or different sex.

sexual politics the link feminists argue exists between sexuality and power, and between sexuality and race, class, and gender oppression.

sexual revolution the widespread changes in men's and women's roles and a greater public acceptance of sexuality as a normal part of social development.

sexual scripts the ideas taught to us about what is appropriate sexual behavior for a person of our gender.

significant others those with whom we have a close affiliation.

social capital *see* cultural capital.

social change the alteration of social interaction, social institutions, stratification systems, and elements of culture over time.

social class the social structural hierarchical position groups hold relative to the economic, social, political, and cultural resources of society.

social construction perspective a theoretical perspective that explains identity and society as created and learned within a cultural, social, and historical context.

social control the process by which groups and individuals within those groups are brought into conformity with dominant social expectations.

social control agents those who regulate and administer the response to deviance, such as the police or mental health workers.

social control theory theory that explains deviance as the result of the weakening of social bonds.

social fact social pattern that is external to individuals.

social institution an established and organized system of social behavior with a recognized purpose.

social interaction behavior between two or more people that is given meaning.

social learning theory a theory of socialization positing that the formation of identity is a learned response to social stimuli.

social media the term used to refer to the vast networks of social interaction that new media have created.

social mobility a person's movement over time from one class to another.

social movement a group that acts with some continuity and organization to promote or resist social change in society.

social network a set of links between individuals or other social units such as groups or organizations.

social organization the order established in social groups.

social sanction a mechanism of social control that enforces norms.

social stratification a relatively fixed hierarchical arrangement in society by which groups have different access to resources, power, and perceived social worth; a system of structured social inequality.

social structure the pattern of social relationships and social institutions that make up society.

socialism an economic institution characterized by state ownership and management of the basic industries.

socialization the process through which people learn the expectations of society.

socialization agents those who pass on social expectations.

society a system of social interaction, typically within geographical boundaries, that includes both culture and social organization.

socioeconomic status (SES) a measure of class standing, typically indicated by income, occupational prestige, and educational attainment.

sociological imagination the ability to see the societal patterns that influence individual and group life.

sociology the study of human behavior in society.

spurious correlation a false correlation between *X* and *Y*, produced by their relationship to some third variable (*Z*) rather than by a true causal relationship to each other.

state the organized system of power and authority in society.

status an established position in a social structure that carries with it a degree of prestige.

status attainment the process by which people end up in a given position in the stratification system.

status inconsistency exists when the different statuses occupied by the individual bring with them significantly different amounts of prestige.

status set the complete set of statuses occupied by a person at a given time.

stereotype an oversimplified set of beliefs about the members of a social group or social stratum that is used to categorize individuals of that group.

stereotype interchangeability the principle that negative stereotypes are often interchangeable from one racial group (or gender or social class) to another.

stereotype threat the effect of a negative stereotype about one's self upon one's own test performance.

stigma an attribute that is socially devalued and discredited.

Stockholm syndrome a process whereby a captured person identifies with the captor as a result of becoming inadvertently dependent upon the captor.

structural strain theory a theory that interprets deviance as originating in the tensions that exist in society between cultural goals and the means people have to achieve those goals.

subculture the culture of groups whose values and norms of behavior are somewhat different from those of the dominant culture.

symbolic interaction theory a theoretical perspective claiming that people act toward things because of the meaning things have for them.

symbol thing or behavior to which people give meaning.

taboo behavior that bring the most serious sanctions.

taking the role of the other the process of imagining oneself from the point of view of another.

teacher expectancy effect the effect of a teacher's expectations on a student's actual performance, independent of the student's ability.

Temporary Assistance for Needy Families (TANF) federal program by which grants are given to states to fund welfare.

terrorism the unlawful use of force or violence against people or property to intimidate or coerce a government or population in furtherance of political or social objectives.

Title IX legislation that prohibits schools that receive federal funds from discriminating based on gender.

total institution an organization cut off from the rest of society in which individuals are subject to strict social control.

totalitarian state an extreme form of authoritarianism where the state has total control over all aspects of public and private life.

totem an object or living thing that a religious group regards with special awe and reverence.

tracking grouping, or stratifying, students in school on the basis of ability test scores.

traditional authority authority stemming from long-established patterns that give certain people or groups legitimate power in society.

transgender those who deviate from the binary (that is, male or female) system of gender.

transnational family families where one parent (or both) lives and works in one country while the children remain in their country of origin.

triad a group consisting of three people.

troubles privately felt problems that come from events or feelings in one individual's life.

underemployment the condition of being employed at a skill level below what would be expected given a person's training, experience, or education.

unemployment rate the percentage of those not working, but officially defined as looking for work.

urban underclass a grouping of people, largely minority and poor, who live at the absolute bottom of the socioeconomic ladder in urban areas.

urbanization the process by which a community acquires the characteristics of city life.

utilitarian organization a profit or nonprofit organization that pays its employees salaries or wages.

validity the degree to which an indicator accurately measures or reflects a concept.

values the abstract standards in a society or group that define ideal principles.

variable something that can have more than one value or score.

verstehen the process of understanding social behavior from the point of view of those engaged in it.

wealth the monetary value of everything one actually owns.

White privilege the ability for Whites to maintain an elevated status in society that masks racial inequality.

work productive human activity that produces something of value, either goods or services.

world cities cities that are closely linked through the system of international commerce.

world systems theory theory that capitalism is a single world economy and that there is a worldwide system of unequal political and economic relationships that benefit the technologically advanced countries at the expense of the less technologically advanced.

xenophobia the fear and hatred of foreigners.

zero population growth stable population growth whereby the birthrate is equal to the death rate, without other influences, such as immigration.

References

4th Estate. 2014. "Silenced: Gender Gap in Election Coverage." **www.4thestate.net/female-voices-in-media-infographic/**

Abel, Ernest L. and Michael L. Kruger. 2007. "Gender Related Naming Practices: Similarities and Differences between People and Their Dogs." *Sex Roles: A Journal of Research* 57 (1-2): 15–19.

Abendroth, Anja-Kristin, Matt L. Huffman, and Judith Treas. 2014. "The Parity Penalty in Life Course Perspective: Motherhood and Occupational Status in 13 European Countries." *American Sociological Review* 79 (5): 993.

Aberle, David F., Albert K. Cohen, A. Kingsley Davis, Marion J. Levy Jr., and Francis X. Sutton. 1950. "The Functional Prerequisites of a Society." *Ethics* 60 (January): 100–111.

Abrahamson, Mark. 2006. *Urban Enclaves: Identity and Place in the World,* 2nd ed. New York: Worth.

Acker, Joan. 1992. "Gendered Institutions: From Sex Roles to Gendered Institutions." *Contemporary Sociology* 21 (September): 565–569.

Acker, Joan, Sandra Morgen, and Lisa Gonzales. 2002. "Welfare Restructuring, Work & Poverty: Policy from Oregon." Working Paper, Center for the Study of Women in Society, Eugene, OR.

Acs, Gregory, and Seth Zimmerman. 2008. *U.S. Intergenerational Mobility from 1984 to 2004.* Washington, DC: Pew Charitable Trust Economic Mobility Project. **www.pewtrusts.org**

Adorno, T. W., Else Frenkel-Brunswik, D. J. Levinson, and R. N. Sanford. 1950. *The Authoritarian Personality.* New York: Harper and Row.

Alba, Richard. 1990. *Ethnic Identity: The Transformation of Ethnicity in the Lives of Americans of European Ancestry.* New Haven, CT: Yale University Press.

Alba, Richard, and Gwen Moore. 1982. "Ethnicity in the American Elite." *American Sociological Review* 47 (June): 373–383.

Aldrich, Howard, and Martin Ruef. 2006. *Organizations Evolving.* Thousand Oaks, CA: Sage.

Alexander, Michelle. 2010. *The New Jim Crow: Mass Incarceration in the Age of Colorblindness.* New York: The New Press.

Alexander, Stephanie A. C., Katherine L. Frohlich, Blake D. Poland, Rebecca J. Haines, and Catherine Maule. 2010. "I'm a Young Student, I'm a Girl ... and for Some Reason They Are Hard on Me for Smoking: The Role of Gender and Social Context for Smoking Behaviour." *Critical Public Health* 20 (3): 323–338.

Algars, Monica, Pekka Santtila, and N. Kenneth Sandnabba. 2010. "Conflicted Gender Identity, Body Dissatisfaction, and Disordered Eating in Adult Men and Women." *Sex Roles* 63 (142): 118–125.

Allen, Kathleen P. 2012. "Off the Radar and Ubiquitous: Text Messaging and Its Relationship to 'Drama' and Cyberbullying in an Affluent Academically Rigorous U.S. High School." *Journal of Youth Studies* 15 (1): 99–117.

Alliance for Board Diversity. 2013. *Missing Pieces: Women and Minorities on Fortune 500 Boards.* **www.theabd.org**

Allison, Anne. 1994. *Nightwork: Sexuality, Pleasure, and Corporate Masculinity in a Tokyo Hostess Club.* Chicago: University of Chicago Press.

Allison, Stuart F. H., Amie M. Schuck, and Kim Michelle Lersch. 2005. "Exploring the Crime of Identity Theft: Prevalence, Clearance Rates, and Victim-Offender Characteristics." *Journal of Criminal Justice* 33 (1): 19–29.

Allport, Gordon W. 1954. *The Nature of Prejudice.* Reading, MA: Addison-Wesley.

Alon, Sigal, and Marta Tienda. 2007. "Diversity, Opportunity, and the Shifting Meritocracy in Higher Education." *American Sociological Review* 72 (August): 487–511.

Altman, Dennis. 2001. *Global Sex.* Chicago: University of Chicago Press.

Amato, Paul R. 2010. "Research on Divorce: Continuing Trends and New Developments." *Journal of Marriage and Family* 72 (3): 650–666.

Amato, Paul R., Alan Booth, Alan R. Johnson, and Stacy J. Rogers. 2007. *Alone Together: How Marriage in America Is Changing.* Cambridge, MA: Harvard University Press.

American Association of University Women. 1998. *Gender Gaps: Where Schools Still Fail Our Children.* Washington, DC: American Association of University Women.

American Association of University Women. 2010. "Back to School after All These Years." *AAUW Outlook* 104 (Fall).

American Psychological Association. 2007. *Report of the APA Task Force on the Sexualization of Girls.* Washington, DC: American Psychological Association.

American Psychological Association. 2008. *Stress in America.* Washington, DC: American Psychological Association. **www.apa.org**

American Sociological Association. 2008. *ASA Code of Ethics.* **www.asanet.org/about/ethics.cfm**

Amott, Teresa L., and Julie A. Matthaei. 1996. *Race, Gender, and Work: A Multicultural History of Women in the United States,* 2nd ed. Boston: South End Press.

Andersen, Margaret L. 2004. "From *Brown* to *Grutter*: The Diverse Beneficiaries of *Brown* vs. *Board of Education.*" *Illinois Law Review* 5 (August): 1073–1097.

Andersen, Margaret L. 2015. *Thinking about Women: Sociological Perspectives on Sex and Gender,* 10th ed. Boston: Allyn and Bacon.

Andersen, Margaret L., and Patricia Hill Collins (eds.). 2013. *Race, Class and Gender: An Anthology,* 8th ed. Belmont, CA: Wadsworth/Cengage.

Andersen, Margaret L., and Patricia Hill Collins (eds.). 2016. *Race, Class, and Gender: An Anthology,* 9th ed. Boston, MA: Cengage.

Anderson, Elijah. 2015. "The White Space." *Sociology of Race and Ethnicity* 1 (1): 10–21.

Anderson, David A., and Mykol Hamilton. 2005. "Gender Role Stereotyping of Parents in Children's Picture Books: The Invisible Father." *Sex Roles: A Journal of Research* 52(3-4): 145–151.

Anderson, Elijah. 1976. *A Place on the Corner.* Chicago: University of Chicago Press.

Anderson, Elijah. 1990. *Streetwise: Race, Class, and Change in an Urban Community.* Chicago: University of Chicago Press.

Anderson, Elijah. 1999. *Code of the Street: Decency. Violence, and the Moral Life of the Inner City.* New York: W. W. Norton.

Anderson, Elijah. 2011. *The Cosmopolitan Canopy: Race and Civility in Everyday Life.* New York: W. W. Norton.

Angwin, Julia. 2010. "The Web's New Gold Mine: Your Secrets." *The Wall Street Journal,* July 30.

Anthony, Dick, Thomas Robbins, and Steven Barrie-Anthony. 2002. "Cult and Anticult Totalism: Reciprocal Escalation and Violence." *Terrorism and Political Violence* 14 (Spring): 211–239.

Apple, Michael W. 1991. "The New Technology: Is It Part of the Solution or Part of the Problem in Education?" *Computers in the Schools* 8 (April–October): 59–81.

Arendt, Hannah. 1963. *Eichmann in Jerusalem: A Report on the Banality of Evil.* New York: Viking Press.

Armstrong, Elizabeth M. 2003. *Conceiving Risk, Bearing Responsibility: Fetal Alcohol Syndrome and the Diagnosis of Moral Disorder.* Baltimore, MD: Johns Hopkins University Press.

Armstrong, E. A., L. Hamilton, and B. Sweeney. 2006. "Sexual Assault on Campus: Multilevel, Integrative Approach to Party Rape." *Social Problems* 53 (4): 483–499.

Armstrong, Elizabeth A., Laura Hamilton, and Paula England. 2010. "Is Hooking Up Bad for Young Women?" *Contexts* 9 (Summer): 23–27.

Arnett, J. 2010. "Emerging Adulthood." *New York Times Magazine* (August 18). **www.nytimes.com**

Arnett, J., and J. Tanner, eds. 2010. *Growing into Adulthood: The Lives and Contexts of Emerging Adults*. Washington, DC: American Psychological Association.

Asch, Solomon. 1951. "Effects of Group Pressure upon the Modification and Distortion of Judgments." In *Groups, Leadership, and Men,* edited by H. Guetzkow. Pittsburgh, PA: Carnegie Press.

Asch, Solomon. 1955. "Opinions and Social Pressure." *Scientific American* 19 (July): 31–35.

Atkins, Celeste. 2011. "Big Black Mammas: The Intersection of Race, Gender and Weight in the U.S." Unpublished manuscript, Pima Community College, AZ.

Austin, Algernon. 2013. *High Unemployment Means Native Americans Are Still Waiting for an Economic Recovery*. Economic Policy Institute, December 17. **www.epi.org**

Babbie, Earl. 2013. *The Practice of Social Research,* 13th ed. Belmont, CA: Wadsworth/Cengage.

Baca Zinn, Maxine. 1995. "Chicano Men and Masculinity." pp. 33–41 in *Men's Lives,* 3rd ed., edited by Michael S. Kimmel and Michael A. Messner. Boston: Allyn and Bacon.

Baca Zinn, Maxine, and Bonnie Thornton Dill. 1996. "Theorizing Difference from Multiracial Feminism." *Feminist Studies* 22 (Summer): 321–331.

Baca Zinn, Maxine, Pierrette Hondagneu-Sotelo, and Michael Messner. 2011. *Gender through the Prism of Difference*. New York: Oxford University Press.

Baca Zinn, Maxine, and Bonnie Thornton Dill. 1996. "Theorizing Difference from Multiracial Feminism." *Feminist Studies* 22 (Summer): 321–331

Bailey, Garrick. 2003. *Humanity: An Introduction to Cultural Anthropology,* 6th ed. Belmont, CA: Wadsworth.

Balbo, Nicoletta, and Nicola Barban. 2014. "Does Fertility Behavior Spread among Friends?" *American Sociological Review* 79 (3): 4312–4431.

Bales, Robert F. 1951. *Interaction Process Analysis*. Cambridge: Addison-Wesley.

Bales, Kevin. 2010. *The Slave Next Door: Human Trafficking and Slavery in America Today*. Berkeley, CA: University of California Press.

Ballantine, Jeanne H., and Joan Z. Spade. 2015. *Schools and Society: A Sociological Approach to Education,* 5th ed. Thousand Oaks, CA: Sage Publications.

Bandura, Albert, and R. H. Walters. 1963. *Social Learning and Personality Development*. New York: Holt, Reinhart, and Winston.

Barak, Gregg, Paul Leighton, and Allison Cotton. 2015. *Class, Race, Gender, & Crime: Social Realities of Justice in America,* 4th ed. Lanham, MD: Rowman & Littlefield.

Barr, Donald A. 2014. *Health Disparities in the United States: Social Class, Race, Ethnicity, and Health, Second Edition*. Baltimore, MD: Johns Hopkins University Press.

Barton, Bernadette. 2006. *Stripped: Inside the Lives of Exotic Dancers*. New York: New York University Press.

Baumeister, Roy F., and Brad J. Bushman. 2008. *Social Psychology and Human Nature*. Belmont, CA: Wadsworth.

Baxter, Jennifer, Ruth Weston, and Lixia Qu. 2011. "Family Structure, Co-Parental Relationship Quality, Post-Separation Paternal Involvement and Children's Emotional Wellbeing." *Journal of Family Studies* 17 (2): 86–109.

Bean, Frank D., and Marta Tienda. 1987. *The Hispanic Population of the United States*. New York: Russell Sage Foundation.

Becker, Elizabeth. 2000. "Harassment in the Military Is Said to Rise." *The New York Times* (March 10): 14.

Becker, Howard S. 1963. *Outsiders: Studies in the Sociology of Deviance*. New York: Free Press.

Becker, L. B., T. Vlad, W. Kazragis, C. Toledo, and P. Desnoes. 2010. *2010 Annual Survey of Journalism and Mass Communication Graduates*. Athens, GA: James M. Cox, Jr. Center for International Mass Communication Training and Research, Grady College of Journalism and Mass Communication, University of Georgia.

Belknap, Joanne. 2001. *The Invisible Woman*. Belmont, CA: Wadsworth.

Bell, Myrtle P. 2011. *Diversity in Organizations*. Stamford, CT: Cengage Learning.

Bell, Leslie C. 2013. *Hard to Get: 20-Something Women and the Paradox of Sexual Freedom*. Berkeley, CA: University of California Press.

Bellah, Robert (ed.). 1973. *Emile Durkheim on Morality and Society: Selected Writings*. Chicago: University of Chicago Press.

Beller, Emily, and Michael Hout. 2006. "Intergenerational Social Mobility: The United States in Comparative Perspective." *The Future of Children* 16 (Fall): 19–36.

Ben-Yehuda, Nachman. 1986. "The European Witch Craze of the Fourteenth-Seventeenth Centuries: A Sociologist's Perspective." *American Journal of Sociology* 86: 1–31.

Benedict, Ruth. 1934. *Patterns of Culture*. Boston: Houghton Mifflin.

Berger, Peter L. 1963. *Invitation to Sociology: A Humanistic Perspective*. Garden City, NY: Doubleday Anchor.

Berger, Peter L., Brigitte Berger, and Hansfried Kellner. 1974. *The Homeless Mind: Modernization and Consciousness*. New York: Vintage Books.

Berger, Peter L., and Thomas Luckmann. 1967. *The Social Construction of Reality: A Treatise in the Sociology of Knowledge*. Garden City, NY: Anchor Books.

Bernstein, Nina. 2002. "Side Effect of Welfare Law: The No-Parent Family." *The New York Times* (July 29): A1.

Best, Joel. 1999. *Random Violence: How We Talk about New Crimes and New Victims*. Berkeley, CA: University of California Press.

Best, Joel. 2006. *Flavor of the Month: Why Smart People Fall for Fads*. Berkeley, CA: University of California Press.

Best, Joel. 2008. *Social Problems*, 2nd ed. New York: W. W. Norton.

Best, Joel. 2011. *Everyone's a Winner: Life in Our Own Congratulatory Culture*. Berkeley, CA: University of California Press.

Bishaw, Alemayehi. 2011. *Areas with Concentrated Poverty, 2006–2010*. Washington, DC: U.S. Census Bureau.

Black, M. C., K. C. Basile, M. J. Breiding, S. G. Smith, M. I. M. L. Walters, M. T. Merrick, J. Chen, and M. R. Stevens. 2011. *The National Intimate Partner and Sexual Violence Survey: 2010 Summary Report*. Atlanta, GA: National Center for Injury Prevention and Control, Centers for Disease Control and Prevention.

Blair-Loy, Mary, and Amy S. Wharton. 2002. "Employees' Use of Work-Family Policies and the Workplace Social Context." *Social Forces* 80 (3): 813–845.

Blake, C. Fred. 1994. "Footbinding in Neo-Confucian China and the Appropriation of Female Labor." *Signs* 19 (Spring): 676–712.

Blankenship, Kim. 1993. "Bringing Gender and Race in U.S. Employment Discrimination Policy." *Gender & Society* 7 (June): 204–226.

Blassingame, John. 1973. *The Slave Community: Plantation Life in the Antebellum South*. New York: Oxford University Press.

Blau, Peter M., and Otis Dudley Duncan. 1967. *The American Occupational Structure*. New York: Wiley.

Blau, Peter M., and W. Richard Scott. 1974. *On the Nature of Organizations*. New York: Wiley.

Bloom, Shelah. 2008. "Violence against Women and Girls: A Compendium of Monitoring and Evaluation Indicators," MEASURE Evaluation. Chapel Hill, NC: Caroline Population Center.

Blumer, Herbert, 1969. *Studies in Symbolic Interaction*. Englewood Cliffs, NJ: Prentice Hall.

Bobo, Lawrence D. 2004. "Inequalities That Endure? Racial Ideology, American Politics, and the Peculiar Role of the Social Sciences." pp. 13–42 in *The Changing Terrain of Race and Ethnicity*, edited by Maria Krysan and Amanda E. Lewis. New York: Russell Sage Foundation.

Bobo, Lawrence D. 2006. "The Color Line, the Dilemma, and the Dream: Race Relations in America at the Close of the Twentieth Century." pp. 87–95 in *Race and Ethnicity in Society: The Changing Landscape*, edited by Elizabeth Higginbotham and Margaret L. Andersen. Belmont, CA: Wadsworth.

Bobo, Lawrence. 2012 (January). "Post Racial Dreams, American Realities." Paper read before the Department of Sociology, Princeton University, Princeton, NJ.

Bocian, Debbie Gruenstein, Wei Li, and Keith S. Ernst. 2010. *Foreclosures by Race and Ethnicity: The Demographics of a Crisis.* Durham, NC: Center for Responsible Lending.

Bogle, Kathleen. 2008. *Hooking Up: Sex, Dating, and Relationships on Campus.* New York: New York University Press.

Bonilla-Silva, Eduardo. 1997. "Rethinking Racism: Toward a Structural Interpretation." *American Sociological Review* 62 (3): 465–480.

Bonilla-Silva, Eduardo. 2013. *Racism without Racists: Colorblind Racism and the Persistence of Inequality in America,* 4th ed. Lanham, MD: Rowman and Littlefield.

Bonilla-Silva, Eduardo, and Gianpaolo Baiocchi. 2001. "Anything but Racism: How Sociologists Limit the Significance of Racism." *Race and Society* 4 (2): 117–131.

Bontemps, Arna (ed.). 1972. *The Harlem Renaissance Remembered.* New York: Dodd, Mead.

Boo, Katherine. 2012. *Behind the Beautiful Forevers: Life, Death, and Hope in a Mumbai Undercity.* New York: Random House.

Boonstra, Heather, Rachel Benson Gold, Cory L. Richards, and Lawrence B. Finer. 2006. *Abortion in Women's Lives.* New York: Guttmacher Institute.

Bourdieu, Pierre. 1984. *Distinction: A Social Critique of the Judgement of Taste,* translated by Richard Nice. Cambridge, MA: Harvard University Press.

Bowles, Samuel, Herbert Gintis, and Melissa Osborn Groves (eds.). 2005. "Introduction." pp. 1–22 in *Unequal Chances: Family Background and Economic Success.* Princeton, NJ: Princeton University Press.

Boyd, Danah. 2014. *It's Complicated: The Social Lives of Networked Teens.* New Haven, CT: Yale University Press.

Boyle, Karen. 2011. "Producing Abuse: Selling the Harms of Pornography." *Women's Studies International Forum* 34 (6): 593–602.

Brame, Robert, Shawn D. Bushway, Ray Paternoster, and Michael G. Turner. 2014. "Demographic Patterns of Cumulative Arrest Prevalence by Ages 18 and 23." *Crime & Delinquency* 60 (3): 471–486.

Branch, Enobong Hannah. 2011. *Opportunity Denied: Limiting Black Women in Devalued Work.* New Brunswick, NJ: Rutgers University Press.

Branch, Taylor. 2006. *At Canaan's Edge: America in the King Years, 1965–1968.* New York: Simon and Schuster.

Brand, Jennie E., and Yu Xie. 2010. "Who Benefits Most from College? Evidence for Negative Selection in Heterogeneous Economic Returns to Higher Education." *American Sociological Review* 75 (2): 273–302.

Braunstein, Ruth. 2011. "Who Are 'We the People'"? *Contexts* 10 (May): 72–73.

Brechin, Steven R. 2003. "Comparative Public Opinion and Global Climatic Change and the Kyoto Protocol: The U.S. versus the World?" *International Journal of Sociology and Social Policy* 23 (10): 106–134.

Bridges, Ana J., Robert Wosnitzer, Erica Scharrer, Chyng Sun, and Rachael Liberman. 2010. "Aggression and Sexual Behavior in Best-Selling Pornography Videos: A Content Analysis Update." *Violence Against Women* 16 (10): 1065–1085.

Brodkin, Karen. 2006. "How Did Jews Become White Folks?" pp. 59–66 in Elizabeth Higginbotham and Margaret L. Andersen, eds., *Race and Ethnicity in Society: The Changing Landscape.* Belmont, CA: Wadsworth.

Brooke, James. 1998. "Utah Struggles with a Revival of Polygamy." *The New York Times* (August 23): A12.

Brooks-Gunn, Jeanne, Wen-Jui Han, and Jane Waldfogel. 2002. "Maternal Employment and Child Cognitive Outcomes in the First Three Years of Life: The NICHD Study of Early Child Care." *Child Development* 73 (July–August): 1052–1072.

Brown, Cynthia. 1993. "The Vanished Native Americans." *The Nation* 257 (October 11): 384–389.

Brown, Elaine. 1992. *A Taste of Power: A Black Woman's Story.* Pantheon: New York.

Brown, Michael K., Martin Carnoy, Troy Duster, Elliott Currie, David B. Oppenheimer, Marjorie Schulz, and David Wellman. 2005. *Whitewashing Race: The Myth of a Color-Blind Society.* Berkeley, CA: University of California Press.

Broyard, Bliss. 2007. *One Drop: My Father's Hidden Life.* New York: Little, Brown.

Brulle, Robert J., and David N. Pellow. 2006. "Environmental Justice: Human Health and Environmental Inequalities." *Annual Review of Public Health* 27 (April): 103–123.

Budig, Michelle J. 2002. "Male Advantage and the Gender Composition of Jobs: Who Rides the Glass Escalator?" *Social Problems* 49 (May): 257–277.

Buddel, Neil. 2011. "Queering the Workplace." *Journal of Gay & Lesbian Social Services* 23 (1): 131–146.

Bullard, Robert D., and Beverly Wright. 2009. *Race, Place, and Environmental Justice after Hurricane Katrina.* Boulder, CO: Westview Press.

Burt, Cyril. 1966. "The Genetic Determination of Differences in Intelligence: A Study of Monozygotic Twins Reared Together and Apart." *British Journal of Psychology* 57 (1–2): 137–153.

Bushman, Brad P. 1998. "Primary Effects of Media Violence on the Accessibility of Aggressive Constructs in Memory." *Personality and Social Psychology Bulletin* 24: 537–545.

Butsch, Richard. 1992. "Class and Gender in Four Decades of Television Situation Comedy: Plus ça Change ..." *Critical Studies in Mass Communication* 9 (4): 387–399.

Buunk, Bram, and Ralph B. Hupka. 1987. "Cross-Cultural Differences in the Elicitation of Sexual Jealousy." *Journal of Sex Research* 23 (February): 12–22.

Cabrera, Nolan Leon. 2014. "Exposing Whiteness in Higher Education: White Male College Students Minimizing Racism, Claiming Victimization, and Recreating White Supremacy." *Race, Ethnicity, and Education* 17 (March): 30–55.

Calarco, Jessica McCrory. 2014. "Coached for the Classroom: Parents' Cultural Transmission and Children's Reproduction of Educational Inequalities." *American Sociological Review* 79 (5): 1015–1037.

Cantor, Joanne. 2000. "Media Violence." *Journal of Adolescent Health* 27 (August): 30–34.

Carlowicz, Michael. 2014. "*Global Temperatures.*" National Aeronautics and Space Administration. **www .earthobservatory.nasa.gov**

Carmichael, Stokely, and Charles V. Hamilton. 1967. *Black Power: The Politics of Liberation in America.* New York: Vintage.

Carr, Deborah. 2010. "Cheating Hearts." *Contexts* 9 (Summer): 58–60.

Carroll, J. B., 1956. *Language, Thought, and Reality: Selected Writings of Benjamin Lee Whorf.* Cambridge, MA: MIT Press.

Carter, J. S., Mamadi Corra, and Shannon K. Carter. 2009. "The Interaction of Race and Gender: Changing Gender-Role Attitudes, 1974–2006." *Social Science Quarterly* 90(1): 196–211.

Carson, E. Ann, and William J. Sabol. 2012. *Prisoners in 2011.* Washington, DC: Bureau of Justice Statistics, U.S. Department of Justice.

Carson, E. Ann. 2014. *Prisoners in 2013.* Washington, DC: Bureau of Justice Statistics, U.S. Department of Justice. **www.bjs.gov**

Cassidy, J., and P. R. Shaver (eds.). 2008. *Handbook of Attachment: Theory, Research, and Critical Applications,* 2nd ed. New York: Guilford.

Catalano, Shannon. 2012. *Intimate Partner Violence, 1993–2010.* Washington, DC: U.S. Bureau of Justice Statistics. **www.bjs.gov**

Cavalier, Elizabeth. 2003. "'I Wear Dresses, I Wear Muscles': Media Images of Women Soccer Players." Paper presented at the annual meeting of the Southern Sociological Society, New Orleans, Louisiana.

Center for Responsive Politics. 2012. *Stats at a Glance.* **www.opensecrets.org**

Center for Responsive Politics. 2014. "Super PACs." **www.opensecrets.org**

Centers for Disease Control and Prevention. 2011. "National Center for Chronic Disease Prevention and Health Promotion, Division of Nutrition, Physical Activity, and Obesity." Atlanta, GA: Centers for Disease Control and Prevention.

Centers for Disease Control and Prevention. 2013. *Health United States 2013*. Atlanta, GA: Centers for Disease Control and Prevention.

Centers, Richard. 1949. *The Psychology of Social Classes*. Princeton, NJ: Princeton University Press.

Chafetz, Janet. 1984. *Sex and Advantage*. Totowa, NJ: Rowman and Allanheld.

Chalfant, H. Paul, Robert E. Beckley, and C. Eddie Palmer. 1987. *Religion in Contemporary Society*, 2nd ed. Palo Alto, CA: Mayfield.

Chang, Jung. 1991. *Wild Swans: Three Daughters of China*. New York: Simon & Schuster.

Chen, Elsa, Y. F. 1991. "Conflict between Korean Greengrocers and Black Americans." Unpublished senior thesis, Princeton University.

Cherlin, Andrew J. 2010. "Demographic Trends in the United States: A Review of Research in the 2000s." *Journal of Marriage and Family* 72 (June): 403–419.

Children's Bureau. 2012. *Child Maltreatment 2012*. Washington, DC: U.S. Department of Health and Human Services. **www.acf.hhs.gov**

Childs, Erica Chito. 2005. *Navigating Interracial Borders: Black-White Couples and Their Social World*. New Brunswick, NJ: Rutgers University Press.

Chopra, Rohit. July 17, 2013. "Student Debt Swells, Federal Loans Now Top a Trillian." Consumer Financial Protection Bureau. **www.consumerfinance.gov**

Chou, Rosalind S., and Joe Feagin. 2008. *The Myth of the Model Minority: Asian Americans Facing Racism*, second edition. New York: Paradigm Publishers.

Clawson, Dan, and Naomi Gerstel. 2002. "Caring for Our Young: Child Care in Europe and the United States." *Contexts* 1 (Fall–Winter): 28–35.

Coale, Ansley. 1986. "Population Trends and Economic Development." pp. 96–104 in *World Population and the U.S. Population Policy: The Choice Ahead*, edited by J. Menken. New York: W. W. Norton.

Cole, David. 1999. "The Color of Justice." *The Nation* (October 11): 12–15.

Coles, Roberta, and Charles Green, eds. 2009. *The Myth of the Missing Black Father*. New York: Columbia University Press.

Colley, Ann, and Zazie Todd. 2002. "Gender-Linked Differences in the Style and Content of E-mails to Friends." *Journal of Language and Social Psychology* 21 (December): 380–393.

Collins, Patricia Hill. 1990. *Black Feminist Theory: Knowledge, Consciousness and the Politics of Empowerment*. Cambridge, MA: Unwin Hyman.

Collins, Patricia Hill. 1998. *Fighting Words: Black Women and the Search for Justice*. Minneapolis, MN: University of Minnesota Press.

Collins, Patricia Hill. 2004. *Black Sexual Politics: African Americans, Gender, and the New Racism*. New York: Routledge.

Collins, Patricia Hill, Lionel A. Maldonado, Dana Y. Takagi, Barrie Thorne, Lynn Weber, and Howard Winant. 1995. "On West and Fenstermaker's 'Doing Difference.'" *Gender & Society* 9 (August): 491–505.

Collins, Randall, and Michael Makowsky. 1972. *The Discovery of Society*. New York: Random House.

Coltrane, Scott, and Melinda Messineo. 2000. "The Perpetuation of Subtle Prejudice: Race and Gender Imagery in 1990s Television Advertising." *Sex Roles* 42 (March): 363–389.

Connell, Catherine. 2014. *School's Out: Gay and Lesbian Teachers in the Classroom*. Berkeley, CA: University of California Press.

Connell, Raewyn. 2009. "Accountable Conduct." *Gender & Society* 24: 104–111.

Connell, Catherine. 2010. "Doing, Undoing, or Redoing Gender? Learning from the Workplace Experiences of Transpeople." *Gender & Society* 24 (1): 31–55.

Connell, Raewyn. 2012. "Transsexual Women and Feminist Thought." *Signs* 37 (4): 857–88.

Conrad, Peter, and Joseph W. Schneider. 1992. *Deviance and Medicalization: From Badness to Sickness*, expanded ed. Philadelphia, PA: Temple University Press.

Cook, S. W. 1988. "The 1954 Social Science Statement and School Segregation: A Reply to Gerard." pp. 237–256 in *Eliminating Racism: Profiles in Controversy*, edited by D. A. Taylor. New York: Plenum.

Cook, Thomas, and John Wall, eds. 2011. *Children and Armed Conflict: Cross-Disciplinary Investigations*. Hampshire, UK: Palgrave Macmillan.

Cookson, Peter W., Jr., and Caroline Hodges Persell. 1985. *Preparing for Power: America's Elite Boarding Schools*. New York: Basic Books.

Cooley, Charles Horton. 1902. *Human Nature and Social Order*. New York: Scribner's.

Cooley, Charles Horton. 1967 [1909]. *Social Organization*. New York: Schocken Books.

Coontz, Stephanie. 1992. *The Way We Never Were*. New York: Basic Books.

Cooper, Marianne. 2014. *Cut Adrift: Families in Insecure Times*. Berkeley, CA: University of California Press.

Copen, Casey E., Kimberly Daniels, Jonathan Vespa, and William D. Mosher. 2012. "First Marriages in the United States: Data from the 2006–2010 National Survey of Family Growth." *National Health Statistics Reports*, no. 29, March 22. **www.cdc.gov**

Cortes, Patricia. 2008. "The Effect of Low-Skilled Immigration on U.S. Prices: Evidence from CPI Data." *Journal of Political Economy* 116 (3): 381–422.

Coser, Lewis. 1977. *Masters of Sociological Thought*. New York: Harcourt Brace Jovanovich.

Coston, Beverly M., and Michael Kimmel. 2012. "Seeing Privilege Where It Isn't Marginalized Masculinities and the Intersectionality of Privilege." *Journal of Social Issues* 68 (March): 97–111.

Coy, Maddy. 2014. "Pornographic Performances: A Review of Sexualisation and Racism in Music Videos." *End Violence against Women*. **endviolenceagainstwomen.org.uk**

Craig, Lyn, and Killian Mullan. 2013. "Parental Leisure Time: A Gender Comparison in Five Countries." *Social Politics* 20 (3): 329–357.

Crane, Diana (ed.). 1994. *The Sociology of Culture: Emerging Theoretical Perspectives*. Cambridge, England: Blackwell.

Cravey, Tiffany, and Aparna Mmitra. 2011. "Demographics of the Sandwich Generation by Race and Ethnicity in the United States." *Journal of Socio-Economics* 40(3): 306–311.

Crittenden, Ann. 2002. *The Price of Motherhood: Why the Most Important Job in the World Is the Least Valued*. New York: Holt.

Cronin, Ann, and Andrew King. 2014. "Only Connect? Older Lesbian, Gay and Bisexual (LGB) Adults and Social Capital." *Ageing & Society* 34 (2): 258–279.

Cunningham, Mick. 2001. "The Influence of Parental Attitudes and Behaviors on Children's Attitudes toward Gender and Household Labor in Early Adulthood." *Journal of Marriage and Family* 63 (February): 111–122.

D'Emilio, John. 1998. *Sexual Politics, Sexual Communities: The Making of a Homosexual Minority in the United States, 1940–1970*, rev. ed. Chicago: University of Chicago Press.

Dahrendorf, Rolf. 1959. *Class and Class Conflict in Industrial Society*. Stanford, CA: Stanford University Press.

Dalmage, Heather M. 2000. *Tripping on the Color Line: Black-White Multiracial Families in a Racially Divided World*. New Brunswick, NJ: Rutgers University Press.

Dalton, Susan E., and Denise D. Bielby. 2000. "'That's Our Kind of Constellation': Lesbian Mothers Negotiate Institutionalized Understandings of Gender within the Family." *Gender & Society* 14 (February): 36–61.

Darling-Hammond, Linda. 2010. *The Flat World and Education: How America's Commitment to Equity Will Determine Our Future*. New York: Teachers College Press.

Davies, James B., Susanna Sandstrom, Anthony Shorrocks, and Edward N. Wolff. 2008. "The World Distribution of Household Wealth." Helsinki, Finland: UNU-WIDER, World Institute for Development Economics Research.

Davies, Michelle, Jennifer Gilston, and Paul Rogers. 2012. "Examining the

Relationship between Male Rape Myth Acceptance, Female Rape Myth Acceptance, Victim Blame, Homophobia, Gender Roles, and Ambivalent Sexism." *Journal of Interpersonal Violence* 27 (14): 2807–2823.

Davis, Kingsley. 1945. "The World Demographic Transition." *Annals of the American Academy of Political and Social Sciences* 237 (April): 1–11.

Davis, Kingsley, and Wilbert E. Moore. 1945. "Some Principles of Stratification." *American Sociological Review* 10 (April): 242–247.

Deegan, Mary Jo. 1988. "W. E. B. DuBois and the Women of Hull-House, 1895–1899." *The American Sociologist* 19 (Winter): 301–311.

Dellinger, Kristen, and Christine L. Williams. 1997. "Makeup at Work: Negotiating Appearance Rules in the Workplace." *Gender & Society* 11 (April): 151–177.

Deming, Michelle E., Eleanor K. Covan, Suzanne C. Swan, and Deborah L. Billings. 2013. "Exploring Rape Myths, Gendered Norms, Group Processing, and the Social Context of Rape among College Women: A Qualitative Analysis." *Violence Against Women* 19 (4): 465–485.

Demos. 2014. *Student Loan Debt by Race and Ethnicity.* **www.demos.org**

DeNavas-Walt, Carmen, and Bernadette D. Proctor. 2014. *Income and Poverty in the United States: 2013.* Washington, DC: U.S. Census Bureau. **www.census.gov**

Denissen, Amy M., and Abigail C. Saguy. 2014. "Gendered Homophobia and the Contradictions of Workplace Discrimination for Women in the Building Trades." *Gender & Society* 28 (3): 381–403.

DeVault, Marjorie. 1991. *Feeding the Family: The Social Organization of Caring as Gendered Work.* Chicago: University of Chicago Press.

Devine-Eller, Audrey. 2012. "Timing Matters: Test Preparation, Race, and Grade Level." *Sociological Forum* 27 (2): 458–480.

Dickinson, Timothy. 2014. "Inside the Koch Brothers Toxic Empire." *Rolling Stone,* September 24.

Dill, Bonnie Thornton. 1988. "Our Mothers' Grief: Racial Ethnic Women and the Maintenance of Families." *Journal of Family History* 13 (October): 415–431.

DiMaggio, Anthony. 2011. *The Rise of the Tea Party: Political Discontent and Corporate Media in the Age of Obama.* New York: Monthly Review Press.

DiMaggio, Paul J., and Walter W. Powell. 1991. "Introduction." pp. 1–38 in *The New Institutionalism in Organizational Analysis,* edited by W. W. Powell and P. J. DiMaggio. Chicago: University of Chicago Press.

Dines, Gail, and Jean M. Humez. 2002. *Gender, Race, and Class in Media,* 2nd ed. Thousand Oaks, CA: Sage.

DiPrete, Thomas A., and Claudia Buchmann. 2014. *The Rise of Women: The Growing Gender Gap in Education and What It Means for American Schools.* New York: Russell Sage Foundation.

Dirks, Danielle, and Jennifer C. Mueller. 2010. "Racism and Polar Culture." pp. 115–129 in *Handbook of the Sociology of Racial and Ethnic Relations,* edited by Hernán Vera and Joe R. Feagin. New York: Springer.

Doan, Long, Annalise, Loehr, and Lisa R. Miller. 2014. "Formal Rights and Informal Privileges for Same-Sex Couples: Evidence from a National Survey Experiment." *American Sociological Review* 79 (December): 1172–1195.

Doermer, Jill K., and Stephen Demuth. 2010. "The Independent and Joint Effects of Race/Ethnicity, Gender, and Age on Sentencing Outcomes in U. S. Federal Courts." *Justice Quarterly* 27 (1): 1–27.

Dolan, Jill. 2006. "Blogging on Queer Connections in the Arts and the Five Lesbian Brothers." *GLQ : A Journal of Lesbian and Gay Studies* 12 (3): 491–506.

Domhoff, G. William. 2002. *Who Rules America?* New York: McGraw-Hill.

Domhoff, G. William. 2013. *Who Rules America? The Triumph of the Corporate Rich.* New York: McGraw Hill.

Dominguez, Silvia, and Celeste Watkins. 2003. "Creating Networks for Survival and Mobility: Social Capital among African-American and Latin-American Low-Income Mothers." *Social Problems* 50 (1): 111–135.

Dowd, James J., and Laura A. Dowd. 2003. "The Center Holds: From Subcultures to Social Worlds." *Teaching Sociology* 31 (January): 20–37.

Drake, Bruce. 2014 (October 29). "Ferguson Highlights Deep Divisions between Blacks and Whites in America." Washington, DC: Pew Research Center. **www.pewresearch.org**

Drier, Peter. 2012. "The Battle for the Republican Soul: Who Is Drinking the Tea Party?" *Contemporary Sociology* 41 (November): 756–762.

DuBois, W. E. B. 1901. "The Freedmen's Bureau." *Atlantic Monthly* 86 (519): 354–365.

DuBois, W. E. B. 1903. *The Souls of Black Folk: Essays and Sketches.* Chicago: A. C. McClurg and Co.

Dubrofsky, Rachel. 2006. "The Bachelor: Whiteness in the Harem." *Critical Studies in Media Communication* 23 (1): 39–56.

Dudley, Carl S., and David A. Roozen. 2001. *Faith Communities Today: A Report on Religion in the United States Today.* Hartford, CT: Hartford Institute for Religion Research, Hartford Seminary.

Due, Linnea. 1995. *Joining the Tribe: Growing Up Gay & Lesbian in the '90s.* New York: Doubleday.

Duffy, Mignon. 2011. *Making Care Count: A Century of Gender, Race, and Paid Care Work.* New Brunswick, NJ: Rutgers University Press.

Duggan, Maeve, and Aaron Smith. September 16, 2013. "Cell Internet Use 2013." Pew Research Center's Internet & American Life Project. Washington, DC: Pew Research Center.

Duneier, Mitchell. 1999. *Sidewalk.* New York: Farrar, Strauss and Giroux.

Dunne, Gillian A. 2000. "Opting into Motherhood: Lesbians Blurring the Boundaries and Transforming the Meaning of Parenthood and Kinship." *Gender & Society* 14 (February): 11–35.

Dunne, Mairead, and Louise Gazely. 2008. "Teachers, Social Class, and Underachievement." *British Journal of Sociology of Education* 29 (5): 451–463.

Durkheim, Emile. 1951 [1897]. *Suicide.* Glencoe, IL : Free Press.

Durkheim, Emile. 1947 [1912]. *Elementary Forms of Religious Life.* Glencoe, IL: Free Press.

Durkheim, Emile. 1950 [1938]. *The Rules of Sociological Method.* Glencoe, IL: Free Press.

Durkheim, Emile. 1964 [1895]. *The Division of Labor in Society.* New York: Free Press.

Eagan, Kevin, et al. 2013. *The American Freshman: National Norms Fall 2011.* Higher Education Research Institute. Los Angeles, CA: University of California, Los Angeles.

Eberhardt, Jennifer L. 2010. "Enduring Racial Associations: African Americans, Crime, and Animal Imagery." pp. 439–457 in *Doing Race: 21 Essays for the Twenty-First Century,* edited by Paula M. Moya and Hazel Rose Markus. New York: Oxford University Press.

Eberhardt, Jennifer L., Phillip Atiba Goff, Valerie J. Purdie, and Paul G. Davies. 2004. "Seeing Black: Race, Crime, and Visual Processing." *Journal of Personality and Social Behavior* 87 (6): 876–893.

Economic Mobility Report. 2012. *Pursuing the American Dream: Economic Mobility across Generations.* Washington, DC: Pew Charitable Trusts. **www.pewtrusts.org**

Edin, Kathryn. 2000. "What Do Low-Income Single Mothers Say about Marriage?" *Social Problems* 47 (February): 112–133.

Edin, Kathyrn, and Maria Kefalas. 2005. *Promises I Can Keep: Why Poor Women Put Motherhood before Marriage.* Berkeley, CA: University of California Press.

Edin, Kathyrn, and Laura Lein. 1997. *Making Ends Meet: How Single Mothers Survive Welfare and Low-Wage Work.* New York: Russell Sage Foundation.

Edin, Kathyrn, and Timothy J. Nelson. 2013. *Doing the Best I Can: Fatherhood in the Inner City.* Berkeley, CA: University of California Press.

Egan. Timothy. 2006. *The Worst Hard Time: The Story of Those Who Survived the Great American Dust Bowl.* Boston: Houghton Mifflin.

Ehrlich, Paul. 1968. *The Population Bomb.* New York: Ballantine Books.

Ehrlich, Paul R., and Jianguo Liu. 2002. "Some Roots of Terrorism." *Population and Environment* 24 (2): 183–192.

Eitzen, D. Stanley. 2009. "Dimensions of Globalization." pp. 37–42 in *Globalization: The Transformation* of Social World*, edited by D. Stanley Eitzen and Maxine Baca Zinn. Belmont, CA: Cengage.

Eitzen, D. Stanley. 2012. *Fair and Foul: Beyond the Myths and Paradoxes of Sport*. Lanham: MD: Rowman and Littlefield.

Eitzen, D. Stanley, and Maxine Baca Zinn. 2012. *In Conflict and Order*, 13th ed. Upper Saddle River: Pearson.

Elizabeth, Vivienne. 2000. "Cohabitation, Marriage, and the Unruly Consequences of Difference." *Gender & Society* 14 (February): 87–110.

Emerson, Michael O., and David Hartman. 2006. "The Rise of Religious Fundamentalism." *Annual Review of Sociology* 32: 127–144.

England, Paula. 2010. "The Gender Revolution: Uneven and Stalled." *Gender & Society* 24 (April): 149–166.

England, Paula, and Jonathan Bearak. 2014. "The Sexual Double Standard and Gender Differences in Attitudes toward Casual Sex among U.S. University Students." *Demographic Research* 30 (46): 1327–1338.

Enloe, Cynthia. 2001. *Bananas, Beaches, and Bases: Making Feminist Sense of International Politics*, updated ed. Berkeley, CA: University of California Press.

Epstein, Marina, and L. Monique Ward. 2011. "Exploring Parent–Adolescent Communication About Gender: Results from Adolescent and Emerging Adult Samples." *Sex Roles* 65 (1–2): 108–118.

Erikson, Eric. 1980. *Identity and the Life Cycle*. New York: W. W. Norton.

Erikson, Kai. 1966. *Wayward Puritans: A Study in the Sociology of Deviance*. New York: Wiley.

Ermann, M. David, and Richard J. Lundman. 2001. *Corporate and Governmental Deviance*. New York: Oxford University Press.

Espeland, Wendy Nelson. 1998. *The Struggle for Water: Politics, Rationality, and Identity in the American Southwest*. Chicago: University of Chicago Press.

Essed, Philomena. 1991. *Understanding Everyday Racism*. Newbury Park, CA: Sage.

Etzioni, Amatai. 1975. *A Comparative Analysis of Complex Organization: On Power, Involvement, and Their Correlates*, rev. ed. New York: Free Press.

Etzioni, Amatai, John Wilson, Bob Edwards, and Michael W. Foley. 2001. "A Symposium on Robert D. Putnam's *Bowling Alone: The Collapse and Revival of American Community*." *Contemporary Sociology* 30 (May): 223–230.

Fausto-Sterling, Anne. 1992. *Myths of Gender: Biological Theories about Women and Men*. New York: Basic Books.

Fausto-Sterling, Anne. 2000. *Sexing the Body: Gender Politics and the Construction of Sexuality*. New York: Basic Books.

Feagin, Joe R. 2000. *Racist America: Roots, Future Realities, and Racial Reparations*. New York: Routledge.

Feagin, Joe R. 2007. *Systemic Racism: A Theory of Oppression*. New York: Routledge.

Federal Bureau of Investigation. 2011. *Uniform Crime Reports*. Washington, DC: U.S. Department of Justice. **www.fbi.gov**

Federal Bureau of Investigation. 2012. *2012 Hate Crime Statistics*. Washington, DC: U.S. Department of Justice. **www.fbi.gov**

Federal Bureau of Investigation. 2013. *Uniform Crime Reports*. Washington, DC: U.S. Department of Justice.

Federal Trade Commission. 2003. *Identity Theft Survey Report*. **www.ftc.gov** (accessed March 15, 2005).

Fein, Melvyn L. 1988. "Resocialization: A Neglected Paradigm." *Clinical Sociology* 6 (1): 88–100.

Ferber, Abby. 1998. *White Man Falling: Race, Gender, and White Supremacy*. Lanham, MD: Rowman and Littlefield.

Ferber, Abby. 1999. "What White Supremacists Taught a Jewish Scholar about Identity." *The Chronicle of Higher Education* (May 7): 86–87.

Ferdinand, Peter. 2000. *The Internet, Democracy, and Democratization*. London: Frank Cass Publishers.

Ferguson, Priscilla Parkhurst. 2014. "Inside the Extreme Sport of Competitive Eating." *Contexts* 13 (Summer): 26–31.

Festinger, Leon, Stanley Schachter, and Kurt Back. 1950. *Social Pressures in Informal Groups: A Study of Human Factors in Housing*. Stanford, CA: Stanford University Press.

Fetner, Tina. 2012. "The Tea Party: Manufactured Dissent or Complex Social Movement?" *Contemporary Sociology* 41 (November): 762–766.

Finkel, David. 2014. *Thank You for Your Service*. New York: Picador.

Finley, Erin P., Mary Bollinger, Polly H. Noël, Megan E. Amuan, Laurel A. Copland, Jacqueline A. Pugh, Albana Dassori, Raymond Palmer, Craig Bryan, and Mary Jo V. Pugh. 2015. "A National Cohort Study of the Association between the Polytrauma Clinical Trial and Suicide-Related Behavior among U.S. Veterans Who Served in Iraq and Afghanistan." *American Journal of Public Health* 105 (2): 380–387.

Fischer, Claude S., Michal Hout, and Martin Sánchez Jankowski. 1996. *Inequality by Design: Cracking the Bell Curve Myth*. Princeton, NJ: Princeton University Press.

Fischer, Nancy L. 2003. "Oedipus Wrecked? The Moral Boundaries of Incest." *Gender & Society* 17 (1): 92–110.

Fishbein, Allan J., and Patrick Woodall. 2006. "Women are Prime Targets for Subprime Lending: Women Are Disproportionately Represented in High-Cost Mortgage Market." Washington, DC: Consumer Federation of America.

Fisher, Bonnie S., Francis T. Cullen, and Michael G. Turner. 2000. *Sexual Victimization of College Women*. Washington, DC: Bureau of Justice Statistics.

Fishman, Charles. 2011. *The Big Thirst: The Secret Life and Turbulent Future of Water*. New York: Free Press.

Flanagan, William G. 1995. *Urban Sociology: Images and Structure*. Boston: Allyn and Bacon.

Fleury-Steiner, Ben. 2012. *Disposable Heroes: The Betrayal of African American Veterans*. Lanham, MD: Rowman and Littlefield.

Flippen, Chenoa A. 2014. "Intersectionality at Work: Determinants of Labor Supply among Immigrant Latinas." *Gender & Society* 28 (3): 404–434.

Folbre, Nancy. 2001. *The Invisible Heart: Economics and Family Values*. New York: New Press.

Fortini, Amanda. 2005. "The Great Flip-Flop Flap." *Slate*, July 22. **www.slate.com**

Foucault, Michael. 1995. *Discipline and Punish*. New York: Knopf Doubleday Publishing Group.

Frankenberg, Erica, and Chungmei Lee. 2002. *Race in American Public Schools: Rapidly Resegregating School Districts*. Los Angeles: Civil Rights Project, UCLA.

Frazier, E. Franklin. 1957. *The Black Bourgeoisie*. New York: Collier Books.

Fredrickson, George M. 2003. *Racism: A Short History*. Princeton, NJ: Princeton University Press.

Freedle, Roy O. 2003. "Correcting the SATs Ethnic and Social Class Bias: A Method of Re-Estimating SAT Scores." *Harvard Educational Review* 73 (Spring): 1–43.

Freedman, Estelle B., and John D'Emilio. 1988. *Intimate Matters: A History of Sexuality in America*. New York: Harper & Row.

Freeman, Jo. 1983. "A Model for Analyzing the Strategic Options of Social Movement Organizations." pp. 193–210 in *Social Movements of the Sixties and Seventies*, edited by Jo Freeman. New York: Longman.

Fried, Amy. 1994. "'It's Hard to Change What We Want to Change': Rape Crisis Centers as Organizations." *Gender & Society* 4 (December): 562–583.

Friedman, Thomas L. 1999. *The Lexus and the Olive Tree*. New York: Farrar, Strauss, and Giraux.

Fry, Richard. 2012. *A Record One-in-Five Households Now Owe Student Loan Debt*. Washington, DC: Pew Research Center. **www.pewsocialtrends.org**

Fryberg, Stephanie, and Alisha Watts. 2010. "We're Honoring You Dude: Myths, Mascots, and American Indians." pp. 458–480 in *Doing Race: 21 Essays for the 21st Century*, edited by Hazel Rose Markus and Paula M. L. Moya. New York: W. W. Norton.

Frye, Marilyn. 1983. *The Politics of Reality*. Trumansburg, NY: The Crossing Press.

Gaertner, Samuel L., and John F. Dovidio. 2005. "Understanding and Addressing Contemporary Racism: From Aversive Racism to the Common In-group Identity Model." *Journal of Social Issues* 61 (3): 615–639.

Gallagher, Charles A. 2013. "Color-Blind Privilege." pp. 91–95 in *Race, Class and Gender: An Anthology,* 7th ed., edited by Margaret L. Andersen and Patricia Hill Collins. Belmont CA: Wadsworth/Cengage.

Gallagher, Sally K. 2003. *Evangelical Identity and Gendered Family Life.* New Brunswick, NJ: Rutgers University Press.

Gallup, George H., Jr. 2003. "Current Views on Premarital, Extramarital Sex." Princeton, NJ: Gallup Organization. **www.gallup.com**

Gallup Poll. 2008. "Homosexual Relations." *The Gallup Poll.* Princeton, NJ: Gallup Organization. **www.gallup.com**

Gallup Poll. 2012. "Guns." *The Gallup Poll.* Princeton, NJ: Gallup Organization. **www.gallup.com**

Gallup Poll. 2014. "Confidence in Institutions." *The Gallup Poll.* Princeton, NJ: Gallup Organization. **www.gallup.com**

Gamson, Joshua. 1998. *Freaks Talk Back: Tabloid Talk Shows and Sexual Nonconformity.* Chicago: University of Chicago Press.

Gans, Herbert. 1982 [1962]. *The Urban Villagers: Group and Class in the Life of Italian Americans.* New York: Free Press.

Garfinkel, Harold. 1967. *Studies in Ethnomethodology.* Englewood Cliffs, NJ: Prentice Hall.

Gates, Gary J., and Frank Newport. 2012 (October 18). "Special Report: 3.4% of U.S. Adults Identify as LGBT." *The Gallup Poll.* Princeton, NJ: Gallup Organization. **www.gallup.com**

Gates, Henry Louis, Jr., and Nellie Y. McKay, eds. 1997. *The Norton Anthology of African American Literature.* New York: W. W. Norton.

Genovese, Eugene. 1972. *Roll, Jordan, Roll: The World the Slaves Made.* New York: Pantheon.

Gentile, Douglas A., Lindsay C. Mathieson, and Nicki R. Crick. 2011. "Media, Violence Associations with the Form and Function of Aggression among Elementary School Children." *Social Development* 20 (2): 213–232.

Gerschick, T. J., and A. S. Miller. 1995. "Coming to Terms: Masculinity and Physical Disability." pp. 183–204 in *Men's Health and Illness: Gender, Power, and the Body,* Research on Men and Masculinities Series, vol. 8, edited by D. Sabo and D. F. Gordon. Thousand Oaks, CA: Sage Publications.

Gerson, Kathleen. 2010. *The Unfinished Revolution: How a New Generation Is Shaping Family, Work, and Gender in America.* New York: Oxford University Press.

Gerstel, Naomi. 2000. "The Third Shift: Gender and Care Work Outside the Home." *Qualitative Sociology* 23 (4): 467–483.

Gerstel, Naomi, and Sally Gallagher. 2001. "Men's Caregiving: Gender and the Contingent Character of Care." *Gender & Society* 15 (April): 197–217.

Gerth, Hans, and C. Wright Mills (eds.). 1946. *From Max Weber: Essays in Sociology.* New York: Oxford University Press.

Giannarelli, Linda, and James Barsimantov. 2000. *Child Care Expenses of America's Families.* Washington. DC: Urban Institute.

Giddings, Paula. 1994. *In Search of Sisterhood: Delta Sigma Theta and the Challenge of the Black Sorority Movement.* New York: William Morrow.

Giddings, Paula. 2008. *A Sword among Lions: Ida B. Wells and the Campaign against Lynching.* New York: Amistad.

Gilbert, D. T., and P. S. Malone. 1995. "The Correspondence Bias." *Psychological Bulletin* 117 (1): 21–38.

Gile, K.G., L.G. Johnston,and M.J. L Salganik. 2015. "Diagnostics for Respondent-driven Sampling." *Journal of the Royal Statistical Society* 178(1): 241–269.

Gilkes, Cheryl Townsend. 2000. *"If It Wasn't for the Women ... Black Women's Experience and Womanist Culture in Church and Community.* Maryknoll, NY: Orbis Books.

Gimlin, Debra. 1996. "Pamela's Place: Power and Negotiation in the Hair Salon." *Gender & Society* 10 (October): 505–526.

Gimlin, Debra. 2002. *Body Work: Beauty and Self-Image in American Culture.* Berkeley, CA: University of California Press.

Gitlin, Todd. 2002. *Media Unlimited: How the Torrent of Images and Sounds Overwhelms Our Lives.* New York: Metropolitan Books.

Glassner, Barry. 1999. *Culture of Fear: Why Americans Are Afraid of the Wrong Things.* New York: Basic Books.

Glaze, Lauren E., and Laura M. Maruschak. 2008. "Parents in Prison and Their Minor Children." Washington, DC: U.S. Bureau of Justice Statistics. **www.ojp.usdoj.gov**

Glenn, Evelyn Nakano. 1986. *Issei, Nisei, War Bride: Three Generations of Japanese American Women in Domestic Service.* Philadelphia, PA: Temple University Press.

Glenn, Evelyn Nakano. 2002. *Unequal Freedom: How Race and Gender Shaped American Citizenship and Labor.* Cambridge, MA: Harvard University Press.

Glock, Charles, and Rodney Stark. 1965. *Religion and Society in Tension.* Chicago: Rand McNally.

Goffman, Alice. 2009. "On the Run: Wanted Men in a Philadelphia Ghetto." *American Sociological Review* 74 (June): 339–357.

Goffman, Erving. 1959. *The Presentation of Self in Everyday Life.* Garden City, NY: Doubleday.

Goffman, Erving. 1961. *Asylums: Essays on the Social Situation of Mental Patients and Other Inmates.* Garden City, NY: Anchor.

Goffman, Erving. 1963. *Stigma: Notes on the Management of Spoiled Identity.* Englewood Cliffs, NJ: Prentice Hall.

Goldberg, Abbie, Deborah Kashy, and JuilAnna Smith. 2012. "Gender-Typed Play Behaviors in Early Childhood: Adopted Children with Lesbian, Gay, and Heterosexual Parents." *Sex Roles* 67 (9–10): 503–515.

Gordon, Jesse, and Knickerbocker Designs. 2001. "The Sweat behind the Shirt." *The Nation* (March 10): 14ff.

Gordon, Linda. 1977. *Woman's Body/Woman's Right.* New York: Penguin.

Gore, Albert. 2006. *An Inconvenient Truth: The Planetary Emergency of Global Warming and What We Can Do about It.* Emmaus, PA: Rodale.

Gottfredson, Michael R., and Travis Hirschi. 1990. *A General Theory of Crime.* Stanford, CA: Stanford University Press.

Gottfredson, Michael R., and Travis Hirschi. 1995. "National Crime Control Policies." *Society* 32 (January–February): 30–36.

Gould, Stephen Jay. 1999. "The Human Difference." *The New York Times* (July 2).

Gramsci, Antonio. 1971. *Selections from the Prison Notebooks of Antonio Gramsci,* edited by Quintin Hoare and Geoffrey Nowell. London: Lawrence and Wishart.

Granovetter, Mark. 1973. "The Strength of Weak Ties." *American Journal of Sociology* 78 (May): 1360–1380.

Granovetter, Mark. 1974. *Getting a Job: A Study of Contacts and Careers.* Cambridge, MA: Harvard University Press.

Granovetter, Mark S. 1995. "Afterward 1994: Reconsiderations and a New Agenda." pp. 139–182 in *Getting a Job,* 2nd ed. Chicago: University of Chicago Press.

Grant, Don Sherman, II, and Ramiro Martínez Jr. 1997. "Crime and the Restructuring of the U.S. Economy: A Reconsideration of the Class Linkages." *Social Forces* 75 (March): 769–799.

Graves, Joseph L. 2004. *The Race Myth: Why We Pretend Race Exists in America.* New York: Dutton.

Greenebaum, Jessica B. 2012. "Managing Impressions: 'Face-Saving' Strategies of Vegetarians and Vegans." *Humanity & Society* 36 (4): 309–325.

Grimal, Pierre (ed.). 1963. *Larousse World Mythology.* New York: Putnam.

Grindstaff, Laura. 2002. *The Money Shot: Trash Class, and the Making of TV Talk Shows.* Chicago: University of Chicago Press.

Guerrero, Laura K., Peter A. Andersen, and Walid A. Afifi. 2010. *Close Encounters: Communication in Relationships,* 3rd ed. Thousand Oaks, CA: Sage.

Gurin, Patricia, E. L. Dey, Sylvia Hurtado, and G. Gurin. 2002. "Diversity and Higher Education: Theory and Impact on Educational Outcomes." *Harvard Educational Review* 72 (3): 330–366.

Guttierez y Muhs, Gabriella, Yolanda Flores Niemann, Carmen G. Gonzalez, and Angela P. Harris (eds.). 2012. *Presumed Incompetent: The Intersections of Race and Class for Women in Academia.* Logan, UT: Utah State University Press.

Guttmacher Institute. 2014. *Facts on American Teens' Sexual and Reproductive Health.* New York: Guttmacher Institute. **www.guttmacher.org**

Habermas, Jürgen. 1970. *Toward a Rational Society: Student Protest, Science, and Politics.* Boston: Beacon Press.

Haddad, Yvonne Yazbeck, Jane I. Smith, and John L. Esposito (eds.). 2003. *Religion and Immigration: Christian, Jewish, and Muslim Experiences in the United States.* Walnut Creek, CA: AltaMira Press.

Haley-Lock, Anna, and Stephanie Ewert. 2011. "Serving Men and Mothers: Workplace Practices and Workforce Composition in Two U.S. Restaurant Chains and States." *Community, Work & Family* 14 (4): 387–404.

Hall, Edward T. 1966. *The Hidden Dimension.* New York: Doubleday.

Hamer, Jennifer. 2001. *What It Means to Be Daddy.* New York: Columbia University Press.

Hamilton, Laura, and Elizabeth A. Armstrong. 2009. "Gendered Sexuality in Young Adulthood: Double Binds and Flawed Options." *Gender & Society* 23 (October): 589–616.

Hamilton, Mykol C. 1988. "Using Masculine Generics: Does Generic He Increase Male Bias in the User's Imagery?" *Sex Roles* 19 (December): 785–799.

Handlin, Oscar. 1951. *The Uprooted.* Boston: Little, Brown.

Haney, C., C. Banks, and P. G. Zimbardo. 1973. "Interpersonal Dynamics in a Simulated Prison." *International Journal of Criminology and Penology* 1: 69–97.

Haney, Lynne. 1996. "Homeboys, Babies, Men in Suits: The State and the Reproduction of Male Dominance." *American Sociological Review* 61 (October): 759–778.

Harden, Blaine. 2012. *Escape from Camp 14: One Man's Remarkable Odyssey from North Korea to Freedom in the West.* New York: Viking.

Hargittai, Eszter, and Miguel Angel Centeno. 2001. "Introduction: Defining a Global Geography." *American Behavioral Scientist* 44 (10): 1545–1560.

Harnois, Catherine E. 2010. "Race, Gender, and the Black Women's Standpoint." *Sociological Forum* 25 (1): 68–85.

Harp, Dustin, and Mark Tremayne. 2006. "The Gendered Blogosphere: Examining Inequality Using Network and Feminist Theory." *Journalism & Mass Communication Quarterly* 83 (2): 247–264.

Harris, Angel L. 2006. "I Don't Hate School: Revisiting 'Oppositional Culture' Theory of Blacks' Resistance to Schooling." *Social Forces* 85 (2): 797–834.

Harris, Marvin. 1974. *Cows, Pigs, Wars, and Witches: The Riddles of Culture.* New York: Vintage.

Harris, Richard J., Juanita M. Firestone, and Mary Bollinger. 2000. "Gender Role Attitudes: Native Americans in Comparative Perspective." *Free Inquiry in Creative Sociology* 28 (2): 63–76.

Harrison, Roderick. 2000. "Inadequacies of Multiple Response Race Data in the Federal Statistical System." Manuscript. Joint Center for Political and Economic Studies and Howard University, Department of Sociology, Washington, DC.

Hartley, D. 2014. "Rural Health Disparities, Population Health, and Rural Culture." *American Journal of Public Health* 94 (10): 1675–1678.

Hauan, Susan M., Nancy S. Landale, and Kevin T. Leicht. 2000. "Poverty and Work Effort among Urban Latino Men." *Work and Occupations* 27 (May): 188–222.

Hayes-Smith, Rebecca. 2011. "Gender Norms in the *Twilight* Series." *Contexts* 10 (Spring): 78–79.

Hays, Sharon. 2003. *Flat Broke with Children: Women in the Age of Welfare Reform.* New York: Oxford University Press.

Hearnshaw, Leslie. 1979. *Cyril Burt: Psychologist.* Ithaca, NY: Cornell University Press.

Hendy, Helen M., Cheryl Gustitus, and Jamie Leitzel-Schwalm. 2001. "Social Cognitive Predictors of Body Images in Preschool Children." *Sex Roles* 44 (May): 557–569.

Henry, Kathy. 2008. "Warrior Woman—Ida B. Wells Barnett." *Ezinearticles.* **ezinearticles.com**

Henslin, James M. 1993. "Doing the Unthinkable." pp. 253–262 in *Down to Earth Sociology,* 7th ed., edited by James M. Henslin. New York: Free Press.

Henson, K. T. 1995. *Curriculum Development for Educational Reform.* New York: Harper Collins.

Herring, Cedric. 2009. "Does Diversity Pay?: Race, Gender, and the Business Case for Diversity." *American Sociological Review* 74 (2): 208–224.

Herrnstein, Richard J., and Charles Murray. 1994. *The Bell Curve: Intelligence and Class Structure in American Life.* New York: Free Press.

Hertz, Rosanna. 2006. *Single by Chance, Mothers by Choice: How Women Are Choosing Parenthood without Marriage and Creating the New American Family.* New York: Oxford University Press.

Hertz, Tom. 2006. *Understanding Mobility in America.* Washington, DC: Center for American Progress. **www.americanprogress.org**

Hesse-Biber, Sharlene Hagy. 2007. *The Cult of Thinness,* 2nd ed. New York: Oxford University Press.

Higginbotham, A. Leon, Jr. 1978. *In the Matter of Color: Race and the American Legal Process.* New York: Oxford University Press.

Higginbotham, Elizabeth, and Margaret L. Andersen, eds. 2012. *Race and Ethnicity in Society: The Changing Landscape,* 3rd edition. Belmont, CA: Wadsworth/Cengage.

Hill, Lori Diane. 2001. "Conceptualizing Educational Attainment Opportunities of Urban Youth: The Effects of School Capacity, Community Context and Social Capital." PhD diss., University of Chicago.

Hirschi, Travis. 1969. *Causes of Delinquency.* Berkeley: University of California Press.

Hirschman, Charles. 1994. "Why Fertility Changes." *Annual Review of Sociology* 20: 203–223.

Hirschman, Charles, and Douglas S. Massey. 2008. "Places and Peoples: The New American Mosaic." pp. 1–21 in *New Faces in New Places: The Changing Geography of American Immigration,* ed. by Douglass Massey. New York: Russell Sage Foundation.

Hochschild, Arlie Russell. 1983. *The Managed Heart: Commercialization of Human Feelings.* Berkeley, CA: University of California Press.

Hochschild, Arlie Russell. 1997. *The Time Bind: When Work Becomes Home and Home Becomes Work.* New York: Metropolitan Books.

Hochschild, Arlie Russell. 2003. *The Commercialization of Intimate Life: Notes from Home and Work.* Berkeley: University of California Press.

Hochschild, Arlie Russell, with Anne Machung. 1989. *The Second Shift: Working Parents and the Revolution at Home.* New York: Viking.

Holt, Thomas J., and Michael G. Turner. 2012. "Examining Risks and Protective Factors of On-Line Identity Theft." *Deviant Behavior* 33: 308–323.

Holyoke, Thomas T. 2009. "Interest Group Competition and Coalition Formation." *American Journal of Political Science* 53 (April): 360–375.

Hondagneu-Sotelo, Pierrette. 2001. *Doméstica: Immigrant Workers Cleaning and Caring in the Shadows of Affluence.* Berkeley, CA: University of California Press.

Horton, Hayward Derrick, Beverlyn Lundy Allen, Cedric Herring, and Melvin E. Thomas. 2000. "Lost in the Storm: The Sociology of the Black Working Class, 1850 to 1990." *American Sociological Review* 65 (February): 128–137.

Howard, Philip H., and Patricia Allen. 2010. "Beyond Organic and Fair Trade? An Analysis of Ecolabel Preferences in the United States." *Rural Sociology* 75 (2): 244–269.

Hoyt, Wendy, and Lori R. Kogan. 2001. "Satisfaction with Body Image and Peer Relationships for Males and Females in a College Environment." *Sex Roles* 45 (August): 199–215.

Huesmann, L. R., J. Moise, C. D. Podolski, and J. D. Eron. 2003. "Longitudinal Relations between Childhood Exposure to Media Violence and Adult Aggression and Violence, 1977–1992." *Developmental Psychology* 39 (2): 201–221.

Hughes, Langston. 1967. *The Big Sea.* New York: Knopf.

Hull, Kathleen E., Ann Meier, and Timothy Ortyl. 2010. "The Changing Landscape of Love and Marriage." *Contexts* 9 (Spring): 32–37.

Hunt, Ruth. 2005. "Subjective Rating by Skin Color Gradation, Gender, and

Other Characteristics." Junior Project, Princeton University, manuscript.

Ignatiev, Noel. 1995. *How the Irish Became White.* New York: Routledge.
Insight. 2013 (August). *The Relationship between Alcohol and Sexual Assault.* **www.nvcc.edu**
Institute of Medicine of the National Academies. 2010. *Returning Home from Iraq and Afghanistan: Preliminary Assessment of Readjustment Needs of Veterans, Service Members, and Their Families.* Committee on the Initial Assessment of Readjustment Needs of Military Personnel, Veterans, and Their Families. Washington, DC: National Academies Press.
Inter-Parliamentary Union. 2012. *Women in National Parliaments 2012.* **www.ipu.org**
International Labour Organization. 2014. *What is Child Labor.* **www.ilo.org**
Irwin, Katherine. 2001. "Legitimating the First Tattoo: Moral Passage through Informal Interaction." *Symbolic Interaction* 24 (March): 49–73.
Itakura, Hiroko. 2014. "Femininity in Mixed-Sex Talk and Intercultural Communication." *Pragmatics and Society* 5 (3): 455–483.

Jakubczyk, A., A. Klimkiewicz, A. Krasowska, M. Kopera, A. Slawinska-Ceran, K. J. Brower, and M. Wojnar. 2014. "History of Sexual Abuse and Suicide Attempts in Alcohol-Dependent Patients." *Child Abuse & Neglect* 38 (9): 1560–1568.
Jack, Anthony Abraham. 2014. "Culture Shock Revisited: The Social and Cultural Contingencies to Class Marginality." *Sociological Forum* 28 (June): 453–475.
Jackson, Pamela Braboy. 2000. "Stress and Coping among Black Elites in Organizational Settings." Unpublished manuscript.
Jackson, Pamela B., Peggy A. Thoits, and Howard F. Taylor. "The Effects of Tokenism on America's Black Elite." Paper read before the American Sociological Association, Los Angeles, CA, August 1994.
Jackson, Pamela B., Peggy A. Thoits, and Howard F. Taylor. 1995. "Composition of the Workplace and Psychological Well-Being: The Effects of Tokenism on America's Black Elite." *Social Forces* 74 (December): 543–557.
Jacobs, David, Zhenchao Qian, Jason T. Charmichael, and Stephanie L. Kent. 2007. "Who Survives Death Row: An Individual and Contextual Analysis." *American Sociological Review* 72 (August): 610–632.
Jacobs, Jerry A., and Kathleen Gerson. 2004. *The Time Divide: Work, Family, and Gender Inequality.* New York: Oxford University Press.
Janis, Irving L. 1982. *Groupthink: Psychological Studies of Policy Decisions and Fiascos,* 2nd ed. Boston: Houghton Mifflin.
Jäntti, Markus et al. 2006. "American Exceptionalism in a New Light: A Comparison of Intergenerational Earnings Mobility in the Nordic Countries, the United Kingdom and the United States." Discussion Paper No. 1938 (January). Institute for the Study of Labor. **http://ftp.iza.org/dp1938.pdf**
Jenness, Valerie, and Sarah Fenstermaker. 2014. "Agnes Goes to Prison: Gender Authenticity, Transgender Inmates in Prisons for Men, and Pursuit of 'The Real Deal.'" *Gender & Society* 28 (February): 5–31.
Jimenez, Tomas. 2009. *Replenished Ethnicity: Mexican Americans, Immigration, and Identity.* Berkeley, CA: University of California Press.
Johnson, Jennifer A. 2011. "Mapping the Feminist Political Economy of the Online Commercial Pornography Industry: A Network Approach." *International Journal of Media and Cultural Politics* 7 (2): 189–208.
Johnstone, Ronald I. 1992. *Religion in Society: A Sociology of Religion,* 4th ed. Englewood Cliffs, NJ: Prentice Hall.
Jones, Diane Carlson. 2001. "Social Comparison and Body Image: Attractiveness Comparison to Models and Peers among Adolescent Girls and Boys." *Sex Roles* 45 (November): 645–664.
Jones, James M., John F. Dovidio, and Deborah L. Vietze. 2013. *The Psychology of Diversity.* Malden, MA: Wiley Blackwell.
Jordan, Catheleen, and David Cory. 2010. "Boomers, Boomerangs, and Bedpans." *National Social Science Journal* 34 (1): 79–84.
Jordon, Winthrop D. 1968. *White over Black: American Attitudes toward the Negro 1550–1812.* Chapel Hill, NC: University of North Carolina Press.
Jussim, Lee, and Kent D. Harber. 2005. "Teacher Expectations and Self-Fulfilling Prophesies: Knowns and Unknowns, Resolved and Unresolved Controversies." *Personality and Social Psychology Review* 9 (2): 131–155.

Kadushin, Charles. 1974. *The American Intellectual Elite.* Boston: Little, Brown.
Kaiser Family Foundation. 2010. *Prescription Drug Trends.* Menlo Park, CA: Kaiser Family Foundation. **www.kff.org**
Kalleberg, Arne. 2013. *Good Jobs, Bad Jobs: The Rise of Polarized and Precarious Employment Systems in the United States, 1970s to 2000s.* New York: Russell Sage Foundation.
Kalof, Linda. 1999. "The Effects of Gender and Music Video Imagery on Sexual Attitudes." *Journal of Social Psychology* 139 (June): 378–385.
Kamin, Leon J. 1974. *The Science and Politics of IQ.* Potomac, MD: Lawrence Erlbaum.
Kane, Emily. 2012. *The Gender Trap: Parents and the Pitfalls of Raising Boys and Girls.* New York: New York University Press.
Kang, Miliann. 2010. *The Managed Hand: Race, Gender and the Body in Beauty Service Work.* Berkeley, CA: University of California Press.
Kanter, Rosabeth Moss. 1977. *Men and Women of the Corporation.* New York: Basic Books.
Kaplan, Elaine Bell. 1996. *Not Our Kind of Girl: Unraveling the Myths of Black Teenage Motherhood.* Berkeley, CA: University of California Press.
Kaplan, Mark S., Nathalie Huguet, Benston H. McFarland, and Jason T. Newsom. 2007. "Suicide among Male Veterans: A Prospective Population-Based Study." *Journal of Epidemiology and Community Health* 61 (July): 619–624.
Kaufman, Gayle. 1999. "The Portrayal of Men's Family Roles in Television Commercials." *Sex Roles* 41 (September): 439–458.
Kendall, Lori. 2002. "'Oh No! I'm a Nerd!': Hegemonic Masculinity on an Online Forum." *Gender & Society* 14 (April): 256–274.
Kennedy, Randall. 2003. *Interracial Intimacies: Sex, Marriage, Identity and Adoption.* New York: Pantheon.
Kephart, W. H. 1993. *Extraordinary Groups: An Examination of Unconventional Life,* rev. ed. New York: Martin's Press.
Kerr, N. L. 1992. "Issue Importance and Group Decision Making." pp. 68–88 in *Group Process and Productivity,* edited by S. Worchel, W. Wood, and J. A. Simpson. Newbury Park, CA: Sage.
Kessler, Suzanne J. 1990. "The Medical Construction of Gender: Case Management of Intersexed Infants." *Signs* 16 (Autumn): 3–26.
Khashan, Hilal, and Lina Kreidie. 2001. "The Social and Economic Correlates of Islamic Religiosity." *World Affairs* 1654 (Fall): 83–96.
Kibria, Nazli, Cara Bowman, and Megan O'Leary. 2014. *Race and Immigration.* Malden, MA: Polity.
Kim, Elaine H. 1993. "Home Is Where the *Han* Is: A Korean American Perspective on the Los Angeles Upheavals." pp. 215–235 in *Reading Rodney King/Reading Urban Uprising,* edited by Robert Gooding-Williams. New York: Routledge.
Kimmel, Michael. 2000. "Saving the Males: The Sociological Implications of Virginia Military Institute and the Citadel." *Gender & Society* 14 (August): 494–516.
Kimmel, Michael. 2008. *Guyland: The Perilous World Where Boys Become Men.* New York: Harper.
Kimmel, Michael S., and Michael A. Messner. 2012. *Men's Lives,* 9th ed. Boston: Allyn and Bacon.
King, Eden B., Jonathan J. Mohr, Chad I. Peddie, Kristen P. Jones, and Matt Kendra. 2014. "Predictors of Identity Management: An Exploratory Experience-Sampling Study of Lesbian, Gay, and Bisexual Workers." *Journal of Management* 20 (June): 1–27.
Kistler, Michelle E., and Moon J. Lee. 2010. "Does Exposure to Sexual Hip-Hop Music Videos Influence the

Sexual Attitudes of College Students?" *Mass Communication and Society* 12 (January): 67–86.

Kitano, Harry. 1976. *Japanese Americans: The Evolution of a Subculture,* 2nd ed. New York: Prentice Hall.

Kleinfeld, Judith. 1999. "Student Performance: Males versus Females." *National Affairs* 134 (Winter): 3–20.

Klinenberg, Eric. 2002. *Heat Wave: A Social Autopsy of Disaster in Chicago.* Chicago: University of Chicago Press.

Klinenberg, Eric. 2012. *Going Solo: The Extraordinary Rise and Surprising Appeal of Living Alone.* New York: The Penguin Press.

Klinger, Lori J., James A. Hamilton, and Peggy J. Cantrell. 2001. "Children's Perceptions of Aggression and Gender-Specific Content in Toy Commercials." *Social Behavior and Personality* 29 (1): 11–20.

Kloer, Amanda. 2010. "Sex Trafficking and HIV/AIDS: A Deadly Junction for Women and Girls." *Human Rights Magazine* 37 (Spring).

Kluegel, J. R., and Lawrence Bobo. 1993. "Dimensions of Whites' Beliefs about the Black-White Socioeconomic Gap." pp. 127–147 in *Race and Politics in American Society,* edited by P. Sniderman, P. Tetlock, and E. Carmines. Stanford, CA: Stanford University Press.

Kneebone, Elizabeth, and Natalie Holmes. 2014. *New Census Data Show Few Metro Areas Made Progress Against Poverty in 2013.* Washington, DC: Brookings. **www.brookings.edu**

Ko, Eunjeong, Sunhee Cho, Ramona Perez, Younsook Yeo, and Helen Palomino. 2013. "Good and Bad Death: Exploring the Perspectives of Older Mexican Americans." *Journal of Gerontological Social Work* 56 (1): 6–25.

Kochen, M. (ed.). 1989. *The Small World.* Norwood, NJ: Ablex Press.

Kochhar, Rakesh, Richard Fry, and Paul Taylor. 2011. "Wealth Gaps Rise to Record Highs between Whites, Blacks, Hispanics." Washington, DC: Pew Research Center. **www.pewsocialtrends.org**

Kocieniewski, David, and Robert Hanley. 2000. "Racial Profiling Was Routine, New Jersey Says." *The New York Times* (November 28): 1.

Kohut, Andrew. 2007. "Muslim Americans: Middle Class and Mostly Mainstream." Washington, DC: Pew Research Center.

Komter, Martha L. 2006. "From Talk to Text: The Interactional Construction of a Police Record." *Research on Language and Social Interaction* 39 (3): 201–228.

Konishi, Hideo, and Debraj Ray. 2003. "Coalition Formation as a Dynamic Process." *Journal of Economic Theory* 110 (May): 1–41.

Koser, Khalid. 2007. *International Migration: A Very Short Introduction.* New York: Oxford University Press.

Kozol, Jonathan. 2006. *The Shame of the Nation: The Restoration of Apartheid Schooling in America.* New York: Broadway.

Kreager, David, and Jeremy Staff. 2009. "The Sexual Double Standard and Adolescent Acceptance." *Social Psychology Quarterly* 72 (June): 143–164.

Kristof, Nicholas D. 2008. "Racism without Racists." *The New York Times* (October 4). **www.nytimes.com**

Kroll, Luisa, and Kerry A. Dolan. 2012. "The Forbes 400: The Richest People in America." *Forbes* (September 19).

Kurz, Demie. 1995. *For Richer for Poorer: Mothers Confront Divorce.* New York: Routledge.

Laakso, Janice. 2013. "Flawed Policy Assumptions and HOPE VI." *Journal of Poverty* 17 (1): 29–46.

Lacy, Karen R. 2007. *Blue-Chip Black: Race, Class and Status in the New Black Middle Class.* Berkeley, CA: University of California Press.

LaFlamme, Darquise, Andree Pomerrleau, and Gerard Malcuit. 2002. "A Comparison of Fathers' and Mothers' Involvement in Childcare and Stimulation Behaviors during Free-Play with Their Infants at 9 and 15 Months." *Sex Roles* 11–12 (December): 507–518.

Lamanna, Mary Ann, and Agnes Riedman. 2012. *Marriage and Families: Making Choices in a Diverse Society,* 11th ed. Belmont, CA: Wadsworth.

Lamont, Michèle. 1992. *Money, Morals, and Manners: The Culture of the French and the American Upper-Middle Class.* Chicago: University of Chicago Press.

Langman, Lauren, and Douglas Morris. 2002. "Internetworked Social Movements: The Promises and Prospects for Global Justice." Paper presented at the International Sociological Association, Brisbane, Australia.

Langton, Lynn, and Sofi Sinozich. 2014. *Rape and Sexual Assault among College-Age Females, 1995–2013.* Washington, DC: U.S. Bureau of Justice Statistics. **www.bjs.gov**

Laumann, Edward O., John H. Gagnon, Robert T. Michael, and Stuart Michaels. 1994. *The Social Organization of Sexuality: Sexual Practices in the United States.* Chicago: University of Chicago Press.

Ledger, Kate. 2009. "Sociology and the Gene." *Contexts* 8 (3): 16–20.

Lee, Donghoon. 2013. *Household Debt and Credit. Student Debt.* New York: Federal Reserve Bank of New York Consumer Credit Panel. **www.newyorkfed.org**

Lee, Hangwoo. 2006. "Privacy, Publicity, and Accountability of Self Presentation in an On-line Discussion Group." *Sociological Inquiry* 76 (1): 1–22.

Lee, M. A., Joachim Singelmann, and Anat Yom-Tov. 2008. "Welfare Myths: The Transmission of Values and Work among TANF Families." *Social Science Research* 37 (2): 516–529.

Lee, Richard M., Harold D. Grotevant, Wendy L. Hellerstedt, and Megan R. Gunnar. 2006. "Cultural Socialization in Families with Internationally Adopted Children." *Journal of Family Psychology* 20 (4): 571–580.

Lee, Sharon M. 1993. "Racial Classification in the U.S. Census: 1890–1990." *Ethnic and Racial Studies* 16 (1): 75–94.

Lee, Stacey J. 2009. *Unraveling the "Model Minority" Stereotype: Listening to Asian American Youth,* 2nd ed. New York: Teacher's College Press.

Legerski, Elizabeth Miklya, and Marie Cornwall. 2010. "Working-Class Job Loss, Gender, and the Negotiation of Household Labor." *Gender & Society* 24 (August): 447–474.

Leidner, Robin. 1993. *Fast Food, Fast Talk: Service Work and the Routinization of Everyday Life.* Berkeley, CA: University of California Press.

Lemert, Edwin M. 1972. *Human Deviance, Social Problems, and Social Control.* Englewood Cliffs, NJ: Prentice Hall.

Lengermann, P. M., and Niebrugge-Brantley, J. 1998. *The Woman Founders: Sociology and Social Theory, 1830–1930.* Boston: McGraw-Hill.

Lenski, Gerhard. 2005. *Ecological-Evolutionary Theory: Principles and Applications.* Boulder, CO: Paradigm Publishers.

Levanon, Asaf, Paula England, and Paul Allison. 2009. "Occupational Feminization and Pay: Assessing Causal Dynamics Using 1950–2000 U.S. Census Data." *Social Forces* 88 (2): 865–892.

Levine, Linda. 2012. *An Analysis of the Distribution of Wealth across Households.* Congressional Research Service. **www.crs.gov**

Levy, Ariel. 2005. *Female Chauvinist Pigs: Women and the Rise of Raunch Culture.* New York: Free Press.

Levy, Becca R., Pil H. Chung, Talya Bedford, and Kristina Navrazhina. 2013. "Facebook as a Site for Negative Age Stereotypes." *The Gerontologist* 54 (2): 172–176.

Lewin, Tamar. 2002. "Study Links Working Mothers to Slower Learning." *The New York Times* (July 17): A14.

Lewin, Tamar. 2010. "Baby Einstein Founder Goes to Court." *The New York Times* (January 13): A15.

Lewis, Amanda. 2003. *Race in the Schoolyard: Negotiating the Color Line in Classrooms and Communities.* New Brunswick, NJ: Rutgers University Press.

Lewis, Oscar. 1960. *Five Families: Mexican Case Studies in the Culture of Poverty.* New York: Basic Books.

Lewis, Oscar. 1966. "The Culture of Poverty." *Scientific American* 215 (October): 19–25.

Lewontin, Richard. 1996. *Human Diversity.* New York: W. H. Freeman.

Limoncelli, Stephanie. 2010. *The Politics of Trafficking: The First International Movement to Combat the Sexual Exploitation of Women.* Palo Alto, CA: Stanford University Press.

Lin, Nan. 1989. "The Small World Technique as a Theory Construction Tool." pp. 231–238 in *The Small World,* edited by M. Kochen. Norwood, NJ: Ablex Press.

Ling, Haping 2009. *Emerging Voices: Experiences of Underrepresented Asian Americans.* New Brunswick, NJ: Rutgers University Press.

Linton, Ralph. 1936. *The Study of Man*. New York: Appleton Century Crofts.
Locklear, Erin M. 1999. "Where Race and Politics Collide: The Federal Acknowledgement Process and Its Effects on Lumsee and Pequot Indians." Unpublished senior thesis, Princeton University.
Loewen, James W. 2007. *Lies My Teacher Told Me: Everything Your American History Textbook Got Wrong.* New York: Touchstone.
Long, George I. 2007. *Employer-Provided "Quality of Life" Benefits for Workers in Private Industry, 2007.* Washington, DC: Bureau of Labor Statistics. **www.bls.gov**
Lopez, Mark. 2009. "Dissecting the 2008 Electorate: Most Diverse in History." Pew Research Center. **www.pewresearch.com**
Lorber, Judith. 1994. *Paradoxes of Gender.* New Haven, CT: Yale University Press.
Lorenz, Conrad. 1966. *On Aggression*. New York: Harcourt Brace Jovanovich.
Lovejoy, Meg. 2001. "Disturbances in the Social Body: Differences in Body Image and Eating Problems among African American and White Women." *Gender & Society* 15 (April): 239–261.
Lu, Melody C. 2005. "Commercially Arranged Marriage Migration: Case Studies of Cross-Border Marriages in Taiwan." *Indian Journal of Gender Studies* 12 (2–3): 275–303.
Lucal, Betsy. 1994. "Class Stratification in Introductory Textbooks: Relational or Distributional Models?" *Teaching Sociology* 22 (April): 139–150.
Luker, Kristin. 1975. *Taking Chances: Abortion and the Decision Not to Contracept.* Berkeley, CA: University of California Press.
Luker, Kristin. 1984. *Abortion and the Politics of Motherhood.* Berkeley, CA: University of California Press.
Luker, Kristin. 1996. *Dubious Conceptions: The Politics of Teenage Pregnancy.* Cambridge, MA: Harvard University Press.
Lyons, Anthony, Marian Pitts, and Jeffrey Grierson. 2013. "Growing Old as a Gay Man: Psychosocial Well-being of a Sexual Minority." *Research on Aging* 35 (3): 275–295.
Lyons, Linda. 2002. "Teen Attitudes Contradict Sex-Crazed Stereotype." *The Gallup Poll.* Princeton, NJ: The Gallup Organization. January 29, **www.gallup.com**
Lyons, Linda. 2004. "Teens: Sex Can Wait." *The Gallup Poll.* Princeton, NJ; The Gallup Organization. **www.gallup.com**

Machel, Graca. 1996. *Impact of Armed Conflict on Children.* New York: UNICEF/United Nations.
MacKinnon, Catherine. 1983. "Feminism, Marxism, Method, and the State: An Agenda for Theory." *Signs* 7 (Spring): 635–658.
MacKinnon, Catherine A. 2006. "Feminism, Marxism, Method, and the State: An Agenda for Theory." pp. 829–868 in *The Canon of American Legal Thought,* edited by D. Kennedy and W. F. Fisher. Princeton: Princeton University Press.
Mackintosh, N. J. 1995. *Cyril Burt: Fraud or Framed?* Oxford, England: Oxford University Press.
Malamuth, Neil M., Gert M. Hald, and Mary Koss. 2012. "Pornography, Individual Differences in Risk and Men's Acceptance of Violence against Women in a Representative Sample." *Sex Roles* 66 (7–8): 427–439.
Malcomson, Scott L. 2000. *One Drop of Blood: The American Misadventure of Race.* New York: Farrar, Strauss, and Giroux.
Maldonado, Lionel, A. 1997. "Mexicans in the American System: A Common Destiny." In *Ethnicity in the United States: An Institutional Approach,* edited by William Velez. Bayside, NY: General Hall.
Malinauskas, Brenda et al. 2006. "Dieting Practices, Weight Perceptions, and Body Composition: A Comparison of Normal Weight, Overweight, and Obese College Females." *Nutrition Journal* 5 (March 31): 5–11.
Maltby, Lauren E., M. Elizabeth L. Hall, Tamara L. Anderson, and Keith Edwards. 2010. "Religion and Sexism: The Moderating Role of Participant Gender." *Sex Roles* 62: 615–622.
Malthus, Thomas Robert. 1926 [1798]. *First Essay on Population 1798.* London: Macmillan.
Mandel, Daniel. 2001. "Muslims on the Silver Screen." *Middle East Quarterly* 8 (Spring): 19–30.
Mantsios, Gregory. 2010. "Media Magic: Making Class Invisible." pp. 386–394 in *Race, Class, and Gender: An Anthology,* edited by Margaret L. Andersen and Patricia Hill Collins. Belmont, CA: Wadsworth.
Marcuse, Herbert. 1964. *One-Dimensional Man.* Boston: Beacon Press.
Marks, Carole. 1989. *Farewell, We're Good and Gone: The Great Black Migration.* Bloomington, IN: Indiana University Press.
Marks, Carole, and Deana Edkins. 1999. *The Power of Pride: Stylemakers and Rulebreakers of the Harlem Renaissance.* New York: Crown.
Martin, Karin A. 2005. "William Wants a Doll. Can He Have One? Feminists, Child Care Advisors, and Gender-Neutral Child Rearing." *Gender & Society* 19 (August): 456–479.
Martin, Karin A., and Emily Kazyak. 2009. "Hetero-romantic Love and Heterosexiness in Children's G-Rated Films." *Gender & Society* 23 (June): 315–336.
Martin, Patricia Yancey, and Robert Hummer. 1989. "Fraternities and Rape on Campus." *Gender & Society* 3 (December): 457–473.
Martineau, Harriet. 1837. *Society in America.* London: Saunders and Otley.
Martineau, Harriet. 1838. *How to Observe Morals and Manners.* London: Charles Knight and Co.
Martinez, Michael E., and Robin A. Cohen. 2014. *Health Insurance Coverage: Early Release of Estimates from National Health Interview Survey, January–June 2014.* Centers for Disease Control and Prevention, Hyattsville, MD: U.S. Department of Health and Human Services, National Center for Health Statistics.
Marx, Anthony. 1997. *Making Race and Nation: A Comparison of the United States, South Africa, and Brazil.* New York: Cambridge University Press.
Marx, Karl. 1967 [1867]. *Capital.* F. Engels (ed.). New York: International Publishers.
Marx, Karl. 1972 [1843]. "Contribution to the Critique of Hegel's *Philosophy of Right.*" pp. 11–23 in *The Marx-Engels Reader,* edited by Robert C. Tucker. New York: W. W. Norton.
Massey, Douglas S. 2005. *Strangers in a Strange Land: Humans in an Urbanizing World.* New York: Norton.
Massey, Douglas S., and Nancy A. Denton. 1993. *American Apartheid: Segregation and the Making of the Underclass.* Cambridge, MA: Harvard University Press.
Mazumder, Bhashkar. 2008. "Intergenerational Economic Mobility in the US: 1940 to 2000." *Journal of Human Resources* 43 (January): 139–172.
McCabe, Janice. 2005. "What's in a Label? The Relationship between Feminist Self-Identification and 'Feminist' Attitudes among U.S. Women and Men." *Gender & Society* 19 (4): 480–505.
McCabe, Janice, Emily Fairchild, Liz Grauerholz, Bernice A. Pescosolido, and Daniel Tope. 2011. "Gender in Twentieth Century Children's Books: Patterns of Disparity in Titles and Central Characters." *Gender & Society* 25 (April): 197–225.
McCarthy, Justin. 2014. "Same-Sex Marriage Support Reaches New High at 55%." *The Gallup Poll.* Princeton, NJ: The Gallup Organization. **www.gallup.com**
McCartney, Suzanne, Alemayehi Bishaw, and Kayla Fontenot. 2013 (February). *Poverty Rates for Selected Detailed Race and Hispanic Groups by State and Place: 2007–2011.* Washington, DC: U.S. Census Bureau. **www.census.gov**
McClelland, Susan. 2003. "A Grim Toll on the Innocent." *Maclean's* (May 12): 20.
McClintock, Elizabeth Aura. 2010. "When Does Race Matter? Race, Sex, and Dating at an Elite University." *Journal of Marriage and the Family* 72 (February): 45–72.
McClintock, Elizabeth Aura. 2011. "Handsome Wants as Handsome Does: Physical Attractiveness and Gender Differences in Revealed Sexual Preferences." *Biodemography and Social Biology* 57 (2): 221–257.
McGrath, Timothy. 2014. "What People around the World Are Saying about Ferguson." *USA Today*, November 25.
McGuire, Meredith. 2008. *Religion: The Social Context*, 5th ed. Long Grove, IL: Waveland Press.
McIntyre, Robert S., Matthew Gardner, and Richard Phillips. 2014 (February).

The Sorry State of Corporate Taxes. Washington, DC: Citizens for Tax Justice and the Institute of Taxation and Economic Policy. **www.itep.org**

McNamee, Stephen, and Robert K. Miller. 2009. *The Meritocracy Myth.* Lanham, MD: Rowman and Littlefield.

McVeigh, Ricky. 2012. "Making Sense of the Tea Party." *Contemporary Sociology* 41 (November): 766–769.

Mead, George Herbert. 1934. *Mind, Self, and Society.* Chicago, IL: University of Chicago Press.

Meier, Barry. 2013. "Maker Hid Data about Design Flaw in Hip Implant." *The New York Times* (January 13): B1 and B6.

Mendes, Elizabeth, Lydia Saad, and Kyley McGeeney. 2012. "Stay at Home Moms Report More Depression, Sadness, Anger." *The Gallup Poll.* Princeton, NJ: The Gallup Organization, May 18. **www.gallup.com**

Meredith, Martin. 2003. *Elephant Destiny: Biography of an Endangered Species in Africa.* New York: HarperCollins.

Mernissi, Fatema. 2011. *Beyond the Veil: Male-Female Dynamics in Modern Muslim Society.* London: Seqi Books.

Merton, Robert K. 1957. *Social Theory and Social Structure.* New York: Free Press.

Merton, Robert K. 1968. "Social Structure and Anomie." *American Sociological Review* 3 (5): 672–682.

Merton, Robert, and Alice K. Rossi. 1950. "Contributions to the Theory of Reference Group Behavior." pp. 279–334 in *Continuities in Social Research Studies, Scope and Method of "The American Soldier,"* edited by Robert K. Merton and Paul F. Lazarsfeld. New York: Free Press.

Messina-Dysert, Gina, and Rosemary Ruether, eds. 2014. *Feminism and Religion in the Twenty-First Century: Technology, Dialogue and Expanding Borders.* New York: Routledge.

Messner, Michael A. 2002. *Taking the Field: Women, Men, and Sports.* Minneapolis, MN: University of Minnesota Press.

Messner, Michael A. 2009. *It's All for the Kids: Gender, Families, and Youth Sports.* Berkeley: University of California Press.

Messner, Steven F. 2011. *Crime and the American Dream.* Belmont, CA: Wadsworth/Cengage.

Meyer, B. D., and J. X. Sullivan. 2012. "Identifying the Disadvantaged: Official Poverty, Consumption Poverty, and the New Supplemental Poverty Measure." *Journal of Economic Perspectives* 26 (3): 111–136.

Mezey, Nancy J. 2008. *New Choices, New Families: How Lesbians Decide about Motherhood.* Baltimore, MD: Johns Hopkins University Press.

Mickelson, Roslyn Arlin (ed.). 2000. *Children on the Streets of the Americas: Globalization, Homelessness, and Education in the United States, Brazil, and Cuba.* New York: Routledge.

Milanovic, Brandon. 2010. *The Haves and Have Nots: A Brief and Idiosyncratic History of Global Inequality.* New York: Basic Books.

Milgram, Stanley. 1974. *Obedience to Authority: An Experimental View.* New York: Harper & Row.

Milkie, Melissa A., Suzanne M. Bianchi, Marybeth J. Mattingly, and John P. Robinson. 2002. "Gendered Division of Childrearing: Ideals, Realities, and the Relationship to Parental Well-being." *Sex Roles* 47 (July): 21–38.

Miller, Susan L. 1997. "The Unintended Consequences of Current Criminal Justice Policy." Talk presented at Research on Women Series, University of Delaware, Newark, DE.

Mills, C. Wright. 1956. *The Power Elite.* New York: Oxford University Press.

Mills, C. Wright. 1959. *The Sociological Imagination.* New York: Oxford University Press.

Miner, Horace. 1956. "Body Ritual among the Nacirema." *American Anthropologist* 58 (3): 503–507.

Mirandé, Alfredo. 1979. "Machismo: A Reinterpretation of Male Dominance in the Chicano Family." *The Family Coordinator* 28 (4): 447–449.

Mirandé, Alfredo. 1985. *The Chicano Experience.* Notre Dame, IN: Notre Dame University Press.

Mishel, Lawrence, Josh Bivens, Elise Gould, and Heidi Shierholz. 2012. *The State of Working America,* 12th ed. Ithaca, NY: Cornell University Press.

Misra, Joy, Stephanie Moller, and Maria Karides. 2003. "Envisioning Dependency: Changing Media Depictions of Welfare in the 20th Century." *Social Problems* 50 (November): 482–504.

Mizruchi, Ephraim H. 1983. *Regulating Society: Marginality and Social Control in Historical Perspective.* New York: Free Press.

Moen, Phyllis. 2003. *It's about Time: Couples and Careers.* Ithaca, NY: Cornell University Press.

Moen, Phyllis, Jungmeen E. Kim, and Heather Hofmeister. 2001. "Couples' Work/Retirement Transitions, Gender, and Marital Quality." *Social Psychology Quarterly* 64 (March): 55–71.

Mohai, Paul, and Robin Saha. 2007. "Racial Inequality in the Distribution of Hazardous Waste: A National-Level Reassessment." *Social Problems* 54 (3): 343–370.

Moore, Gwen. 1979. "The Structure of a National Elite Network." *American Sociological Review* 44 (October): 673–692.

Moore, Joan. 1976. *Hispanics in the United States.* Englewood Cliffs, NJ: Prentice Hall.

Moore, Robert B. 1992. "Racist Stereotyping in the English Language." pp. 317–328 in *Race, Class, and Gender: An Anthology,* 2nd ed., edited by Margaret L. Andersen and Patricia Hill Collins. Belmont, CA: Wadsworth.

Moore, Valerie A. 2001. "'Doing' Racialized and Gendered Age to Organize Peer Relations: Observing Kids in Summer Camp." *Gender & Society* 15 (December): 835–858.

Mora, Christina. 2009. *DeMuchos, Uno: The Institutionalization of Latino Panethnicity in the United States, 1960–1990.* PhD diss., Princeton University, Princeton, NJ.

Morin, Rich, and Seth Motel. 2012. "A Third of Americans Now Say They Are in the Lower Classes." *Pew Social & Demographic Trends.* Washington, DC: Pew Research Center.

Morgan, Marcyliena. 2009. *The Real Hiphop: Battling for Knowledge, Power, and Respect in the LA Underground.* Durham, NC: Duke University Press.

Morgan, Marcyliena. 2010. (With Dawn-Elissa Fischer). "Hiphop and Race: Blackness, Language and Creativity" pp. 509–527 in *Doing Race: 21 Essays for the 21st Century,* edited by Hazel Rose Markus and Paula M. L. Moya. New York: W. W. Norton.

Morning, Ann. 2011. *The Nature of Race: How Scientists Think and Teach about Human Differences.* New York: New York University Press.

Morris, Aldon. 1984. *The Origins of the Civil Rights Movement: Black Communities Organizing for Change.* New York: Free Press.

Morris, Aldon D. 1999. "A Retrospective on the Civil Rights Movement: Political and Intellectual Landmarks." *Annual Review of Sociology* 25: 517–539.

Morris, Aldon. 2015. *The Scholar Denied: W. E. B. DuBois and the Birth of Modern Sociology.* Berkeley, CA: University of California Press.

Moskos, Peter. 2008. *Cop in the Hood: My Year in Policing Baltimore's Eastern District.* Princeton, NJ: Princeton University Press.

Mullen, Ann L. 2010. *Degrees of Inequality: Culture, Class, and Gender in American Higher Education.* Baltimore, MD: Johns Hopkins University Press.

Murphy, Joseph. 2014. "The Social and Educational Outcomes of Homeschooling." *Sociological Spectrum* 34 (May): 244–272.

Myers, Steven Lee. 2000. "Survey of Troops Finds Antigay Bias Common in Service." *The New York Times* (March 24): 1.

Myers, Walter D. 1998. *Amistad Affair.* New York: NAL/Dutton.

Nagel, Joane. 1996. *American Indian Ethnic Renewal: Red Power and the Resurgence of Identity and Culture.* New York: Oxford University Press.

Nagel, Joane. 2003. *Race, Ethnicity, and Sexuality: Intimate Intersections, Forbidden Frontiers.* New York: Oxford University Press.

Nanda, Serena. 1998. *Neither Man Nor Woman: The Hijras of India.* Belmont, CA: Wadsworth.

Nargiso, Jessica, Erica L. Ballard, and Margie R. Skeet. 2015. "A Systematic Review of Risk and Protective Factors Associated with Nonmedical Use of Prescription Drugs among Youth in the United States: A Social Ecological Perspective." *Journal of Studies in Alcohol & Drugs* 76 (1): 5–20.

National Center for Education Statistics. 2013. *The Condition of Education, 2012.* Washington, DC: U.S. Department of Education.

National Center for Education Statistics. 2013. *Private School Universe Survey 1995–1996 through 2011–2012.* Washington, DC: National Center for Education Statistics, U.S. Department of Education. Table 205.20.

National Center for Health Statistics. 2012. "Health, United States 2011." Hyattsville, MD: U.S. Department of Health and Human Services.

National Center for Health Statistics. 2013. *Health United States 2013.* Washington, DC: National Center for Health Statistics.

National Center on Elder Abuse. 2014. **www.ncea.aoa.gov**

National Coalition against Domestic Violence. 2001. **www.ncadv.org**

National Coalition for the Homeless. 2014. "Fact Sheet." **www.nationalhomeless.org**

National Human Genome Research Institute. *Fact Sheet: The Human Genome Project.* National Institutes of Health. **www.genome.gov**

National Institutes of Health. 2007. *National Center for Complementary and Alternative Medicine.* Hyattsville, MD: U.S. Department of Health and Human Services.

National Research Council. 2012. *Advancing the Science of Climate Change.* Washington, DC: The National Academies Press. **www.nas-sites.org**

Nawaz, Maajid. 2013. *Radical: My Journey Out of Islamist Extremism.* Guilford, CT: Lyons Press.

Nee, Victor. 1973. *Longtime Californ': A Documentary Study of an American Chinatown.* New York: Pantheon Books.

Negrón-Muntaner, Frances with Chelsea Abbos, Luis Figueroa, and Samuel Robson. 2014. *The Latino Media Gap: A Report on the Status of U.S. Latinos in U.S. Media.* New York: Columbia University Center for the Study of Race and Ethnicity. **www.columbia.edu**

Neighbors, L. A., and J. Sobal. 2007. "Prevalence and Magnitude of Body Weight and Shape Dissatisfaction among University Students." *Eating Behaviors* 8 (December): 429–439.

Nielsen, Søren Beck. 2014. "Medical Record Keeping as Interactional Accomplishment." *Pragmatics and Society* 5 (2): 221–242.

Nielsen Report. 2014. *Shifts in Viewing: The Cross-Platform Report.* **www.nielsen.com**

Newman, Katherine. 1999. *No Shame in My Game: The Working Poor in the Inner City.* New York: Russell Sage Foundation/Vintage Books.

Newman, Katherine S. 2012. *The Accordion Family: Boomerang Kids, Anxious Parents, and the Private Toll of Global Competition.* Boston: Beacon Press.

Newman, Katherine S., Cybelle Fox, David Harding, Jal Mehta, and Wendy Roth. 2006. *Rampage: Social Roots of School Shootings.* New York: Basic Books.

Newport, Frank. 2012a. "Bias against Mormon Presidential Candidate Same as in 1967." *The Gallup Poll.* Princeton, NJ: The Gallup Organization. **www.gallup.com**

Newport, Frank, 2012b. "Seven in 10 Americans Are Moderately or Very Religious." *The Gallup Poll.* Princeton, NJ: The Gallup Organization. **www.gallup.com**

Newport, Frank. 2012c. "Young Adults Admit Too Much Time on Cell Phones, Internet." *The Gallup Poll.* Princeton, NJ: The Gallup Organization. **www.gallup.com**

Newport, Frank, 2013. "In U.S. 4 in 10 Report Attending Church in Last Week." Princeton, NJ: *The Gallup Poll.* Princeton, NJ: The Gallup Organization. **www.gallup.com**

Newport, Frank, and Joseph Carroll. 2005. "Another Look at Evangelicals in America Today." *The Gallup Poll.* Princeton, NJ: The Gallup Organization. **www.gallup.com**

Ngai, Mae M. 2012. "Impossible Subjects: Illegal Aliens and the Making of Modern America." pp. 192–196 in *Race and Ethnicity in Society: The Changing Landscape*, 3rd ed., edited by Elizabeth Higginbotham and Margaret L. Andersen. Belmont, CA: Wadsworth/Cengage.

Noah, Timothy. 2012. *The Great Divergence: America's Growing Inequality Crisis and What We Can Do about It.* New York: Bloomsbury.

Nolan, Patrick, and Gerhard Lenski. 2014. *Human Societies.* New York: Oxford University Press.

Norgaard, Kari M. 2006. "'People Want to Protect Themselves a Little Bit': Emotions, Denial, and Social Movement Nonparticipation." *Sociological Inquiry* 76 (3): 372–396.

Norris, Pippa, and Ronald Inglehart. 2002. "Islamic Culture and Democracy: Testing the 'Clash of Civilizations' Thesis." *Comparative Sociology* 1: 44 (1) 235–263.

Oakes, Jeannie. 2005. *Keeping Track: How Schools Structure Inequality,* 2nd ed. New Haven, CT: Yale University Press.

Oakes, Jeannie, Karen Hunter Quartz, Steve Ryan, and Martin Lipton. 2000. *Becoming Good American Schools. The Struggle for Civic Virtue in Education Reform.* San Francisco: Jossey-Bass.

Offer, Shira. 2014. "Time with Children and Employed Parents' Emotional Well-being." *Social Science Research* 47 (September): 192–203.

Ogburn, William F. 1922. *Social Change with Respect to Cultural and Original Nature.* New York: B. W. Huebsch.

Ogden, Cynthia L., Margaret D. Carrol, Brian K. Kit, and Katherine Flegal. 2014. "Prevalence of Childhood and Adult Obesity in the United States, 2011–2012." *JAMA* 311 (8): 806–815.

Oliver, Melvin L., and Thomas M. Shapiro. 2006. *Black Wealth/White Wealth: A New Perspective on Racial Inequality,* 10th ed. New York: Routledge.

Oliver, Melvin L., and Thomas M. Shapiro. 2008. "Sub-Prime as Black Catastrophe." *The American Prospect,* September 22.

Ollivier, Michele. 2000. "'Too Much Money off Other People's Backs': Status in Late Modern Societies." *Canadian Journal of Sociology* 25 (Fall): 441–470.

Omi, Michael, and Howard Winant. 2014. *Racial Formation in the United States,* 3rd ed. New York: Routledge.

O'Neil, John. 2002. "Parent Smoking and Teenage Sex." *The New York Times,* Sept. 3, p. F7.

Orfield, Gary, Erica Frankenberg, and Laurie Russman. 2014. "School Resegregation and Civil Rights Challenges for the Obama Administration." Los Angeles, CA: The Civil Rights Project/Proyecto Derechos Civiles, University of California Los Angeles. **www.civilrightsproject.ucla.edu**

Organization for Economic Co-operation and Development (OECD). 2014. *Society at a Glance—OECD Social Indicators.* **www.oecd.org/els/social/indicators/SAG**

Ortiz, Isabel, and Matthew Cummins. 2011. "Global Inequality: Beyond the Bottom Billion." *Social and Economic Policy Working Paper.* UNICEF, April. **www.unicef.org**

Padavic, Irene, and Barbara Reskin. 2002. *Women and Men at Work,* 2nd ed. Thousand Oaks, CA: Sage.

Padgett, Tim, Anthony Esposito, and Aaron Nelson. 2011. "The Chilean Miners." *Time* (January 3): 105–111.

Page, Charles H. 1946. "Bureaucracy's Other Face." *Social Forces* 25 (October): 89–94.

Page, Scott E. 2007. *The Difference: How the Power of Diversity Creates Better Groups, Firms, Schools, and Societies.* Princeton, NJ: Princeton University Press.

Pager, Devah. 2007. *Marked: Race, Crime, and Finding Work.* Chicago: University of Chicago Press.

Pain, Emil. 2002. "The Social Nature of Extremism and Terrorism." *Social Sciences* 33: 55–68.

Painter, Nell Irvin. 2011. *The History of White People.* New York: W. W. Norton.

Park, Robert E., and Ernest W. Burgess. 1921. *Introduction to the Science of Society.* Chicago: University of Chicago Press.

Parker, Kim. 2012. *The Boomerang Generation: Feeling OK about Living with Mom and Dad.* Washington, DC: Pew Research Center. **ww.pewsocialtrends.org**

Parker-Pope, Tara. 2008. "Love, Sex, and the Changing Landscape of Infidelity." *The New York Times,* October 27.

Parreñas, Rhacel Salazar. 2001. *Servants of Globalization: Women, Migration, and Domestic Work.* Stanford, CA: Stanford University Press.

Parreñas, Rhacel. 2005. *Children of Global Migration: Transnational Families and Gendered Woes.* Stanford, CA: Stanford University Press.

Parsons, Talcott (ed.). 1947. *Max Weber: The Theory of Social and Economic Organization.* New York: Free Press.
Parsons, Talcott. 1951a. *The Social System.* Glencoe, IL: Free Press.
Parsons, Talcott. 1951b. *Toward a General Theory of Action.* Cambridge, MA: Harvard University Press.
Parsons, Talcott. 1966. *Societies: Evolutionary and Comparative Perspectives.* Englewood Cliffs, NJ: Prentice Hall.
Pascoe, C. J. 2011. *Dude, You're a Fag: Masculinity and Sexuality in High School.* Berkeley, CA: University of California Press.
Paternoster, Ray, Jean Marie McGloin, Holly Nguyen, and Kyle J. Thomas. 2013. "The Causal Impact of Exposure to Deviant Peers: An Experimental Investigation." *Journal of Research in Crime and Delinquency* 50 (4): 476–503.
Pathways to Peace. 2009. "Impact on War on Poverty." **www.icrc.org**
Pattillo-McCoy, Mary. 2013. *Black Picket Fences: Privilege and Peril among the Black Middle Class,* 2nd ed. Chicago: University of Chicago Press.
Paulus, P. B., T. S. Larey, and M. T. Dzindolet. 2001. "Creativity in Groups and Teams." pp. 319–338 in *Groups at Work: Theory and Research*, edited by M. E. Turner. Mahwah, NJ: Erlbaum.
Pedraza, Silvia. 1996. "Cuba's Refugees: Manifold Migrations." pp. 263–279 in *Origins and Destinies: Immigration, Race, and Ethnicity in America,* edited by Silvia Pedraza and Rubén Rumbaut. Belmont, CA: Wadsworth.
Pelham, Brett, and Steve Crabtree. 2009. "Religiosity and Perceived Intolerance of Gays and Lesbians." Princeton, NJ: Gallup Poll, March 10. **www.gallup.com**
Pellow, David N. 2004. "The Politics of Illegal Dumping: An Environmental Justice Framework." *Qualitative Sociology* 27 (4): 511–525.
Peri, Giovanni. 2014. "Does Immigration Hurt the Poor?" *Pathways* (Summer): 15–18.
Peri, Giovanni, and Chad Sparber. 2009. "Task Specialization, Immigration, and Wages." *American Economic Journal: Applied Economics* 1 (July): 135–169.
Perlmutter, David. 2008. *Blogwars: The New Political Battleground.* New York: Oxford University Press.
Perrow, Charles. 1986. *Complex Organization: A Critical Essay,* 3rd ed. New York: Random House.
Perrow, Charles. 1994. "The Limit of Safety: The Enhancement of a Theory of Accidents." *Journal of Contingencies and Crisis Management* 22 (4): 212–220.
Perrow, Charles. 2007. *Organizing America: Wealth, Power, and Origins of American Capitalism.* Princeton: Princeton University Press.
Pershing, Jana L. 2003. "Why Women Don't Report Sexual Harassment: A Case Study of an Elite Military Institution." *Gender Issues* 21 (4): 3–30.
Pescosolido, Bernice A., Elizabeth Grauerholz, and Melissa A. Milkie. 1997. "Culture and Conflict: The Portrayal of Blacks in U.S. Children's Picture Books through the Mid- and Late-Twentieth Century." *American Sociological Review* 62 (June): 443–464.
Pettigrew, Thomas F. 1992. "The Ultimate Attribution Error: Extending Allport's Cognitive Analysis of Prejudice." pp. 401–419 in *Readings about the Social Animal,* edited by Elliott Aronson. New York: Freeman.
Pettit, Becky, and Stephanie Ewert. 2009. "Employment Gains and Wage Declines: The Erosion of Black Women's Relative Wages since 1980." *Demography* 46 (3): 469–492.
Pew Forum on Religion & Public Life. 2012. *Asian Americans: A Mosaic of Faiths.* Washington, DC: Pew Research Center. **www.pewforum.org**
Pew Research Center. 2007. "Muslim Americans: Middle Class and Mostly Mainstream." Washington, DC: Pew Research Center. **www.pewresearch.org**
Pew Research Center. 2009. *Dissecting the 2008 Election: Most Diverse in U.S. History*. Washington, DC: Pew Research Center. **www.pewresearch.org**
Pew Research Center. 2010. *Marrying Out: One-in-Seven New U.S. Marriages Is Interracial or Interethnic.* Washington, DC: Pew Research Center. **www.pewsocialtrends.org**
Pew Research Center. 2014 (December 8). "Sharp Divisions in Reaction to Brown, Garner Decisions." Washington, DC: Pew Research Center. **www.pewresearch.org**
Pew Research Center. 2015 (June 8). "Changing Attitudes on Gay Marriage." Washington, DC: Pew Research Center. **www.pewresearch.org**
Pew Research Center for the People and the Press. 2014. "For 2016 Hopefuls, Washington Experience Could Do More Harm Than Good." Washington, DC: Pew Research Center, May 14. **www.pewresearch.org**
Pew Research Global Attitudes Project. 2014. *Emerging and Developing Economies Much More Optimistic than Rich Countries about the Future.* Washington, DC: Pew Research Center. **www.pewglobal.org**
Pew Research Internet Project. 2014. *Mobile Technology Fact Sheet.* **www.pewinternet.org**
Pew Research Religion & Public Life Project. 2014. "The Shifting Religious Identity of Latinos in the United States." Washington, DC: Pew Research Center. **www.pewforum.org**
Pfeffer, Fabian T., Sheldon Danziger, and Robert F. Shoeni. 2014. *Wealth Levels, Wealth Inequality, and the Great Recession.* New York: Russell Sage Foundation. **www.russellsage.org**
Piketty, Thomas. 2014. *Capital in the Twenty-First Century*. Cambridge, MA: Harvard University Press.
Piontak, Joy Rayanne, and Michael D. Schulman. 2014. "Food Insecurity in Rural America." *Contexts* 13 (Summer): 75–77.
Polce-Lynch, Mary, Barbara J. Myers, Wendy Kliewer, and Christopher Kilmartin. 2001. "Adolescent Self-Esteem and Gender: Exploring Relations to Sexual Harassment, Body Image, Media Influence, and Emotional Expression." *Journal of Youth and Adolescence* 30 (April): 225–244.
Popenoe, David. 2001. "Today's Dads: A New Breed?" *The New York Times* (June 19): A22.
Portes, Alejandro. 2002. "English-Only Triumphs, but the Costs Are High." *Contexts* 1 (February): 10–15.
Portes, Alejandro, and Rubén G. Rumbaut. 1996. *Immigrant America: A Portrait,* 2nd ed. Berkeley, CA: University of California Press.
Portes, Alejandro, and Rubén G. Rumbaut. 2001. *Legacies: The Story of the Immigrant Second Generation.* Berkeley, CA: University of California Press.
Powell, Brian, Catherine Bolzendahl, Claudia Geist, and Lala Carr Steelman. 2010. *Counted Out: Same-Sex Relations and Americans' Definitions of Family.* New York: Russell Sage Foundation.
Press, Andrea. 2002. "The Paradox of Talk." *Contexts* 1 (Fall–Winter): 69–70.
Press, Eyal. 1996. "Barbie's Betrayal." *The Nation* (December 30): 11–16.
Preves, Sharon E. 2003. *Intersex and Identity: The Contested Self.* New Brunswick, NJ: Rutgers University Press.
Price-Glynn, Kim. 2010. *Strip Club: Gender, Power, and Sex Work.* New York: New York University Press.
Prokos, Anastasia. 2011. "An Unfinished Revolution." *Gender & Society* 25 (1): 75–80.
Prostitutes Education Network. 2009. **www.bayswan.org**
Puffer, Phyllis. 2009. "Durkheim Did Not Say 'Normlessness': The Concept of Anomic Suicide for Introductory Sociology Courses." *Southern Rural Sociology* 24 (2): 200–222.

Quadagno, Jill. 2005. *One Nation, Uninsured: Why the U.S. Has No National Health Insurance.* New York: Oxford University Press.

Rajan, Ramkishen S., and Sadhana Srivastava. 2007. *Harvard Asia Pacific Review* 9 (Winter): n.p.
Raley, Sara B., Marybeth J. Mattingly, and Suzanne M. Bianchi. 2006. "How Dual Are Dual-Income Couples? Documenting Change from 1970 to 2001." *Journal of Marriage and Family* 68 (1): 11–28.
Rampersad, Arnold. 1986. *The Life of Langston Hughes: Vol. I: 1902–1941. I, Too, Sing America.* New York: Oxford University Press.
Rampersad, Arnold. 1988. *The Life of Langston Hughes: Vol. II: 1941–1967. I Dream A World.* New York: Oxford University Press.
Rashid, Ahmed. 2000. *Taliban: Militant Islam, Oil, and Fundamentalism in Central Asia.* New Haven, CT: Yale University Press.

Rasmussen Reports. 2009. "62% Say Today's Children Will Not Be Better Off than Their Parents." **www.rasmussenreports.com**

Read, Jen'nan Ghazal. 2003. "The Sources of Gender Role Attitudes among Christian and Muslim Arab-American Women." *Sociology of Religion* 64 (Summer): 207–222.

Read, Piers Paul. 1974. *Alive: The Story of the Andes Survivors.* Philadelphia, PA: Lippincott.

Reich, Robert. 2010. "The Root of Economic Fragility and Political Anger." **ww.robertreich.org**

Reid, T. R. 2010. *The Healing of America: A Global Quest for Better, Cheaper, and Fairer Health Care.* New York: Penguin Books.

Reiman, Jeffrey H. 2012. *The Rich Get Richer and the Poor Get Prison,* 10th ed. Boston: Allyn and Bacon.

Reiman, Jeffrey, and Paul Leighton. 2012. *The Rich Get Richer and the Poor Get Prison: A Reader,* 10th ed. Upper Saddle River, NJ: Pearson.

Renzetti, Claire, Jeffrey L. Edieson, and Raquel K. Bergen. 2010. *Sourcebook on Violence against Women.* Thousand Oaks, CA: Sage Publications.

Reskin, Barbara. 1988. "Bringing the Men Back In: Sex Differentiation and the Devaluation of Women's Work." *Gender & Society* 2 (March): 58–81.

Reyns, Bradford W. 2013. "Online Routines and Identity Theft Victimization: Further Expanding Routine Activity Theory beyond Direct-Contact Offenses." *Journal of Research in Crime and Delinquency* 50 (2), 216–238.

Rheault, Magail, and Dalia Mogahed. 2008. "Moral Issues Divide Westerners from Muslims in the West." *The Gallup Poll.* Princeton, NJ: Gallup Organization. **www.gallup.com**

Rich, Adrienne. 1980. "Compulsory Heterosexuality and Lesbian Existence." *Signs* 5 (Summer): 631–660.

Ridgeway. Cecilia L. 2011. *Framed by Gender: How Gender Inequality Persists in the Modern World.* New York: Oxford University Press.

Riegle-Crumb, Catherine, and Melissa Humphries. 2012. "Exploring Bias in Math Teachers' Perceptions of Students' Ability by Gender and Race/Ethnicity." *Gender & Society* 26 (April): 290–322.

Riffkin, Rebecca. 2014. "Americans Favor Ban on Smoking in Public, but not Total Ban." *The Gallup Poll,* July 30. Princeton, NJ: The Gallup Organization. **www.gallup.com**

Risman, Barbara, and Pepper Schwartz. 2002. "After the Sexual Revolution: Gender Politics in Teen Dating." *Contexts* 1 (Spring): 16–24.

Ritzer, George. 2010. *The McDonaldization of Society,* 6th ed. Thousand Oaks, CA: Sage Publications.

Rivas-Drake, Deborah. 2011. "Ethnic-Racial Socialization and Adjustment among Latino College Students: The Mediating Roles of Ethnic Centrality, Public Regard, and Perceived Barriers to Opportunity." *Journal of Youth and Adolescence* 40 (May): 606–619.

Roberts, Dorothy. 1997. *Killing the Black Body: Race, Reproduction and the Meaning of Liberty.* New York: Vintage Books.

Roberts, Dorothy. 2012. *Fatal Invention: How Science, Politics, and Big Business Re-Create Race in the Twenty-First Century.* New York: New Press.

Robertson, Tatsha, and Garrance Burke. 2001. "Fighting Terror: Concerned Family;" "Shock, Worry for Family of U.S. Man Captured with Taliban." *The Boston Globe* (December 4): A1.

Robinson, Jo Ann Gibson. 1987. *The Montgomery Bus Boycott and the Women Who Started It.* Knoxville, TN: The University of Tennessee Press.

Roda, Allison, and Amy Stuart Wells. 2013. "School Choice Policies and Racial Segregation: Where Parents' Good Intensions, Anxiety, and Privilege Collide." *American Journal of Education* 119 (February): 261–293.

Rodriguez, Clara E. 1989. *Puerto Ricans: Born in the U.S.A.* Boston: Unwin Hyman.

Rodriguez, Clara. 2000. *Changing Race: Latinos, the Census, and the History of Ethnicity.* New York: New York University Press.

Rodriguez, Clara E. 2009. "Changing Race." pp. 22–25 in *Race and Ethnicity in Society: The Changing Landscape,* 3rd ed., edited by Elizabeth Higginbotham and Margaret L. Andersen. Belmont, CA: Wadsworth.

Romain, Suzanne. 1999. *Communicating Gender.* Mahwah, NJ: Erlbaum.

Ropelato, Jerry. 2007. "Internet Pornography Statistics." *Top Ten Reviews.* **http://internet-filter-review.toptenreviews.com/internet-pornography-statistics-pg4.html**

Rosenbaum, Janet Elise. 2009. "Patient Teenagers? A Comparison of the Sexual Behavior of Virginity Pledgers and Matched Nonpledgers." *Pediatrics* 123 (January): 110–120.

Rosenfeld, Michael J., and Reuben J. Thomas. 2012. "Searching for a Mate: The Rise of the Internet as a Social Intermediary." *American Sociological Review* 77 (August): 523–547.

Rosengarten, Danielle. 2000. "Modern Times." *Dollars & Sense* (September): 4.

Rosenhan, David L. 1973. "On Being Sane in Insane Places." *Science* 179 (January 19): 250–258.

Rosenthal, Robert, and Lenore Jacobson. 1968. *Pygmalion in the Classroom: Teacher Expectations and Pupils' Intellectual Development.* New York: Holt, Rinehart and Winston.

Rospenda, Kathleen M., Judith A. Richman, and Candice A. Shannon. 2009. "Prevalence and Mental Health Correlates of Harassment and Discrimination in the Workplace. Results from a National Study." *Journal of Interpersonal Violence* 24 (5): 819–843.

Rossi, Alice S., and Peter H. Rossi. 1990. *Of Human Bonding: Parent-Child Relations across the Life Course.* New York: Aldine de Gruyter.

Royster, Deirdre. 2003. *The Invisible Hand: How White Networks Exclude Black Men from Blue-Collar Jobs.* Berkeley, CA: University of California Press.

Rudski, Jeffrey M. 2014. "Treatment Acceptability, Stigma, and Legal Concerns of Medical Marijuana Are Affected by Method of Administration." *Journal of Drug Issues* 44 (3): 308–320.

Ruef, Martin, Howard Aldrich, and N. Carter. 2003. "The Structure of Foundation Teams: Homophily, Strong Ties, and Isolation among U.S. Entrepreneurs." *American Sociological Review* 68 (April): 195–222.

Rueschmeyer, Dietrich, and Theda Skocpol. 1996. *States, Social Knowledge, and the Origins of Modern Social Policies.* Princeton, NJ: Princeton University Press.

Rugh, Jacob S., and Douglas S. Massey. 2010. "Racial Segregation and the American Foreclosure Crisis." *American Sociological Review* 75 (October): 629–651.

Rupp, Leila J., and Verta Taylor. 2003. *Drag Queens at the 801 Cabaret.* Chicago: University of Chicago Press.

Rupp, Leila J., and Verta Taylor. 2010. "Straight Girls Kissing." *Contexts* 9 (Summer): 28–33.

Rust, Paula. 1995. *Bisexuality and the Challenge to Lesbian Politics.* New York: New York University Press.

Rust, Paula C. 1993. "'Coming Out' in the Age of Social Constructionism: Sexual Identity Formation among Lesbian and Bisexual Women." *Gender & Society* 7 (March): 50–77.

Rutter, Virginia, and Pepper Schwartz. 2011. *The Gender of Sexuality: Exploring Sexual Possibilities.* Lanham, MD: Rowman and Littlefield.

Ruvolo, Julie. 2011. "How Much of the Internet is Actually for Porn?" *Forbes,* September 7. **www.forbes.com**

Ryan, Kathryn M. 2011. "The Relationship between Rape Myths and Sexual Scripts: The Social Construction of Rape." *Sex Roles: A Journal of Research* 65 (11–12): 774–782.

Ryan, William. 1971. *Blaming the Victim.* New York: Pantheon.

Saad, Lydia. 2011 (August 8). "Plenty of Common Ground Found in Abortion Debate." *The Gallup Poll.* Princeton, NJ: The Gallup Organization. **www.gallup.com**

Saad, Lydia. 2012a. "In U.S., Half of Women Prefer a Job Outside the Home." *The Gallup Poll.* Princeton, NJ: The Gallup Organization, September 7. **www.gallup.com**

Saad, Lydia. 2012c. (May 14). "U.S. Acceptance of Gay/Lesbian Relations is the New Normal." *The Gallup Poll.* Princeton, NJ: The Gallup Organization. **www.gallup.com**

Saad, Lydia. 2014. "U.S. Still Split on Abortion: 47% Pro-Choice, 46% Pro-Life." *The Gallup Poll.* Princeton,

NJ: The Gallup Organization. **www.gallup.com**
Sabol, William J., and Heather Couture. 2008. "Prison Inmates at Mid Year 2007." Washington, DC: U.S. Bureau of Justice Statistics. **www.ojp.usdoj.gov**
Sadker, Myra, and David Sadker. 1994. *Failing at Fairness: How America's Schools Cheat Girls.* New York: Scribner's.
Sadker, David, and Karen R. Zittleman. 2009. *Still Failing at Fairness: How Gender Bias Cheats Girls and Boys in School and What We Can Do about It.* New York: Scribner.
Saez, Emmanuel, and Gabriel Zucman. 2014. "Exploding Wealth Inequality in the United States." Washington, DC: Washington Center for Equitable Growth. **www.equitablegrowth.org**
Saguy, Abigail. 2003. *What Is Sexual Harassment? From Capitol Hill to the Sorbonne.* Berkeley, CA: University of California Press.
Sampson, Robert J. 1987. "Urban Black Violence: The Effect of Male Joblessness and Family Disruption." *American Journal of Sociology* 93 (September): 348–382.
Sanchez-Jankowski, Martin. 1991. *Islands in the Street: Gangs and American Urban Society.* Berkeley: University of California Press.
Sanday, Peggy. 2002. *Women at the Center: Life in a Modern Matriarchy.* Ithaca, NY: Cornell University Press.
Sandberg, Sheryl. 2013. *Lean In: Women, Work, and the Will to Lead.* New York: Knopf.
Sandnabba, N. Kenneth, and Christian Ahlberg. 1999. "Parents' Attitudes and Expectations about Children's Cross-Gender Behavior." *Sex Roles* 40 (February): 249–263.
Santelli, John, et al. 2007. "Explaining Recent Declines in Adolescent Pregnancy in the United States: The Contribution of Abstinence and Increased Contraceptive Use." *American Journal of Public Health* 97 (1): 150–156.
Sapir, Edward. 1921. *Language: An Introduction to the Study of Speech.* New York: Harcourt Brace.
Sara, Siddharth. 2010. *Sex Trafficking: Inside the Business of Modern Slavery.* New York: Columbia University Press.
Sarkasian, Natalia, and Naomi Gerstel. 2012. *Nuclear Family Values, Extended Family Lives: The Power of Race, Class, and Gender.* New York: Russell Sage Foundation.
Saulny, Susan. 2011. "Black? White? Asian? More Younger Americans Choose All of the Above." *The New York Times* (January 29): 1 and 18.
Sayer, Liana C., and Leigh Fine. 2011. "Racial-Ethnic Differences in U.S. Married Women's and Men's Housework." *Social Indicators Research* 101 (2): 259–265.
Schacht, Steven P. 1996. "Misogyny on and off the 'Pitch': The Gendered World of Male Rugby Players." *Gender & Society* 10 (October): 550–565.
Shaefer, H. Luke, and Kathyrn Edin. 2014. "The Rise of Extreme Poverty in the United States." *Pathways* (Summer): 28–32.
Schaffer, Kay, and Song Xianlin. 2007. "Unruly Spaces: Gender, Women's Writing and Indigenous Feminism in China." *Journal of Gender Studies* 16 (1): 17–30.
Schalet, Amy. 2010. "Sex, Love, and Autonomy in the Teenage Sleepover." *Contexts* 9 (Summer): 16–21.
Schilt, Kristen. 2011. *Just One of the Guys? Transgender Men and the Persistence of Gender Inequality.* Chicago: University of Chicago Press.
Schlosser, Eric. 2001. *Fast Food Nation: The Dark Side of the All-American Meal.* New York: Houghton Mifflin.
Schmitt, Eric. 2001. "Segregation Growing among U.S. Children." *The New York Times* (May 6): 28.
Schmitt, Frederika E., and Patricia Yancey Martin. 1999. "Unobtrusive Mobilization by an Institutionalized Rape Crisis Center: 'It Comes From the Victims.'" *Gender & Society* 13 (3): 364–384.
Schumock, Glen T., Edward C. Li, Katie J. Suda, Linda M. Matusiak, Robert J. Hunkler, Lee C. Vermeulen, and James M. Hoffman. 2014. "National Trends in Prescription Drug Expenditures and Projections for 2014." *American Journal of Health-System Pharmacy* 71: e6–23.
Schwartz, Christine R., and Hongyun Han. 2014. "The Reversal of the Gender Gap in Education and Trends in Marital Dissolution." *American Sociological Review* 79 (August): 605–629.
Schwartz, John, and Matthew L. Wald. 2003. "NASA's Failings Go Far beyond Foam Hitting Shuttle, Panel Says." *The New York Times* (June 7): 1 and 12.
Scott, Ellen K., Andrew S. London, and Nancy A. Myers. 2002. "Dangerous Dependencies: The Intersection of Welfare Reform and Domestic Violence." *Gender & Society* 16 (December): 878–897.
Segal, David R., and Mady Wechsler Segal. 2004. "America's Military Population." *Population Bulletin* 59 (December): 1–44.
Seidman, Steven. 2014. *The Social Construction of Sexuality*, 3rd ed. New York: W. W. Norton.
Sen, Amartya. 2000. "Population and Gender Equity." *The Nation* (July 24–31): 16–18.
Sanders, Clinton R., and D. Angus Vail. 2008. *The Art and Culture of Tattooing.* Philadelphia, PA: Temple University Press.
Shelley, Louise. 2010. *Human Trafficking: A Global Perspective.* Cambridge, MA: Cambridge University Press.
Sherman, Jennifer. 2009. "Bend to Avoid Breaking: Job Loss, Gender Norms, and Family Stability in Rural America." *Social Problems* 56 (November): 599–620.
Shibutani, Tomatsu. 1961. *Society and Personality: An Interactionist Approach to Social Psychology.* Englewood Cliffs, NJ: Prentice Hall.
Shieman, Scott. 2010. "Socioeconomic Status and Beliefs about God's Influence in Everyday Life." *Sociology of Religion* 71 (1): 25–51.
Sikora, Joanna. 2014. "Gender Gap in School Science: Are Single-Sex Schools Important?" *Sex Roles* 70: 400–415.
Silva, Jennifer M. 2014. "Working Class Growing Pains." *Contexts* 13 (2): 26–31.
Silver, Alexandra. 2010. "Brief History of the U.S. Census." *Time* (February 8): 16.
Silverthorne, Zebulon A., and Vernon L Quinsey. 2000. "Sexual Partner Age Preferences of Homosexual and Heterosexual Men and Women." *Archives of Sexual Behavior* 29 (1): 67–76.
Simmel, Georg. 1950 [1902]. "The Number of Members as Determining the Sociological Form of the Group." *The American Journal of Sociology* 8 (July): 1–46.
Simon, David R. 2011. *Elite Deviance,* 10th ed. Boston: Allyn and Bacon.
Singer, Peter W. 2007. *Corporate Warriors: The Rise of the Privatized Military Industry.* Ithaca, NY: Cornell University Press.
Sirvananadan, A. 1995. "La trahision des clercs. (Racism)." *New Statesman and Society* 8: 20–22.
Sjøberg, Gideon. 1965. *The Preindustrial City: Past and Present.* New York: Free Press.
Skocpol, Theda. 1992. *Protecting Soldiers and Mothers: The Origins of Social Policy in the United States.* Cambridge, MA: Belknap Press.
Skocpol, Theda, and Vanessa Williamson. 2012. *The Tea Party and the Remaking of Republican Conservatism.* New York: Oxford University Press.
Smelser, Neil J. 1992. "Culture: Coherent or Incoherent." pp. 3–28 in *Theory of Culture,* edited by R. Münch and N. J. Smelser. Berkeley, CA: University of California Press.
Smith, Brad W., and Malcolm D. Holmes. 2014. "Police Use of Excessive Force in Minority Communities: A Test of the Minority Threat, Place, and Community Accountability Hypotheses." *Social Problems* 61 (February): 83–104.
Smith, Aaron. 2012. "The Best (and Worst) of Mobile Connectivity." Pew Research Internet Project. **www.pewinternet.org**
Smith, Andrew. 2009. "Nigerian Scam E-Mails and the Charms of Capital." *Cultural Studies* 23 (1): 27–47.
Smith, Brad W., and Malcolm D. Holmes. 2014. "Police use of Excessive Force in Minority Communities: A Test of the Minority Threat, Place, and Community Accountability Hypotheses." *Social Problems* 61 (1): 83–104.
Smith, M. Dwayne, Joel A. Devine, and Joseph F. Sheley. 1992. "Crime and Unemployment: Effects Across Age and Race Categories." *Sociological Perspectives* 35 (Winter): 551–572.
Smith, Sandra. 2007. *Lone Pursuit: Distrust and Defensive Individualism among the Black Poor.* New York: Russell Sage Foundation.

Smith, Stacy L., Marc Choueiti, Ashley Prescott, and Katherine Pieper. 2013. *Gender Roles & Occupation: A Look at Character Attributes and Job-Related Aspirations in Film and Television.* **www.seejane.org**

Smith, Tom W., Peter V. Marsden, Michael Hout, and Jibum Kim. *General Social Surveys, 1972–2012.* Chicago: National Opinion Research Center.

Smith, Tom, and Jaesok Son 2013. *Trends in Public Attitudes about Sexual Morality.* Chicago: National Opinion Research Center.

Smith, R. Tyson, and Gala True. 2014. "Warring Identities: Identity Conflict and the Mental Distress of American Veterans of the Wars in Iraq and Afghanistan." *Society and Mental Health* 4 (2): 147–161.

Smolak, L, and M. Levine, eds., 1996. *The Developmental Psychopathology of Eating Disorders: Implications for Research, Prevention, and Treatment.* Hillsdale, NJ: Lawrence Erlbaum Associates Inc.

Snipp, C. Matthew. 1989. *American Indians: The First of This Land.* New York: Russell Sage Foundation.

Snipp, C. Matthew. 1996. "The First Americans: American Indians." pp. 390–404 in *Origins and Destinies: Immigration, Race, and Ethnicity in America,* edited by Sylvia Pedraza and Rubén G. Rumbaut. Belmont, CA: Wadsworth.

Snipp, Matthew. 2007. "An Overview of American Indian Populations." pp. 38–48 in *American Indian Nations: Yesterday, Today, and Tomorrow,* edited by George Horse Capture, Duane Champaign, and Chandler Jackson. Walnut Creek, CA: Altamira Press.

Solomon, Brittany C., and Simine Vazire. 2014. "You Are So Beautiful . . . To Me: Seeing beyond Biases and Achieving Accuracy in Romantic Relationships." *Journal of Personality and Social Psychology* 107 (3): 516–528.

Spencer, Herbert. 1882. *The Study of Sociology.* London: Routledge.

Spencer, Ranier. 2011. *Reproducing Race: The Paradox of Generation Mix.* Las Vegas, Nevada: Lynne Rienner.

Spencer, Ranier. 2012. "Mixed-Race Chic." pp. 67–70 in Elizabeth Higginbotham and Margaret L. Andersen, eds., *Race and Ethnicity in Society: The Changing Landscape,* 3rd ed. Belmont, CA: Wadsworth/Cengage.

Spitzer, Brenda L., Katherine A. Henderson, and Marilyn T. Zivian. 1999. "A Comparison of Population and Media Body Sizes for American and Canadian Women." *Sex Roles* 700 (7/8): 545–565.

Spitzer, Steven. 1975. "Toward a Marxian Theory of Deviance." *Social Problems* 22 (June): 638–651.

Squires, Gregory. 2007. "Demobilization of the Individualistic Bias: Housing Market Discrimination as a Contributor to Labor Market and Economic Inequality." *The Annals of the American Academy of Political and Social Science* 609 (January): 200–214.

Stacey, Judith, and Timothy J. Bibliarz. 2001. "(How) Does the Sexual Orientation of Parents Matter?" *American Sociological Review* 66 (April): 159–183.

Stack, Carol. 1974. *All Our Kin: Strategies for Survival in a Black Community.* New York: Harper Colophon Books.

Steele, Claude M. 1997. "A Threat in the Air: How Stereotypes Shape Intellectual Identity and Performance." *American Psychologist* 52 (6): 613–629.

Steele, Claude. M. 2010. *Whistling Vivaldi and Other Clues to How Stereotypes Affect Us.* New York: W. W. Norton.

Steele, Claude M., and Joshua Aronson. 1995. "Stereotype Threat and the Intellectual Test Performance of African Americans." *Journal of Personality and Social Psychology* 69 (5): 797–811.

Steger, Manfred B. 2009. *Globalization: A Very Short Introduction.* New York: Oxford University Press.

Stein, Arlene, and Marcy Westerling. 2012. "The Politics of Broken Dreams." *Contexts* 11 (Summer): 8–10.

Stern, Jessica. 2003. *Terror in the Name of God: Why Religious Militants Kill.* New York: Ecco.

Sternheimer, Karen. 2007. "Do Video Games Kill?" *Contexts* 6 (Winter): 13–17.

Sternheimer , Karen. 2011. *Celebrity Culture and the American Dream: Stardom and Social Mobility.* New York: Routledge.

Stewart, Susan D. 2001. "Contemporary American Stepparenthood: Integrating Cohabiting and Nonresident Stepparents." *Population Research and Policy* 20 (August): 345–364.

Stombler, Mindy, and Irene Padavic. 1997. "Sister Acts: Resisting Men's Domination in Black and White Fraternity Little Sister Programs." *Social Problems* 44 (May): 257–275.

Stone, Pamela. 2007. *Opting Out: Why Women Really Quit Careers and Head Home.* Berkeley, CA: University of California Press.

Stone, Michael E. 1993. *Shelter Poverty: New Ideas on Housing Affordability.* Philadelphia: Temple University Press.

Stoner, J. A. F. 1961. "A Comparison of Individual and Group Decisions Involving Risk." Unpublished M. A. thesis, MIT.

Stryker, Susan, and Stephen Whittle (eds.). 2006. *The Transgender Studies Reader.* New York: Routledge.

Su, Dejun, Chad Richardson, and Guangzhen Wang. 2010. "Assessing Cultural Assimilation of Mexican Americans: How Rapidly Do Their Gender-Role Attitudes Converge to the U.S. Mainstream?" *Social Science Quarterly* 91 (3): 762–776.

Sullivan, Maureen. 1996. "Rozzie and Harriet? Gender and Family Patterns of Lesbian Coparents." *Gender & Society* 12 (December): 747–767.

Sullivan, Nikki. 2003. *A Critical Introduction to Queer Theory.* New York: New York University Press.

Sullivan, Patrick F. 1995. "Mortality in Anorexia Nervosa." *American Journal of Psychiatry* 152 (July): 1073–1074.

Sumner, William Graham. 1906. *Folkways: A Study of the Sociological Importance of Usages, Manners, Customs, Mores, and Morals.* Boston: Ginn.

Sussman, N. M., and D. H. Tyson. 2000. "Sex and Power: Gender Differences in Computer-Mediated Interactions." *Computers in Human Behavior* 16 (July): 381–394.

Sutherland, Edwin H. 1940. "White Collar Criminality." *American Sociological Review* 5 (February): 1–12.

Sutherland, Edwin H., and Donald R. Cressey. 1978. *Criminology,* 10th ed. New York: Lippincott.

Swidler, Ann. 1986. "Culture in Action: Symbols and Strategies." *American Sociological Review* 51 (April): 273–286.

Switzer, J. Y. 1990. "The Impact of Generic Word Choices: An Empirical Investigation of Age- and Sex-Related Differences." *Sex Roles* 22 (1): 69–82.

Tach, Laura, and Sarah Halpern-Meekin. 2009. "How Does Premarital Cohabitation Affect Trajectories of Marital Quality?" *Journal of Marriage and Family* 71 (May): 298–317.

Takaki, Ronald. 1989. *Strangers from a Different Shore: A History of Asian Americans.* New York: Penguin.

Tanner, Lindsay. 2014. "New Study Suggests Genetic Link for Male Homosexuality." *The Huffington Post,* November 17.

Tatum, Beverly. 1997. *Why Are All the Black Kids Sitting Together at the Cafeteria*? New York: Basic Books.

Taylor, Howard F. 1980. *The IQ Game: A Methodological Inquiry into the Heredity-Environment Controversy.* New Brunswick, NJ: Rutgers University Press.

Taylor, Howard F. 1992. "The Structure of a National Black Leadership Network: Preliminary Findings." Unpublished manuscript, Princeton University.

Taylor, Howard F. 2012. "Defining Race." pp. 7–13 in *Race and Ethnicity in Society: The Changing Landscape,* 3rd ed., edited by in Elizabeth Higginbotham and Margaret L. Andersen. Belmont, CA: Wadsworth/Cengage.

Taylor, Paul. 2012. "The Growing Electoral Clout of Blacks Is Driven by Turnout, Not Demographics." *Pew Social & Demographic Trends.* Washington, DC: Pew Research Center. **www.pewresearch.org**

Taylor, Shelley E. 2010. "Mechanisms Linking Early Life Stress to Adult Health Outcomes." *Proceedings of the National Academy of Sciences* 107(19): 8507–8512.

Taylor, Shelley E., Letitia Anne Peplau, and David O. Sears. 2006. *Social Psychology,* 12th ed. Upper Saddle River, NJ: Prentice-Hall.

Taylor, Shelley E., Letitia Ann Peplau, and David O. Sears. 2013. *Social Psychology,*

13th ed. Upper Saddle River, NJ: Prentice Hall.

Teaster, Pamela B. 2000. "A Response to the Abuse of Vulnerable Adults: The 200 Survey of State Adult Protective Services." Washington, DC: National Center on Elder Abuse. **www.elderabusecenter.org**

Telles, Edward E. 1994. "Residential Segregation and Skin Color in Brazil." *American Sociological Review* 57 (April): 186–197.

Telles, Edward E. 2004. *Race in Another America: The Significance of Skin Color in Brazil.* Princeton, NJ: Princeton University Press.

Telles, Edward E., and Vilma Ortiz. 2008. *Generations of Exclusion: Mexican Americans, Assimilation, and Race.* New York: Russell Sage Foundation.

Telles, Edward, Mark Q. Sawyer, and Gaspar Rivera-Salgado, eds. 2011. *Just Neighbors? Research on African American and Latino Relations in the United States.* New York: Russell Sage Foundation.

Tenenbaum, Harriet R. 2009. "'You'd Be Good at That': Gender Patterns in Parent–Child Talk about Courses." *Social Development* 18 (2): 447–463.

Thoits, Peggy A. 2009. "Sociological Approaches to Mental Illness." In *A Handbook for the Study of Mental Health*, edited by Teresa L. Scheid and Tony L. Brown. Cambridge: Cambridge University Press.

Thomas, William I. 1931. *The Unadjusted Girl.* Boston: Little, Brown.

Thomas, William I., with Dorothy Swaine Thomas. 1928. *The Child in America.* New York: Knopf.

Thompson, Robert Farris. 1993. *Face of the Gods: Art and Alters of Africa and the African Americas.* New York: Random House.

Thornton, Russell. 1987. *American Indian Holocaust and Survival: A Population History.* Norman, OK: University of Oklahoma Press.

Thornton, Russell. 2001. "Trends among American Indians in the United States." pp. 135–169 in *America Becoming: Racial Trends and Their Consequences*, vol. I, edited by Neil J. Smelser, William Julius Wilson, and Faith Mitchell. Washington, DC: National Academies Press.

Tichenor, Veronica. 2005. "Maintaining Men's Dominance: Negotiating Identity and Power When She Earns More." *Sex Roles* 53 (3–4): 191–205.

Tienda, Marta, and Haya Stier. 1996. "Generating Labor Market Inequality: Employment Opportunities and the Accumulation of Disadvantage." *Social Problems* 43 (May): 147–165.

Tierney, Kathleen J. 2007. "From the Margins to the Mainstream? Disaster Research at the Crossroads." *Annual Review of Sociology* 33: 503–525.

Tierney, Kathleen. 2012 (November 11). "After Hurricane Sandy: Understanding a Hurricane's Worst Impacts: Q&A." *Star-Ledger.* **blog.nj.com**

Tierney, Kathleen J., Michael K. Lindell, and Ronald W. Perry. 2001. *Facing the Unexpected: Disaster Preparedness and Response in the United States.* Washington, DC: National Academies Press.

Tierney, Kathleen, Christine Bevc, and Erica Kuglikowski. 2006. "Metaphors Matter: Disaster Frames, Media Myths, and Their Consequences in Hurricane Katrina." *Annals of the American Association of Political and Social Science* 604 (March): 57–81.

Tierney, Patrick. 2000. *Darkness in El Dorado: How Scientists and Journalists Devastated the Amazon.* New York: W. W. Norton.

Timmerman, Kelsey. 2012. *Where Am I Wearing: A Global Tour to the Countries, Factories, and People that Make our Clothes.* Hoboken, NJ: Wiley.

Tjaden, Patricia, and Nancy Thoennes. 2000. *Extent, Nature, and Consequences of Intimate Partner Violence.* Washington, DC: National Institute of Justice and the Centers for Disease Control and Prevention.

Tönnies, Ferdinand. 1963 [1887]. *Community and Society (Gemeinschaft and Gesellschaft).* New York: Harper & Row.

Toossi, Mitra. 2012. "Labor Force Projections to 2020: A More Slowly Growing Workforce." *Monthly Labor Review* (January): 43–64.

Toynbee, Arnold J., and Jane Caplan. 1972. *A Study of History.* New York: Oxford University Press.

TransAtlantic Slave Trade Database. 2014. **www.slavevoyages.org**

Travers, Jeffrey, and Stanley Milgram. 1969. "An Experimental Study of the Small World Problem." *Sociometry* 32 (4): 425–443.

Treiman, Donald J. 2001. "Occupations, Stratification and Mobility." pp. 297–313 in *The Blackwell Companion to Sociology*, edited by Judith R. Blau. Malden, MA: Blackwell.

Tsoukalas, I. 2007. "Exploring the Microfoundations of Consciousness." *Culture and Psychology* 13 (1): 39–81.

Tuan, Yi-Fu. 1984. *Dominance and Affection: The Making of Pets.* New Haven, CT: Yale University Press.

Tuchman, Gaye. 1979. "Women's Depiction by the Mass Media." *Signs* 4 (Spring): 528–542.

Tumin, Melvin M. 1953. "Some Principles of Stratification." *American Sociological Review* 18 (August): 387–393.

Turner, Ralph, and Lewis Killian. 1993. *Collective Behavior*, 4th ed. Englewood Cliffs, NJ: Prentice Hall.

Turner, Terence. 1969. "Tchikrin: A Central Brazilian Tribe and Its Symbolic Language of Body Adornment." *Natural History* 78 (October): 50–59.

Uleman, J. S., L. S. Newman, and G. B. Moskowitz. 1996. "People as Flexible Interpreters: Evidence and Issues from Spontaneous Trait Inference." pp. 211–279 in *Advances in Experimental Social Psychology*, vol. 28, edited by Mark Zanna. Boston: Academic Press.

UNICEF. 2000. *Child Poverty in Rich Nations.* Florence, Italy: United Nations Children's Fund. New York: United Nations. **www.unicef-icdc.org**

United Nations. 2006. *In Depth Study on All Forms of Violence against Women.* New York: United Nations.

United Nations. 2010. *The Gender Inequality Index.* New York: United Nations. **www.undp.org**

United Nations. 2012a. *WomenWatch.* New York: United Nations. **www.un.org/womenwatch**

United Nations. 2012b. "World Population to Increase by 2.6 Billion over Next 45 Years, with All Growth Occurring in Less Developed Regions." Press Release, POP918. **www.un.org**

United Nations. 2013. *World Population Prospects: The 2012 Revision.* New York: United Nations. **www.unpopulation.org**

United Nations. 2014. *Statistics on Literacy.* **www.unesco.org**

United Nations World Food Programme. 2014. **www.wfp.org**

U.S. Bureau of Justice Statistics. 2012a. *Criminal Victimization in the United States 2012, Statistical Tables.* Washington, DC: U.S. Bureau of Justice Statistics.

U.S. Bureau of Justice Statistics. 2012b. *National Crime Victimization Survey.* Department of Justice. **www.bjs.gov**

U.S. Bureau of Justice Statistics. 2013. *Arrest Data.* Department of Justice. **www.bjs.gov**

U.S. Bureau of Labor Statistics. 2010. "Labor Force Participation Rates among Mothers." Washington, DC: U.S. Department of Labor. **www.bls.gov**

U.S. Bureau of Labor Statistics. 2012a. *American Time Use Survey Summary.* Washington, DC: U.S. Department of Labor.

U.S. Bureau of Labor Statistics. 2012b. Employment Projections: Civilian Labor Force Participation Rates by Age, Sex, Race, and Ethnicity. Washington, DC: U.S. Department of Labor. **www.bls.gov**

U.S. Bureau of Labor Statistics. 2012c. *Employment and Earnings.* Washington, DC: U.S. Department of Labor. **www.bls.gov**

U.S. Bureau of Labor Statistics. 2012d. "A Profile of the Working Poor, 2008." Washington, DC: U.S. Department of Labor. **www.bls.gov**

U.S. Bureau of Labor Statistics. 2014a. *Consumer Expenditures–2013.* Washington, DC: U.S. Dpeartment of Labor. **www.bls.gov**

U.S. Bureau of Labor Statistics. 2014b. *Employment and Earnings.* Washington, DC: U.S. Department of Labor. **www.bls.gov**

U.S. Bureau of Labor Statistics. 2014c. *Highlights of Women's Earnings in 2013,* Report 1051. BLS Reports. **www.bls.gov**

U.S. Bureau of Labor Statistics. 2014d. *A Profile of the Working Poor.* Washington, DC: Bureau of Labor Statistics Report, March. **www.bls.gov**

U.S. Census Bureau. 2003. "Racial and Ethnic Classification Used in Census

2000 and Beyond." Washington, DC: U.S. Census Bureau. **www.census.gov**

U.S. Census Bureau. 2012a. *The 2012 Statistical Abstract.* Washington, DC: U.S. Census Bureau. **www.census.gov**

U.S. Census Bureau. 2012b. *America's Families and Living Arrangements: 2012, Detailed Tables.* Current Population Reports. Washington, DC: U.S. Census Bureau. **www.census.gov**

U.S. Census Bureau. 2012c. *Census Bureau Releases Estimate of Undercount and Overcount in the 2010 Census.* Washington, DC: U.S. Census Bureau. **www.census.gov**

U.S. Census Bureau. 2012d. *Current Population Reports: Annual Social and Economic Supplement.* Washington, DC: U.S. Census Bureau. **www.census.gov**

U.S. Census Bureau. 2013. Current Population Survey, School Enrollment Supplement. Washington, DC: U.S. Census Bureau. **www.census.gov**

U.S. Census Bureau. 2014. *The 2012 Statistical Abstract.* Washington, DC: U.S. Census Bureau. **www.census.gov**

U.S. Department of Defense. 2012. *2012 Demographics: A Profile of the Military Community.* Washington, DC: U.S. Department of Defense.

U.S. Department of Justice. 2009. "Child Sex Tourism." Washington, DC: U.S. Department of Justice. **www.justice.gov**

U.S. Department of Justice. 2015. *Investigation of the Ferguson Police Department.* Washington, DC: U.S. Department of Justice. **www.justice.gov**

U.S. Department of Labor. 2014. *Employment and Earnings.* Washington, DC: U.S. Department of Labor. **www.bls.gov**

U.S. Department of State. 2011. *Annual Report on Intercountry Adoption.* Washington, DC: U.S. Department of State.

U.S. Department of State. 2014. *Trafficking in Persons Report: June 2014.* Washington, DC: U.S. Department of State. **www.state.gov**

U.S. Energy Information Administration. 2012. *U.S. Energy Related Carbon Dioxide Emissions 2011.* Washington, DC: U.S. Department of Energy. **www.eia.gov**

U.S. Energy Information Administration. 2012. Washington, DC: U.S. Department of Energy. **www.eia.gov**

U.S. State Department. 2012. *Trafficking in Persons Report, June 2012.* Washington, DC: U.S. State Department. **www.state.gov**

Valentine, Kathyrn, Mary Prentice, Monica F. Torres, and Eduardo Arellano. 2012. "The Importance of Cross-Racial Interactions as Part of College Education: Perceptions of Faculty." *Journal of Diversity in Higher Education* 5 (December): 191–206.

Valian, Virginia. 1999. *Why So Slow? The Advancement of Women.* Cambridge, MA: MIT Press.

Valles, Mike. 2014. "Alternative Medicine and Your Health Insurance." *The Simple Dollar*, September 16. **thesimpledollar.com**

Van Ausdale, Debra, and Joe R. Feagin. 1996. "The Use of Racial and Ethnic Concepts by Very Young Children." *American Sociological Review* 61 (October): 779–793.

van Dijck, José. 2013. "'You Are One Identity': Performing the Self on Facebook and LinkedIn." *Media, Culture, & Society* 35 (2): 199–215.

Vanneman, Reeve, and Lynn Weber Cannon. 1987. *The American Perception of Class.* Philadelphia: Temple University Press.

Vaughan, Diane. 1996. *The Challenger Launch Decision: Risky Technology, Culture, and Deviance at NASA.* Chicago: The University of Chicago Press.

Veblen, Thorstein. 1994 [1899]. *Theory of the Leisure Class.* New York: Penguin.

Veblen, Thorstein. 1953 [1899]. *The Theory of the Leisure Class: An Economic Study of Institutions.* New York: The New American Library.

Veliz, Philip, and Sohaila Shakib. 2012. "Interscholastic Sports Participation and School Based Delinquency: Does Participation in Sport Foster a Positive High School Environment?" *Sociological Spectrum* 32 (6): 558–580.

Vespa, Jonathan. 2009. "Gender Ideology Construction." *Gender & Society* 23 (June): 363–387.

Vidmar, Neil, and Valerie P. Hans. 2007. *American Juries: The Verdict.* Amherst, NY: Prometheus Books.

Vijayasiri, Ganga. 2008. "Reporting Sexual Harassment: The Importance of Organizational Culture and Trust." *Gender Issues* 25 (1): 43–61.

Villareal, Andres. 2010. "Stratification by Skin Color in Contemporary Mexico." *American Sociological Review* 75 (6): 652–678.

Volscho, Thomas W., and Nathan J. Kelly. 2012. "The Rise of the Super-Rich: Power Resources, Taxes, Financial Markets, and the Dynamics of the Top 1 Percent, 1949–2008." *American Sociological Review* 77 (October): 679–699.

Voss, Georgina. 2012. "'Treating It as a Normal Business': Researching the Pornography Industry." *Sexualities* 15 (3–4): 391–410.

Waldfogel, Jane, Wen-Jui Han, and Jeanne Brooks-Gunn. 2002. "The Effects of Early Maternal Employment in Child Cognitive Development." *Demography* 39 (May): 369–392.

Wallace, Walter L. 1971. *The Logic of Science in Sociology.* Chicago: Aldine-Atherton.

Wallace, Walter L. 1983. *Principles of Scientific Sociology.* New York: Aldine Publishing Co.

Wallerstein, Immanuel M. 1974. *The Modern World System: Capitalist Agriculture and the Origins of the European World Economy in the Sixteenth Century.* New York: Academic Press.

Wallerstein, Immanuel M. 1980. *The Modern World-System II.* New York: Academic Press.

Walters, Suzanna Danuta. 1999. "Sex, Text, and Context: (In) Between Feminism and Cultural Studies." pp. 222–260 in *Revisioning Gender,* edited by Myra Marx Ferree, Judith Lorber, and Beth B. Hess. Thousand Oaks, CA: Sage.

Warren, Cortney S., Andrea Schoen, and Kerri J. Schafer. 2010. "Media Internalization and Social Comparison as Predictors of Eating Pathology among Latino Adolescents: The Moderating Effect of Gender and Generational Status." *Sex Roles* 63 (9–10): 712–724.

Washington, Scott. 2011. "Who Isn't Black? The History of the One-Drop Rule." PhD dissertation, Department of Sociology, Princeton University, Princeton, NJ.

Wasserman, Stanley, and Katherine Faust (eds.). 1994. *Social Network Analysis: Methods and Applications.* Cambridge, MA: Cambridge University Press.

Waters, Mary C. 1990. *Ethnic Options: Choosing Identities in America.* Berkeley, CA: University of California Press.

Waters, Mary C., and Peggy Levitt (eds.). 2002. *The Changing Face of Home: The Transnational Lives of the Second Generation.* New York: Russell Sage Foundation.

Watkins, Susan. 1987. "The Fertility Transition: Europe and the Third World Compared." *Sociological Forum* 2 (Fall): 645–673.

Watts, Duncan J. 1999. "Networks, Dynamics, and the Small-World Phenomenon." *American Journal of Sociology* 105 (September): 493–527.

Watts, Duncan J., and Stephen H. Strogatz. 1998. "Collective Dynamics of 'Small World' Networks." *Nature* 393: 440–442. **www.sentencingproject.org**

Weber, Max. 1958 [1904]. *The Protestant Ethic and the Spirit of Capitalism.* New York: Scribner's.

Weber, Max. 1962 [1913]. *Basic Concepts in Sociology.* New York: Greenwood.

Weber, Max. 1978 [1921]. *Economy and Society: An Outline of Interpretive Sociology,* edited by Guenther Roth and Claus Wittich. Berkeley, CA: University of California Press.

Weeks, John R. 2012. *Population: An Introduction to Concepts and Issues,* 12th ed. Belmont, CA: Wadsworth Publishing Co.

Weisburd, David, Cynthia M. Lum, and Anthony Petrosino. 2001. "Does Research Design Affect Study Outcomes in Criminal Justice?" *The Annals of the American Academy of Political and Social Science* 578 (November): 50–70.

Weisburd, David, Stanton Wheeler, Elin Waring, and Nancy Bode. 1991. *Crimes of the Middle Class: White Collar Defenders in the Courts.* New Haven, CT: Yale University Press.

Weitzer, Ronald. 2009. "Sociology of Sex Work." *Annual Review of Sociology* 35 (February): 213–234.

West, Candace, and Sarah Fenstermaker. 1995. "Doing Difference." *Gender & Society* 9 (February): 8–37.

West, Candace, and Don Zimmerman. 1987. "Doing Gender." *Gender & Society* 1 (June): 125–151.

West, Cornel. 1994. *Race Matters.* New York: Vintage.

West, Cornel R. 2004. *Democracy Matters: Winning the Fight against Imperialism.* New York: The Penguin Press.
Westbrook, Laurel, and Kristen Schilt. 2014. "Doing Gender, Determining Gender: Transgender People, Gender Panics, and the Maintenance of the Sex/Gender/Sexuality System." *Gender & Society* 28 (February): 32–57.
Western, Bruce. 2007. *Punishment and Inequality in America.* New York: Russell Sage Foundation.
Western, B. (2014). "Incarceration, Inequality, and Imagining Alternatives." *The Annals of the American Academy of Political and Social Science* 651 (1): 302–306.
White, Deborah Gray. 1999. *Ar'n't I a Woman? Female Slaves in the Plantation South.* New York: W. W. Norton.
White, Jonathan R. 2002. *Terrorism: An Introduction.* Belmont, CA: Wadsworth.
Whitehead, Andrew L., and Joseph O. Baker. 2012. "Homosexuality, Religion, and Science: Moral Authority and the Persistence of Negative Attitudes." *Sociological Inquiry* 82 (4): 487–509.
White House. 2012. *Critical Issues Facing Asian Americans and Pacific Islanders.* Washington DC: Initiative on Asian Americans and Pacific Islanders. **www.whitehouse.gov**
Whiten, A., J. Goodall, W. C. McGrew, T. Nishida, V. Reynolds, Y. Sugiyama, C. E. G. Tutin, R. W. Wrangham, and C. Boesch. 1999. "Cultures in Chimpanzees." *Nature* 399 (June 17): 682–684.
Whorf, Benjamin. 1956. *Language, Thought, and Reality: Selected Writings.* Cambridge, MA: Technology Press, MIT.
Whyte, William F. 1943. *Street Corner Society.* Chicago: University of Chicago Press.
Wilchins, Riki. 2014. *Queer Theory/Gender Theory.* New York: Riverdale.
Wilder, Esther I., and Toni Terling Watt. 2002. "Risky Parental Behavior and Adolescent Sexual Activity at First Coitus." *The Milbank Quarterly* 80 (September): 481–524.
Wilkins, Amy. 2012. "Stigma and Status: Interracial Intimacy and Intersectional Identities among Black College Men." *Gender & Society* 26 (April): 165–189.
Williams, Christine L. 1992. "The Glass Escalator: Hidden Advantages for Men in the 'Female' Professions." *Social Problems* 39 (August): 253–267.
Williams, Christine L. 1995. *Still a Man's World: Men Who Do Women's Work.* Berkeley, CA: University of California Press.
Williams, Christine L., Patti A. Giuffre, and Kirsten Dellinger. 2009. "The Gay-Friendly Closet." *Sexuality Research and Social Policy: Journal of NSRC* 6 (1): 29–45.
Williams, Timothy. 2012. "Suicides Outpacing War Deaths for Troops." *The New York Times*, July 8, p. A10.
Willie, Charles Vert. 1979. *The Caste and Class Controversy.* Bayside, NY: General Hall.
Wilson, Barbara J., Stacy L. Smith, W. James Potter, Dale Kunkel, Daniel Linz, Carolyn M. Colvin, and Edward Donnerstein. 2002. "Violence in Children's Television Programming: Assessing the Risks." *Journal of Communication* 52 (March): 5–35.
Wilson, John F. 1978. *Religion in American Society: The Effective Presence.* Englewood Cliffs, NJ: Prentice Hall.
Wilson, William Julius. 1978. *The Declining Significance of Race: Blacks and Changing American Institutions.* Chicago: University of Chicago Press.
Wilson, William Julius. 1987. *The Truly Disadvantaged: The Inner City, the Underclass and Public Policy.* Chicago: University of Chicago Press.
Wilson, William Julius. 1996. *When Work Disappears: The World of the New Urban Poor.* New York: Knopf.
Wilson, William Julius. 2009. *More Than Just Race: Being Black and Poor in the Inner City.* New York: W. W. Norton.
Winnick, Louis. 1990. "America's Model Minority." *Commentary* 90 (August): 222–229.
Wirth, Louis. 1928. *The Ghetto.* Chicago: University of Chicago Press.
Wolcott, Harry F. 1996. "Peripheral Participation and the Kwakiutl Potlatch." *Anthropology and Education Quarterly* 27 (December): 467–492.
Wolf, Stephen. 2014. "Your Guide to 2014 Election's Results and the 114th Congress Members and Their Districts." *DailyKos,* December 4.
Wolff, Edward N. 2014. "Wealth Inequality." *Pathways*: Special Issue: 34–41.
Woo, Deborah. 1998. "The Gap between Striving and Achieving." pp. 247–256 in *Race, Class, and Gender: An Anthology,* 3rd ed., edited by Margaret L. Andersen and Patricia Hill Collins. Belmont, CA: Wadsworth.
Wood, Julia. 2013. *Gendered Lives: Communication, Culture, and Gender.* Belmont, CA: Wadsworth/Cengage.
Woodhams, Jessica, Claire Cooke, Leigh Harkins, and Teresa da Silva. 2012. "Leadership in Multiple Perpetrator Stranger Rape." *Journal of Interpersonal Violence* 27 (4): 728–752.
Worchel, Stephen, Joel Cooper, George R. Goethals, and James L. Olsen. 2000. *Social Psychology.* Belmont, CA: Wadsworth.
World Bank. 2010. *Independent Evaluation Group, An Evaluation of World Bank Support 1997–2007: Water and Development.* New York: World Bank. **www.worldbank.org**
The World Bank. 2014a. *Health Expenditure, Total (% of GDP).* New York: The World Bank. **www.worldbank.org**
World Bank. 2014b. *Prosperity for All/ Ending Extreme Poverty.* New York: World Bank. **www.worldbank.org**
Wright, Erik Olin. 1979. *Class Structure and Income Determination.* New York: Academic Press.
Wright, Erik Olin. 1985. *Classes.* London: Verso.
Wuthnow, Robert (ed.). 1994. *I Come Away Stronger: How Small Groups Are Shaping American Religion.* Grand Rapids, MI: Eerdmans.
Wuthnow, Robert. 1998. *After Heaven: Spirituality in America since the 1950s.* Berkeley, CA: University of California Press.
Wuthnow, Robert. 2010. *Boundless Faith: The Outreach of American Churches.* Berkeley: University of California Press.

Yardi, Sarita, and Dana Boyd. 2010. "Dynamic Debates: An Analysis of Group Polarization over Time on Twitter." *Bulletin of Science, Technology and Society* 30 (5): 316–327.
Young, Alford A. 2003. *The Minds of Marginalized Black Men: Making Sense of Mobility, Opportunity, and Future Life Chances.* Princeton, NJ: Princeton University Press.

Zhou, Min, and Carl L. Bankston. 2000. "Immigrant and Native Minority Groups, School Performance, and the Problem of Self-Esteem." Paper presented at the Southern Sociological Society, April, New Orleans, LA.
Zimbardo, Phillip G., Ebbe B. Ebbesen, and Christina Maslach. 1977. *Influencing Attitudes and Changing Behavior.* Reading, MA: Addison-Wesley.
Zippel, Kathrin S. 2006. *The Politics of Sexual Harassment: A Comparative Study of the United States, the European Union, and Germany.* Cambridge: Cambridge University Press.
Zola, Irving Kenneth. 1989. "Toward the Necessary Universalizing of a Disability Policy." *Milbank Quarterly* 67 (Suppl. 2): 401–428.
Zola, Irving Kenneth. 1993. "Self, Identity and the Naming Question: Reflections on the Language of Disability." *Social Science and Medicine* 36 (January): 167–173.
Zoroya, Gregg. 2012. "Homeless, At-Risk Vets Double." *USA Today* (December 27).
Zorza, J. 1991. "Woman Battering: A New Cause of Homelessness." *Clearinghouse Review* 25 (4).
Zuckerman, Phil. 2002. "The Sociology of Religion of W. E. B. DuBois." *Sociology of Religion* 63 (July): 239–253.
Zweigenhaft, Richard L., and G. William Domhoff. 2006. *Diversity in the Power Elite: How It Happened, Why It Matters.* Lanham, MD: Rowman and Littlefield.
Zweigenhaft, Richard L., and G. William Domhoff. 2011. *The New CEOs: Women, African Americans, Latino, and Asian American Leaders of Fortune 500 Companies.* Lanham, MD: Rowman and Littlefield.
Zwick, Rebecca (ed.). 2004. *Rethinking the SAT: The Future of Standardized Testing in University Admissions.* New York: Routledge.

Name Index

Subject Index

Note: Page numbers in **bold** indicate definitions/explanations of key terms; page numbers in *italics* indicate boxed material or illustrations.